Ecological Studies

Analysis and Synthesis

Edited by

J. Jacobs, München · O. L. Lange, Würzburg
J. S. Olson, Oak Ridge · W. Wieser, Innsbruck

Volume 6

Klaus Stern   Laurence Roche

Genetics of Forest Ecosystems

With 70 Figures

Springer-Verlag   New York Heidelberg Berlin 1974

Professor Dr. KLAUS STERN †, formerly: Lehrstuhl für Forstgenetik und Forstpflanzenzüchtung, 8400 Göttingen-Weende, Büsgenweg 2, Federal Republic of Germany

LAURENCE ROCHE, B. Agr., M. A., M. F., Ph. D., Professor of Forestry and Head, Department of Forestry, University of Ibadan, Ibadan, Nigeria

ISBN 0-387-06095-2 Springer-Verlag New York · Heidelberg · Berlin
ISBN 3-540-06095-2 Springer-Verlag Berlin · Heidelberg · New York
ISBN 0-412-12980-9 Chapman & Hall Limited London

Dedicated to Olof Langlet

Preface

Throughout the world natural forest ecosystems have been, and are being massively disrupted or destroyed. The boreal forests of Canada are no more immune to man's intervention than the tropical rain forests of Africa, and the day is rapidly approaching when natural forest ecosystems, undisturbed by man, will be found only as remnants in national parks and other protected areas. Yet where they continue to exist these ecosystems are an extraordinarily rich, though relatively neglected source of data that illuminate many aspects of the classic theory of evolution.

The subject matter of this book is not, however, confined to natural forest ecosystems. Forest ecosystems under varying degrees of management, and man-made forests are also a rich source of information on ecological genetics. In general, however, it can be said that the published evidence of this fact has not yet significantly penetrated the botanical literature. All too frequently it is confined to what might be termed forestry journals.

It is hoped that this book will to some extent redress the balance, and draw attention to a body of published work which not only provides a basis for the rational management and conservation of forest ecosystems, but also complements the literature of ecological genetics and evolution.

The first draft of Chapters I to V was written in German by the senior author and translated by E. K. Morgenstern of the Canadian Forestry Service. Dr. Morgenstern's indispensable assistance in this regard is gratefully acknowledged. Dr. Gregorius (Göttingen) helped in developing chapters I and IV. J. Antoine and G. Müller (Göttingen) kindly took care of the subject index.

A book of this nature cannot be published without an acknowledgement of the profound influence of the work of Olof Langlet of Sweden. It is with respect and admiration that this book is dedicated to him.

The tragic death of Klaus Stern, just as the book was ready for press, and when he was at the height of his powers, has impoverished the lives of all who knew him, and left a gap in scientific forestry which will not easily be filled.

Ibadan, February 1974   Laurence Roche

Contents

Introductory Remarks

Forest ecosystems, whether natural or artificial, although greatly influenced by man's activities, continue to be composed predominantly of wild species that have never been subjected to the processes of artificial selection and breeding. Therefore, forest ecologists and geneticists have been more concerned than agronomists and agricultural crop breeders with the inner mechanisms of natural and artificial ecosystems, of which many transitory forms exist. Only by knowing these mechanisms intimately could they hope to improve the economics of forestry; not only through control of stand structure and species composition, maintenance of soil fertility, and choice of varieties, but also through reduction of ecological risks. The latter is regarded as being particulary important as the yields of forestry per unit area are likely to remain substantially below those of agriculture and horticulture. Direct control of diseases and parasites, which is commonly practiced and required in field and garden crops, amounts to an interference into the balance of tree species with these organisms, and this must be kept to a minimum.

Such differences in the objectives of ecological and genetic research in agriculture and horticulture on the one hand, and forestry on the other, become very apparent when comparing the literature of related disciplines in both fields. Forest entomologists, for example, have concentrated on ecological problems and agricultural entomologists on direct parasite control. In forest genetics, actual breeding has received less attention than the study of natural tree populations, their relationship to habitats, and their adaptive strategies.

For the same reasons, foresters and forest biologists have always recognized the need for conservation; i.e., a sustained effort for the protection of the part of the human environment designated as the "landscape." Only recently has the general public become aware of this problem. The control of erosion and other measures for the maintenance of soil fertility, and the protection of water resources, of wildlife, and of nature in general, have always been the concern of forestry in countries where this profession is well developed.

As an integrating discipline of ecosystem biology, forest genetics has a particular role to play — the study of genetic and evolutionary relationships in forest ecosystems — and the aim to understand such systems better for the purposes of management. Forest trees, as the principal objects of genetics research, present certain difficulties that are a result of their long life, their generation intervals, etc. Yet, forest trees are of general interest. They possess characteristics different in character to organisms of classical population genetics (Libby, Seitz and Stettler, 1969). The relationship of populations to their ecosystems is the forest geneticist's central area of inquiry, and, if he is to gain a fairly complete understanding of these relationships, he must regard the intricate patterns of evolution and coevolution. Therefore, forest genetics engages in ecosystems biology, or, more narrowly, in population biology.

Complex systems that are the means or the objects of research invite the danger of descending into holistic categories of thought. Forest science, particulary in central Europe, presents many vivid and deterrent examples. The holistic model of the "organism of the forest," and its dialectic exploitation by holistic forest ecologists, has caused problems and unnecessary difficulties for many forest geneticists. Probably the periodic prevalence of reductionist ideas and methods in forest genetics is a result of this. This is particularly true in the field of forest tree breeding, which is the practice of biological engineering in forest biology. Today, following the spread of TANSLEY's (1935) ideas on ecosystems, this danger is no longer with us to the same degree. With this development the task of systematic illumination of forest ecosystems has been recognized more clearly and experimental approaches using suitable techniques, methods, and models have been replacing the dialectic models.

To achieve a better understanding of the aims of this book, the tasks of forest genetics within the ecosystem's biology need to be outlined more clearly. These tasks are vividly portrayed through the title of HUTCHINSON's (1965) well-known treatise, which supports our own endeavors: "The Ecological Theater and the Evolutionary Play." Thus, the interpretation of evolutionary games in forest ecosystems is the function of physiologists and geneticists, and the stage and the direction are the ecologists' concern. From this, readers may gain the impression that physiology and ecology will receive little attention in this book. However, they can hardly be avoided in a work dedicated to genetic problems. The same view will also facilitate understanding of the structure of this book. Chapter I treats the concept of the ecological niche relatively briefly, which is supplemented in Chapter II by the description of habitat-related adaptive patterns in forest trees in recognition of their evolutionary history. Chapters III and IV present a formal treatment of the genetic systems of populations and of the adaptive strategies that are most likely to be important in forest trees. Chapters V and VI discuss the major forest ecosystems from the geneticist's viewpoint. A chapter on the diverse effects of man's activities on forest ecosystems, including the major aspects of modern forestry practice, is included, although this subject matter might be more logically dealt with in a separate book.

All chapters attempt to limit discussion to problems relevant to forestry. Since attention to methodology is essential when dealing with such unwieldy material as forest trees, emphasis is placed on the methods required to solve these investigational problems. Therefore, methods are included that produce indirect or less reliable results, or results that are interpretable in several different ways. The methods of classical genetics, particularly the analysis of crosses, were not helpful in most cases. With a few exceptions, we have considered field observations and supplementary experiments in the laboratory, greenhouse, growth chamber, and nursery. The reader is directed to FORD's (1964) work on ecological genetics. Our difficulties were similar to those of FORD, and often were greater because of the nature of the material.

There is no intention to present a summary or overall view of the results of forest genetics research concerned with our subject. The examples given are nearly always representative of many other works and have been chosen solely, because of their illustrative or instructive merit. Consequently, the literature cited will not contain all the titles of interest to the reader and the available evidence is correspondingly larger.

I. The Ecological Niche

1. Formal Concept of the Niche

The environmental conditions that permit a population to survive permanently, and with which this population interacts, today are usually designated as its "ecological niche." Originally, the term "niche" did not have the same meaning for all ecologists but this need not be considered here. We above define and use it in the sense of HUTCHINSON (1958).

Let us consider two environmental factors X_1 and X_2. These may assume different values in different areas. Thus every specific environment that is considered only on the basis of these two factors is represented by the pair of values X_1, X_2; i.e., a point in a rectangular coordinate system. In nature upper and lower values of these factors limit specific organisms in their survival and reproduction. Assume a species S_1 existing within the boundaries X_1' (lower limit) and X_1'' (upper limit), and X_2' and X_2'', respectively. Thus, in a system of rectangular coordinates the four corner points, $X_1'X_2'$, $X_1''X_2'$, $X_1'X_2''$, and $X_1''X_2''$, mark the boundaries of the niche for S_1. If relations between X_1 and S_1 are independent of X_2, — i.e., if it is immaterial for S_1 what value X_1 takes at a given value of X_2 (on condition only that both are located within X_1'; X_1'' and X_2'; X_2'') — then the rectangle between the four corner points describes the niche itself. Alternatively, the niche will be bounded by any other areas within the rectangle, its form and size depending upon interaction between X_1 and X_2 (Fig. 1).

With generalization to n environmental factors, the niche of the species S_1 becomes a hyperspace in an n-dimensional system of coordinates. This hyperspace represents the "fundamental niche" $\theta_1^{(f)}$ of the species, a theoretical and abstract

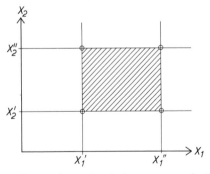

Fig. 1. The ecological niche of a population in relation to two ecological factors X_1 and X_2. X_1' and X_1'', X_2' and X_2'' designate the lower and upper limiting values, respectively. The population is not capable of existing below and above these limits. The hatched rectangle then demarcates the niche if no interactions occur. If there are interactions between X_1 and X_2, the niche is located within the rectangle

definition of the ecological niche discussed in the literature. To give a concrete example, one would have to know all the more relevant physical and biotic factors and be able to depict their interactions.

The fundamental niche may also be perceived as a set of points in an n-dimensional space θ^n. (For an explanation of symbols from set theory see footnote[1].) In a restricted subset of points $R \subset \mathbb{R}^3$ ($\mathbb{R}^3$ designates the physical space) representing particular biotopes, one finds that any point $p_k \in \theta^{(f)}$ corresponds to a number of points $p_k \in R$; i.e., it corresponds to those points that satisfy the conditions for p_{k_i}. The environmental factors $X_1, X_2, \ldots, X_n$ will normally be continuous. Therefore, every small subset of points $\Delta\theta \subseteq \theta^n$ with appropriate $\Delta R \subseteq R$ will correspond probably to larger quantities of small volumes that are scattered in R. Every partial space $\Delta R \subseteq R$, perhaps of the dimensions typical of the migration distance of an animal or the flight of a plant's seed, will therefore contain points that can be delegated to different fundamental niches contained in θ^n, with a definite probability.

Because R is a limited physical space, i.e., the biotope of definite species $S_1; S_2; \ldots; S_n$, there is no reason to assume that a definite point in $\theta_1^{(f)}$ should correspond to any points in R. If there are no points in R sufficient for the existence of S_1 as given by $\theta_1^{(f)}$, then R will be designated as "incomplete for S_1." If some of the points in $\theta_1^{(f)}$ also occur in R, then R is "partially complete for S_1." Finally, R is "complete for S_1" if all points in $\theta_1^{(f)}$ also occur in R.

In all points of the fundamental niche, S_1 will survive permanently, and its probability of surviving at points outside of $\theta_1^{(f)}$ is zero, but within the fundamental niche there ought to be areas with optimal as well as suboptimal or marginal conditions. Here the weakness of the model becomes apparent. Additional parameters could compensate, at least theoretically. Furthermore, the assumption that all environmental factors can be expressed in linear terms is not necessarily valid. Another assumption, that $\theta_1^{(f)}$ is always described by a single hyperspace, is also not convincing. The fundamental niche of trees can differ for different stages of development, and could then be described by two or more discrete hyperspaces in θ^n. This is probably most obvious in insect species, where environments of the larval and the adult stages differ markedly. Finally, competing species or their influences should be measurable in terms of the scale of the coordinate system.

Now let us consider two fundamental niches $\theta_1^{(f)}$ and $\theta_2^{(f)}$ of the species S_1 and S_2. Both may have no points in common so that they can be regarded as separate. If they overlap, then $(\theta_1^{(f)} - \theta_2^{(f)})$ is the subset in $\theta_1^{(f)}$ not contained in $\theta_2^{(f)}$ and $(\theta_2^{(f)} - \theta_1^{(f)})$ the subset in $\theta_2^{(f)}$ not contained in $\theta_1^{(f)}$. $\theta_1^{(f)} \cap \theta_2^{(f)}$ designates the subset of points present in both $\theta_1^{(f)}$ and $\theta_2^{(f)}$, called the intersection.[2] (See Fig. 2).

Niches of Competing Species

The Volterra-Gause principle states that in a limited environment, two competing species with the same requirements will not reach equilibrium, and therefore only one will survive (GAUSE, 1934). Let a_1 and a_2 be the birth rates of two species

---

1 $\subset$: The symbol in set theory for one set properly included in another set. $\in$: The symbol in set theory for membership of an object in a certain set. $\subseteq$: The symbol in set theory for one set properly included in another set and possibly equal to the set.

2 $\cap$: The symbol in set theory for the intersection of two sets that is the set of all objects that are members of both sets.

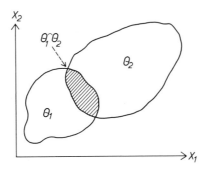

Fig. 2. The ecological niches of two species in relation to two ecological factors. The hatched part indicates the intersection set

competing in the same ecosystem, b_1 and b_2 their death rates, and N_1 and N_2 their population numbers. Then changes in population size at time (t) are given by the following equations:

$$dN_1/dt = a_1N_1 - b_1N_1 f(N_1, N_2)$$

$$dN_2/dt = a_2N_2 - b_2N_2 f(N_1, N_2)$$

Thus the death rates are both a function of N_1 as well as of N_2. S_1 will survive if $a_1/b_1 > a_2/b_2$, and S_2 will survive if $a_2/b_2 > a_1/b_1$. A permanent coexistence of both species is only possible if $a_1/b_1 = a_2/b_2$, but this is an improbable case, because no stable equilibrium exists under this condition (VOLTERRA, 1926).

Yet species competing in the same community are frequently found in nature, and this has led to the common assumption that they occupy different niches. HUTCHINSON (1958) considers this general conclusion to be valid unless it is opposed by contradicting experimental results. Besides, the case of two identical niches, namely, an assumption that for S_1 and S_2, $\theta_1^{(f)} = \theta_2^{(f)}$, is so unlikely that it can be neglected if S_1 and S_2 are really distinguishable species. (We shall see later that our model of competing species has to be enlarged.)

Translated into terms of set theory, the Volterra-Gause principle would mean that there are no small elements in the intersection set $\theta_1^{(f)} \cap \theta_2^{(f)}$ that possess corresponding small elements in R, some of these being permanently occupied by S_1 and others by S_2.

Neglecting the improbable case $\theta_1^{(f)} = \theta_2^{(f)}$, the following situations may be distinguished for competing species:

1. $\theta_2^{(f)}$ is a separate subset of $\theta_1^{(f)}$; ($\theta_2^{(f)}$ is enclosed by $\theta_1^{(f)}$; $\theta_2^{(f)} \subset \theta_1^{(f)}$.

(a) Competition favors S_1 everywhere in R, where elements corresponding to those of $\theta_2^{(f)} \subset \theta_1^{(f)}$ exist. After a sufficiently long period of time, only S_1 will survive.

(b) Competition favors S_2 in all elements of R that possess corresponding elements in $\theta_2^{(f)}$. Both species survive.

2. $\theta_1^{(f)} \cap \theta_2^{(f)}$ is a subset of $\theta_1^{(f)}$ as well as of $\theta_2^{(f)}$. Then S_1 will survive in those parts of R that possess corresponding elements in $(\theta_1^{(f)} - \theta_2^{(f)})$; S_2 will survive in those with corresponding elements in $(\theta_2^{(f)} - \theta_1^{(f)})$. These two subsets are the refugia of the two species.

The realized niche $\theta_1^{(r)}$ of S_1 in the presence of S_2 now consists of $(\theta_1^{(f)} - \theta_2^{(f)})$ plus those additional elements in $\theta_1^{(f)}$. $\theta_2^{(f)}$ that are occupied in R by individuals of S_1. The same argument holds for the realized niche $\theta_2^{(r)}$ and S_2. Expressed in this way, the Volterra-Gause principle is an empirical law stating that realized niches do not overlap. The law can be verified or rejected. General rejection would mean that in nature resources are never limited, because the alternative would be elimination of one of the two species by competition. The niche model derived from set theory allows distinct definitions and refinement, but supplies no evidence for or against the validity of the Volterra-Gause principle. Yet it indicates where experiments testing this principle may be initiated. Two principal methods have been applied here.

Laboratory populations have been employed as "analog computers" to solve the equations resulting from competition. However, this is sensible only if all environmental factors and genetic systems of the populations can be controlled; and since this is not possible, these experiments leave much to be desired. Usually they indicate that as a consequence of partly incomplete R's and forced competition under conditions corresponding to a small part of the intersection set, only one species survives. Furthermore, the experiments indicate that the environmental factors determine the survival of the species, as well as the outcome of competition in that part of the interaction set considered in the experiment. If such experiments contain at least one nonintersecting subset in R, then both species can coexist for any length of time.

The second method consists of the examination of communities in which several related species coexist. It was found that the case where $\theta_1^{(f)} = \theta_2^{(f)}$, rejected here earlier, never really occurred. Where sympatric species of the same genus or subfamily occur, their ecological differences are pronounced. In assessing these differences, niches must be identified as much as possible in today's world of probabilities. Occasionally indirect arguments can be applied; for example, evidence that the differences found are, at least, important for evolution. It is extremely difficult to identify and to quantify the components of competition, as their role is usually demonstrated and measured only indirectly through the outcome of the competitional process.

There are also cases where application of the Volterra-Gause principle is difficult or impossible. This situation occurs, for example, when the reproductive potential of S_1 is less than that of S_2. Annual plants that flower and die at the end of the growing season may distribute their seeds at random over the biotope. In such a case, S_1 may only be capable of occupying elements of the biotope that are vacant by chance. This species is then close to what has been called a fugitive species (HUTCHINSON, 1951, 1953). SKELLAM (1951) and BRIAN (1956b) have discussed a similar situation, where living space became available following a high death rate. The former situation reflects near-catastrophes and accidents of transitory significance in forest communities near or at the climax stage (HUTCHINSON, 1951, 1953), but the latter, seen in relation to forest ecosystems, could initiate successions (SKELLAM, 1951; BRIAN, 1956b). In other respects both cases are identical.

The application of the Volterra-Gause principle is problematic in another case. In the animal kingdom competition is more often a contest than a fight. If the suitable territory is occupied while many individuals exist without territory, size of territory, but not food supply, is the limiting factor. The founding of territories by species competing for food may proceed independently; competition for territory is strictly intraspecific. Although the area is the resource and the object of the contest, each species claiming this area is independent of any other species. With plants the situation is similar, for example, in deciduous forests where vegetation changes with the season. A spring vegetation, before leafing out of the trees forming the canopy, is followed by a summer vegetation adapted to shade.

Changing environmental conditions in a given biotope may reverse the trend of competition over longer or shorter periods. If the periods of environmental change are long compared with the time leading to the final outcome of competition, elimination of the species with certain disadvantages is to be expected. However, if these periods are short, then the mean parameter of climate is decisive for the outcome. Presumably long-term trends would develop, leading inevitably to the final result unless opposing events occur.

BRIAN (1956a) has shown that elimination of certain species trough competition is less likely in the first stages of succession than in climax stages. His rationale is that initially the heterogeneity of environment is more pronounced (possibly this must be understood in a relative sense, for in highly complex climax forests the applicable environmental heterogeneity results solely from highly developed specialization of its components). Numerous intermediate situations may exist between these two extremes, e.g., where the direction of the competition depends on the environment and where a certain environment did not last long enough in time to achieve elimination.

The frequency of species in relation to nonoverlapping portions of their realized niches can be measured in various ways. MACARTHUR (1957) has given the basis for a rational treatment of the problem in his formal theory of the niche. We have already stated that the Volterra-Gause principle is equivalent to the claim that realized niches of coexisting species are without interest.

Let us consider now a biotope R with an equilibrium community consisting of n species $S_1, S_2, \ldots, S_n$ with population sizes $N_1, N_2, \ldots, N_n$. It should be possible for a species S_k to identify within R a number of elements each of which corresponds completely or partially to the realized niche $\theta_k^{(r)}$ but to no other space in θ^n. Suppose that at any time every one of these elements is occupied by an individual of S_k: the total volume of R is the specific biotope of S_k, which is now designated by $N_k \cdot \Delta R(S_k)$ where $\Delta R(S_k)$ is the mean volume in R occupied by an individual of S_k.

The ecosystem (community) in R is in equilibrium for all n species. Consequently all localities suitable for S_k are occupied, namely,

$$R = \sum_1^n N_k \Delta R(S_k).$$

There is no *a priori* information on the distribution of the $N_k \cdot \Delta R(S_k)$ except that these specific biotopes have volumes proportional to $N_1, N_2, \ldots, N_n$. This first approximation is justified if the species are comparable in physiology and size of individuals.

In general, some species will be more frequent than others. The simplest hypothesis would be to assume a random distribution of R among species.

MacArthur (1957) has given a model for the division of R into n parts: $(n-1)$ random dots are tossed upon a straight line of limited length. Alternatively, the line could be divived by n pairs of dots that are randomly thrown. The former technique results in nonoverlapping niches (sections of the straight line), and the latter in overlapping niches. The related mathematical theory will not be discussed.

The authors discovered that several multispecies communities are well described by a distribution for nonoverlapping specific biotopes, i.e., nonoverlapping realized niches. The manner of distribution is independent of the number of dimensions in θ. In comparison, the distribution of overlapping realized niches resulted in few species with moderate frequencies, and in more species with low frequencies than were found in nature. The discrepancy becomes even greater if one extends the study from the straight line to the area or space.

The theory imposes certain limits by stating that the ratio of the total number of individuals $\left(m = \sum_1^n N_k\right)$ to the total number of species remains constant in all larger subdivisions of R. This is highly probable in any "homogeneous-diverse" habitat where the elements of the environmental mosaic are small in proportion to the mobility of the organisms, including their seed dispersal; e.g., where dead and living trees, shrubs, rocks, special conditions for ingrowth into a certain stratum, etc., exist. But in a "heterogeneous-diverse" area, this ratio is expected to become constant only in a portion of the environment: where conditions change from forest to meadow and field, where natural catastrophes have occurred, or in a mosaic of sites in climax forests. MacArthur finds sufficient agreement with his hypothesis if some of the contentious distributions reported in the literature are properly stratified.

Saturation of a biotope is of special interest to the ecologist. How is optimal utilization achieved by a climax community? Many genetic and evolutionary considerations, including the evolution of whole ecosystems, which will be dealt with later, are involved here. It will become apparent that for population genetics the problem must be seen in a different light.

An important characteristic of the niche is the "grain size" of environmental factors. For a population, environments can be fine-grained or coarse-grained in relation to one or more factors. An environmental patch is fine-grained if the population utilizes all components in the proportion of occurrence; it is coarse-grained if utilization is disproportional, i.e., one resource or component is used more than the other (Levins and MacArthur, 1966). Note that a population may treat its environment as fine-grained even if it is coarse-grained for the individual.

No environment is fine-grained in the ideal sense. A measure of deviation has been suggested by MacArthur (1966) for an individual case. Let $p_1, p_2, \ldots, p_n$ be the proportion of components; $t_1, t_2, \ldots, t_n$ the time spent in each component, the sum of t_i being T; then the quantity

$$\sum_{i=1}^n \frac{(t_i - Tp_i)^2}{Tp_i}$$

is distributed like χ^2 with mean $n-1$ and standard deviation $\sigma = 2(n-1)$.

A measure of deviation that is nearly independent of n is the quantity

$$\frac{\dfrac{(t_i - Tp_i)^2}{Tp_i} - (n-1)}{2(n-1)} .$$

Coexistence of different species in a fine-grained mixture or a homogeneous and structureless environment is only possible if the resources are subdivided. Yet the resources are often correlated (as the strata in a multilayered forest) and the natural environment may or will consist of larger blocks of fine-grained patches. In a coarse-grained environment there may be as many species as grain types, although generalists among the species may utilize them like a fine-grained environment.

We are particularly interested in the width of a population's niche, the number of its dimensions, and the degree of overlap with niches of populations of other similar species in the same ecosystem. Of course, it is difficult to find useful measures, and every one of those suggested depends upon certain assumptions (LEVINS, 1968).

The width of a niche may often be determined from investigations of one or more environmental components. Studies of reaction to temperature, mineral nutrition, light requirements in different stages, etc., will indicate whether a population occupies a broad or a narrow niche or is specialized in relation to a certain environmental factor. Yet this procedure depends upon the components chosen for measurement; e.g., they may be poor indicators of total fitness. This procedure is also unsuitable where competition determines realized niches. For example, a comparison of the distribution of two species on different site types and of the diversity of these sites permits only statements about realized niches. A comparison of site types on which field experiments succeeded allows only conclusions concerning fitness in the conditions chosen, not concerning natural conditions. Nevertheless, continuous distribution of a species across a mosaic of different soils should indicate a broad niche and its restriction to a few decrete sites a narrow niche. Similarly, occurrence across a large altitudinal range reveals a broad niche and across a small range a narrow niche. However, adaptation to such conditions through genetic flexibility (especially in such immobile organisms as trees) shows how carefully such observations must be interpreted. In such cases it is hardly possible to discuss the niche of a species without including its adaptive strategy (Chapter IV); genetically differentiated population systems may be involved as have been discovered in all species so far investigated. Thus a species like *Liriodendron tulipifera,* growing over a large range with very diverse soils and climates, may occupy only a narrow niche in a given ecosystem, but may be adapted to so many environments (through evolution of local races, etc.) that the species as a whole must be judged very differently than the individual sub-population.

The measures suggested for the width of a niche are (LEVINS, 1968):

$$\log \theta^{(B)} = - \Sigma\, p_i \log p_i$$

$$1/\theta^{(B)} = \Sigma\, p_i^2$$

where $\theta^{(B)}$ is the width of the niche and p_i the proportion of the species in the ith environment. A measure of viability could also be introduced, perhaps v_i, and used for a meaningful definition of p_i:

$$p_i = v_i / \Sigma \, v_i \, .$$

Again it must be pointed out that the application of such measures requires careful consideration of their suitability for the numerical problem being studied. The relevant dimensions of the niche of a population, too, can only be obtained from a comparison of niches of different populations. The number of dimensions important to us is not the actual, large number of dimensions of the fundamental niche but rather the number of dimensions that is relevant to distinguish niches.

Measures for overlap will be discussed in the chapter dealing with ecosystems. Other properties of the niche that are of interest here will be considered together with examples in the next chapter.

2. Main Characteristics of Ecological Niches of Forest Tree Species

A survey of vegetation types on this planet indicates that forests, however they may be defined, are not ubiquitous. A large amount of literature on plant sociology and plant geography deals with this subject and it is neither planned nor possible to present an account here of the available facts and relationships. Rather, the sole intention is to describe the essential characteristics of niches of forest trees — characteristics that have created the genetic architecture of the whole species or of subpopulations.

A difficulty is that botanists have made little use of the ecological niche concept; the principal literature comes from zoology. However, we have decided to use it as our basic model because it describes the subject of interest in a general way and because it can be profitably used for the discussion of complex relationships.

In one of the few attempts to relate the niche concept to the ecology of higher plants, HARPER (1967) concludes that these plants are situated in the same position of an ecosystem's food chain. Therefore, niches of different species in the same ecosystem overlap on a broad front, but in spite of this, it is often easy to establish where niche divergence of ecologically similar species should be sought (PUTWAIN and HARPER, 1968). Correlations of species frequency with ecological factors such as soil moisture and shade tolerance of seedlings are often demonstrated with simple methods. If the more relevant niche differences are related to competition, and especially to competition at different stages of development, greater methodical obstacles exist. "Selection for mutual avoidance" (ASHTON, 1969 and others), which is more accurately defined as "selection for mutual exclusion by competition," is nothing but niche specialization, and could be the explanation in all those cases where closely related species grow next to each other in ecosystems with many species. These situations have been especially investigated by population biology in recent years (e.g., HARPER et al., 1961; CLATHWORTHY and HARPER, 1962), yet it has become apparent that established theories are sufficient to explain observed equilibria among such species, as well as to describe and explain the evolution of whole ecosystems or populations (for forest ecosystems refer to FEDOROV, 1966, and ASHTON, 1969). We shall return to these problems in Chapter V.

Ecosystems with many species and great stability are generally found in evironments with little fluctuation in time of the relevant niche dimensions; i.e., in stable environments, and where environmental stability has been preserved long enough to permit evolution of stable ecosystems. Such conditions have prevailed especially in the humid tropics and subtropics and also in such regions as parts of the Appalachian-Allegheny Mountains in eastern North America with their mixed forests and many species. Here species occupy microniches that may be created, for example, through mortality of old trees, offering opportunities for growth into the crown layer (or into one of the several strata). At least theoretically, niches may differ in a few factors only, as in the light requirements of young plants (DOBZHANSKY, 1950). In addition, there may be zones with different frequencies of the microniches, such as zones of unlike soil types, or trends in mean values of climate along mountain slopes, leading to a zonation in distribution of the species or to trends in species frequencies.

RICHARDS (1969) directs attention to some difficulties in the application of the niche concept, especially to tropical and subtropical rain forests with their numerous species. In his view differences in adaptation (Chapter II) are insufficient to create niches commensurate in number and kind. The principle of "competitive exclusion" is inadequate to account for the multitude of species inasmuch as character complexes of species with a similar ecology, (e.g., in the same stratum) are adaptively correlated. Our view is that, theoretically, a satisfactory explanation could be provided by a relatively small number of niche differences. (See WILLIAMSON, 1957, who deals with this problem using a simplified method and has designated the relevant niche differences as "control factors." His results indicate that the number of control factors should be at least as large as the number of competing species. Control factors can be complexes of correlated character or parts of such complexes.) Furthermore, other reasons given by RICHARDS for the great number of species, such as competition for minerals or defense mechanisms against diseases and parasites, must also be regarded as niche dimensions. Phenotype and niche are exchangeable notions, as indicated in the preceding section and as will be discussed in detail in Chapter IV. Thus the differences in phenotype emphasized by RICHARDS reflect niche differences. In addition, it is not necessary to limit discussions exclusively to characters related to competition, although this is often done in the literature for reasons of a simpler presentation, e.g., for elucidation of the community matrix (Chapter V).

Trends in the mean values of climatic factors are distinct for mountain slopes, and particularly for large geographic areas. Characters that are subject to direct or correlated selection often reflect the trends in environmental factors — they vary clinally. In most cases, long-term averages of climate constitute inadequate measures of environmental influences along such gradients. Substantial year-to-year deviations from the mean are often of considerable significance for the success of the individual during a short developmental phase, determining its survival and reproduction. For this reason it would be better to regard this problem in relation to niche frequency along the gradient. An example is provided by the length of the growing season, which is taken to begin when the temperature sum reaches a certain level in the spring and to finish when a critical day length is reached in the autumn (see SARVAS, 1965, and ROCHE, 1969; our presentation here is simplified).

In many cases adaptation to the length of the growing season also depends upon early and late frosts, which can do much harm to young plants before they escape the dangers of occasional low temperatures near ground level. Thus four niche factors instead of only one, should be considered: date and distribution of the temperature sum needed to initiate growth, frequency and distribution in time of spring frosts, date of the critical day length in the fall, and frequency and distribution in time of early frost in the fall. Only one of these factors is constant; the remaining three vary from year to year. The effects of the first two are not independent of each other, because the damage resulting from a spring frost at a certain date is also determined by the date when the necessary temperature sum was reached. Therefore, to describe this part of the environment adequately, in addition to mean values of climatic factors, attention should be given to their local heterogeneity, and better still, to the geographic trends to this heterogeneity.

The importance of soil properties as niche factors is much less well known than that of climatic factors, if one disregards obvious cases such as niche limitation through extremes of moisture, ion concentrations, etc. Correlations between soil properties and adaptive characters or characters associated with adaptation have been much less frequently described than correlations between climatic factors and adaptive characters. Most available information comes from extreme situations (Chapter II, Section 1b). Yet soil properties must contribute decisively to the success of a population in a given biotope, as competition experiments indicate (summary given by STERN, 1969). The limits of a species area are often boundaries determined by competition. The species might be very well capable of living and reproducing outside of these boundaries but is prevented from doing so by more competitive populations of other species. This is valid also for closely related species, as MOORE's (1959) experiments with two species of eucalypts indicate. The arrangement of soil properties (as determined by bedrock, etc.) is the most frequent reason for a parapatric or neighboring-sympatric distribution of tree species (JAIN and BRADSHOW, 1966; METTLER and GREGG, 1969). This contrasts with the biotic sympatric distribution in space that is found in complex ecosystems with different microniches that locally reoccur with frequencies determined by soils and/or climatic zones. The consequences of such situations for the adaptation of populations will be discussed in the next chapters. Surely this is one of the most important fields of research in ecological genetics of the near future.

Catastrophic events are typical for many niches of tree species and occur more or less frequently. Thus in some years extreme values are reached in one (or more) climatic factor that is one of the niche dimensions (e.g., temperature, moisture, wind, etc.), or insect epidemics and forest fires occur. Such catastrophes may determine a species' range, their frequency often being the criterion of first importance. For instance, extremely cold winters occurring periodically in central Europe are said to control the eastern limit of white fir (*Abies alba* Mill.). GILBERT (1959) reports a particularly instructive example. On certain sites in Tasmania *Eucalyptus-Nothofagus* stands rejuvenate after forest fires. Both groups of species regenerate in a shelter of *Eucalyptus* species. If the next fire is delayed 350–400 years, the eucalypts will disappear, being unable to survive underneath *Nothofagus* because of insufficient light. The climax forest (or subclimax forest) consists of *Nothofagus* accompanied by few other shade-tolerant species, mostly in the second stratum. If the next fire occurs within less than 350 years, a second story of trees made up of eucalypts (up to 75 meters in height) at wide spacing towers above *Nothofagus* (40 meters). The eucalypts will be nearly evenly aged. If fires become still more frequent, with one or two within a century, *Nothofagus* no longer regenerates and is replaced by other species. With five to ten fires each century, pure *Eucalyptus* forests without some fire-susceptible species develop. Reoccurence of fires at closer intervals leads to forms of vegetation other than forest.

Regular developments take place after periodic catastrophes in other ecosystems. There may be typical successions with a species appearing, disappearing, or increasing in frequency. A succession may lead to a stable equilibrium; i.e., stable in a special sense — species composition and frequencies remain unchanged until the next catastrophe occurs. There may also be cases where a stable climax is not reached (subclimax).

It will be seen that the requirements upon a population in regard to adaptations and adaptive strategies change with position of a species in the succession, i.e., are codetermined by it. Succession begins with pioneer species and ends with prevalence of climax species or with a subclimax vegetation. Both types of species colonize or recolonize, respectively, in different ways and with different requirements on their respective environments. It is possible to group the forests of this planet into categories of the most diverse systems, or vegetational concepts. Plant sociologists and plant geographers have done this frequently and with much success (for example WALTER, 1962, 1966 presents a compendium of this kind that is ecologically and physiologically oriented). Such attempts can be weighted in different ways, depending upon the special interests of the author, and the results will vary. In this section we have tried so set forth the salient characteristics of the ecological niche from the viewpoint of ecological genetics of forest trees. The systematics of niche dimensions is quite independent of the geographic distribution of the numerical value that niche dimensions can take, and the latter could be, and have been often enough (together with historical considerations), the basis of the systematics of vegetation types.

II. Adaptations

Adaptations are defined for our purpose as character combinations, metabolic processes, and developmental pathways that enable an organism to survive in a given niche; and to occupy an optimal position in relation to the specific physiological possibilities set by its evolutionary history. Every adaptation then reflects characteristics of the niche. Now part of this complex are genetic adaptations of populations to environmental heterogeneity, which will be discussed in Chapter IV under the heading "Adaptive Strategies." Adaptive strategies often lead to the development of polymorphic populations. The adaptive value of these populations over all niches of the heterogeneous environment consists in being the source of a variety of phenotypes adapted to certain niches, not of an optimum phenotype for all niches. Thus, adaptations, in the sense of this chapter, are concerned only with "pure" and not with "mixed" strategies. The evolutionary possibilities of a population are set in each case by past experience stored as genetic information and by the value of this past experience under the special conditions of a given niche. Genetic information is modified through reactions to selection in the course of generations. Existing adaptations may be improved and new adaptations may be developed.

The principle of the origin of adaptations has been understood since DARWIN's time. Through experiments, modern population genetics has rounded out many important details and provided better explanations. Yet difficulties are experienced when attempting to explain certain cases in nature that result from the complexity of physiologic, biochemical, genetic, and ecological systems. Therefore, the best known adaptations are those that depend upon simple physiological mechanisms, are simply inherited, and the ecological importance of which is easily recognized and demonstrated. Examples are the development of camouflage by color and pattern (including mimicry, etc.), of special organs or parts of the body in animals, and of color and form of the flower in the higher plants. The following description is limited to a brief account of the major adaptations of forest trees. A general presentation is given by GRANT (1963).

1. Vegetative Cycle

a) Adaptation to Climate

The effect of climate upon vegetation is so strong that any worldwide classification of vegetation must be primarily a climatic classification (WALTER, 1962, and others). Although there are distinct differences in species composition of regions in different continents with similar climates, the vegetation types are so similar to each other that a climatic classification can be superimposed upon the regional divisions. Thus the niche of every population of trees is determined first of all by its macroclimate.

It is probably useful to contrast quantitatively and qualitatively the special conditions of diverse climatic zones as environments of tree populations against the climate of one of the possible extremes, perhaps that of the tropical rain forests. In the extreme, weather remains practically unchanged during the whole year; no seasonal differences exist. There is daily fluctuation in temperature and moisture (and, of course, light intensity) that is associated with day and night, but these changes never reach critical values such as frost, drought, etc. Solar radiation may vary over shorter or longer time periods but these periods are not correlated with seasons. Day length, too, remains practically constant. In these forests, developmental cycles exist in vegetative and generative activities that are similar in some respects to those in forests growing in areas with pronounced seasons but differ in their switch mechanisms (see BÜNNING, 1953 for examples). However, in certain areas catastrophic events such as storms and fires occur that lead to certain specific adaptations and/or adaptive strategies where the events are frequent or influence the composition of vegetation. However, the main characteristics of humid tropical forests are not affected: high average temperatures (at least in lower elevations), high humidity, and precipitation are observed constantly throughout the year, with only small differences between mean annual values.

In tropical or subtropical forests with alternating dry and rainy seasons (semievergreen rain forests and moist to dry deciduous forests), physiological periodicity is often pronounced. Trees in such rain forests usually flush prior to the wet season and often flower before flushing. Flushing may be triggered by a rise in temperature. Leaves are shed earlier in dry years than in wet years. In areas with winter rains one finds evergreen species with shoot extensions in spring. The associated deciduous species flush in spring as well, i.e., bear their leaves during the dry period (BÜNNING, 1953). In dry areas there are only a few shrubs that will flush following the onset of the rainy period and will shed their leaves in succeeding weeks. Such independence from temperature as a trigger mechanism is the exception rather than the rule.

We will be satisfied with these remarks inasmuch as there are few detailed investigations of adaptations to seasonal weather rhythms in these areas. The physiological switch mechanisms of the annual cycle in trees of temperate and boreal climates have been more fully explored. The onset of the dormant period is primarily affected by the diminishing day length in fall; apparently there is a critical day length below which the growth ceases (LYR et al., 1967, ROCHE, 1969), but a drought in late summer or early fall may also influence dormancy. The dormant period in winter can be subdivided into several stages. Criteria used to designate these stages are usually related to possibilities of terminating dormancy or to responses to certain environmental conditions (PERRY, 1971). New information has been derived recently from the enzyme patterns that are typical of these stages.

In temperate and boreal areas the seasonal rhythm is, of course, an adaptation to unfavorable winter weather, particularly to low temperatures. Indeed, resistance against strong winter frosts is a prerequisite for survival in severe winters. The physiological mechanisms that enable a tree to survive extremely low temperatures are highly complex and are not yet fully understood (LYR et al., 1967); they are complex only on account of the complicated stress situation of the plant at such low temperatures (damage to cell plasma, winter drought, parch blight, etc.).

Initiation of shoot growth in spring is primarily controlled by the "temperature sum," i.e., the duration of preceding temperatures above a certain critical point

(Sarvas, 1966). Naturally, the tree's conditions of winter dormancy must also be considered. Even the conditions of the previous fall and winter may have some bearing (preconditioning). Yet the close correlation between temperature sum and shoot extension alone is often surprising (Sarvas, 1966).

Even a change in temperature from day to night may lead to specific adaptations as demonstrated in many phytotron experiments. For example, Kramer (1957) found that optimal growth of *Pinus taeda* seedlings is not achieved merely by favorable temperatures, but by optimal combinations of day and night temperatures.

The climate of regions with seasonal rhythms is characterized by large deviations from long-term means. At the same location, differences from year to year in precipitation, winter temperatures, length of the growing season, and the occurrence of early and late frosts are often much larger than between long-term averages of these variables observed in widely different locations. Thus in contrast to the climate of the wet tropics (and also subtropics) there is pronounced heterogeneity of the environment, even after abatement of the effects of seasonal cycles through special adaptations.

Such heterogeneity is reflected partly in adaptive strategies (Chapter IV), and partly, also, in special characteristics of the adaptations themselves. The frequent occurrence of late frosts at certain localities may lead to the accumulation of late flushing types or the evolution of late flushing local races. These may be limited in distribution to small areas, for example, to frost pockets. Probably many characteristics of forest tree populations in such localities will only be fully understood when annual variation of weather can be considered.

There are, of course, correlations between adaptations and long-term means of climate. Figure 3 depicts a correlation system of this kind where the clinal variation pattern of the dry matter content of Scots pine needles, an adaptive character, was investigated by Langlet (1934) for the first time in a large sample of provenances. Langlet interprets high dry-matter content as an adaptation to low winter temperatures.

Unfortunately, when interpreting conditions in nature, the relationship between climate (expressed through long-term averages) and adaptations has often been taken too far. Langlet (1935) has argued this at an early period. Admittedly, Lundegardh's (1931) statement is still valid: the primary task of experimental ecology is to discover those characters that were "chosen" by natural selection so that these may be investigated quantitatively to determine how adaptation works. In modern terms MacArtur (1967) explains this as follows:

"In fact, the parallel between niche and phenotypes is more than formal. Phenotype embodies all the measurements that can be made on an individual during its lifetime, including those measurements that constitute its niche. Thus phenotype

---

Fig. 3. Relationship of latitude and dry-matter content of one-year-old seedlings of 580 Swedish seed sources of Scots pine determined in 1931 in the nursery of the Swedish Forest Research Institute at Stockholm. The increase in the east–west direction depicted in the northern part is a consequence of topography. (After Langlet, 1934)

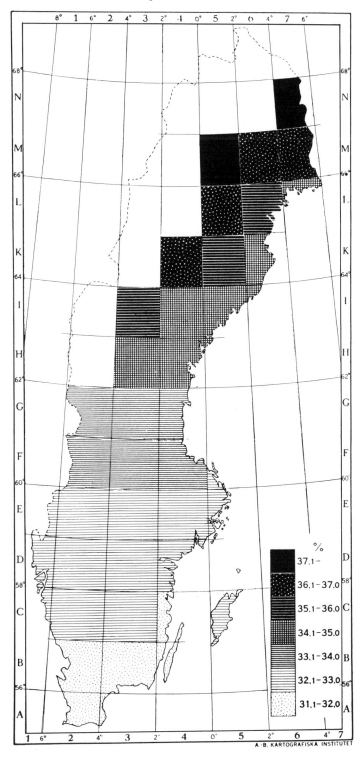

includes niche. And to the extent that all relevant phenotype parameters affect fitness, niche almost includes phenotype."

Clearly, factors other than long-term means of climate must be taken into account and the annual deviations from these are among them.

The length of the growing season is a common climatic factor. Depending upon initiation and cessation of growth, it varies a great deal geographically. In northern latitudes, the more northerly a location and the greater its elevation above sea level, the shorter will be its growing season. For a local race adapted to a shorter growing season this would imply a requirement of a lower – temperature sum for flushing and longer days for triggering the cessation in fall – in order to utilize the short growing season on the one hand but avoid the risk of early fall frosts on the other. These rules are valid indeed and geographic clines have developed, but locally divergent races exist that are adapted to the greater risk of late frosts. The dilemma of marginal populations in the far north has been effectively recorded by SARVAS (1966). Conditions in the high mountains should be similar. In the section on marginal populations (Chapter IV) we return to this subject. The necessity to adapt to the length of the period within which activity is possible at all is also observed in animals. To cite but one example: PENGELLY (1966) found that ground squirrels from the tundra of Alaska develop much faster than those from the southwestern United States.

What we have considered here is that adaptation depends upon the development of mechanisms controlling the initiation and cessation of annual growth. In forest trees a third phenological character is often important: the shift from early to late wood formation. Presumably its mechanism, too, is controlled through or by day length. Thus, in a stand of Norway spruce, LANGNER and STERN (1966) found a negative correlation between date of shoot extension and late wood development: the earlier the growth initiation, the smaller was the percentage of late wood. With independent inheritance of the three phenological characters — growth initiation, late wood development, and growth cessation — this could be expected. In fact, the first and the last character exhibited large and independent genetic variation. The adaptive significance of growth initiation has long been established; the role of cessation particularly in determining wood structure was first demonstrated by DIETRICHSON (1964).

There are limited possibilities to substantiate specific adaptations and thereby correlations among single climatic factors on the one hand and individual characters of the phenotype on the other. Such correlations have often been found (LANGLET, 1964), but their clear interpretations have frequently been impossible. Reasons are, first, the climatic differences among areas are complex and not measurable using a single niche dimension. Consequently, the adaptive significance of character differences becomes evident only within the relevant set of niche dimensions and not when related to a single dimension. Second, correlations exist among the characters subject to natural selection and to other characters that are independent of selection. The application of special techniques is required to separate, at least, the underlying causes of variation from the character complex, making it possible to recognize in broad outlines the relationships that may exist (STERN, 1964). Among studies of forest trees, MORGENSTERN's (1969) investigation of *Picea mariana* probably constitutes one of the most prominent examples. To cite another example: PHARIS and FERRELL (1966) found that coastal provenances of Douglas fir when subjected to artificial drought died before interior provenances.

Extremes of weather such as drought, snow breakage, and ice breakage are of outstanding importance in selection. The existence of the plant formation "forest" is only possible if a minimum of water is regularly available. Below this minimum, or with intermittent water supply, forest is replaced by other vegetation types such as the forest steppe or steppe. Populations of forest trees near such boundary areas often exhibit a special drought resistance that ensures their survival. The special adaptations that permit them to survive permanently, or at least enable them to reoccupy such marginal sites, have (like drought resistance) not yet been fully investigated (VAN BUIJTENEN, 1966).

Resistance against snow and ice breakage is a prerequisite for permanent survival of tree populations in many areas of the temperate and boreal zones. The basis of greater than average resistance may be found in the development of crown and branch shapes that prevent the accumulation of large amounts of snow and ice. In this way particularly, earlier writers explain the narrow growth habit of conifers in the far north and in high mountains. Thus the columnar type of *Picea abies* may be frequent in such areas, with branches drooping along the stem to let snow slide down. Yet wood properties may also play a role. Breaking strength and elasticity of wood probably determine to some extent breakage frequency of branches, crowns, and stems. There are no distinct differences between crown and branch types of *Pinus taeda* provenances from the Piedmont and the Coastal Plain. However, following an ice storm, in a seed orchard only 52% of the Coastal ramets remained undamaged, and 89% of those from the Piedmont (ANONYMOUS, 1971).

Generally, the adaptive significance of wood structure has been little investigated. DIETRICHSON'S (1964) study, cited earlier, is one of the exceptions. BRAUN (1963) could show that the different, known types of wood structure reflect the phylogenetic family tree of shrubs and trees. Evolution has resulted in more complex structures, partly through origin of new elements, partly through more purposeful arrangement of those already available (fibers, vessels, etc.). This is seen in relation to the many functions performed by the wood of trees: conduction of water and nutrients, support of the assimilating components, and storage of reserve materials. Against the background of these many functions BRAUN'S interpretation is intelligible.

Similarly diverse are the requirements that must be met by the form of crowns and branches. The assimilating leaf surface must be held in a favorable position to incoming light, making optimal utilization possible. In addition, maximum resistance against extremes of weather must be achieved, a competitive position must be maintained within crown space, flowers and fruits must be presented in an advantageous manner, etc. MITSCHERLICH (1970) has indicated optimum conditions for some of the relevant factors. But selection operates in a multiplicity of ways and opposing trends may develop. Therefore, a shape meeting all requirements cannot be attained. As a result, different provenances of the same species exhibit extremely different branch and crown types, as may be seen in Scots pine, which has now been studied in this regard for 200 years. Provenances from the Rhein-Main Plains in Germany have extremely wide crowns, are coarse-branched, and have crooked stems as a result of their strong phototropic reaction. In provenances from mountainous and northern and northeastern regions, stems are straight and crowns are narrow and fine-branched with weakly developed phototropism. Presumably the weights of individual niche dimensions under the conditions of the individual source are important. There may be convergence of crown types (and of other adaptations) among different species.

Thus, in extreme cases, spruces and pines of mountain areas look similar, and the species and genera of the upper stratum in tropical rain forests (RICHARDS, 1969) are so much alike that they can be distinguished only upon close examination. Figure 4 depicts character convergence in different species (and genera) using an

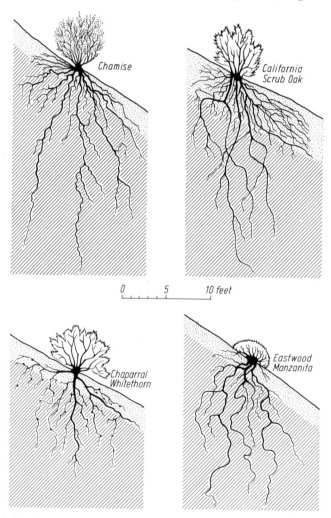

Fig. 4. Growth-form homology of characteristic dominant plants of the Californian chaparral. Chamise (*Adenostoma fasciculatum*), scrub oak (*Quercus dumosa*), chaparral whitehorn (*Ceanothus leucodermis*), and eastwood manzanita (*Arctostaphylos glandulosa*) all belong to different families. Figure taken from HELLMERS et al. (1955). (After MOONEY and DUNN, 1970)

example from an arid region. Striking cases of adaptation in arid regions have been known for a considerable period of time (WENT, 1949, and others).

Adaptations to climate have been the primary concern of classical genecology, which is considered to be 40 years old by HESLOP-HARRISON (1964) and 200 years old by LANGLET (1964; a detailed historical account by LANGLET is in press). TURESSON's (1923) ecotypes are climatic ecotypes if one wishes to retain this term. In their original description these types are morphologically distinct enough to classify them as subspecies. Naturally, there are also less conspicuous adaptations, for instance, the physiological adaptations denoted by LINNAEUS. Forestry science has made notable contributions in this field, which also constitutes one of the primary areas of concern of ecologically orientated forest genetics.

b) Adaptations to Soil

The second environmental complex making up the "site" of European forest ecologists is the soil. This category includes, for example, the sustained supply of minerals, hydrology (including ground-water, level, drainage, storage, capacity, etc.), aeration, concentration of hydrogen ions (pH value), and many other characteristics. That the components of the soil complex must be correlated with one another is recognized at once. Beyond this there are strong correlations between soil and climatic factors, for soil genesis, with all its effects upon individual soil factors, is significantly influenced by climate, i.e., by precipitation, temperature, and their seasonal variation. The strong intercorrelation of individual soil factors is the reason for the failure of many attempts to relate them with characters of tree populations that grow on different soils (VUKOREP, 1970, describes a particularly instructive experiment). Of course, the relevant adaptive characters are correlated among themselves.

Presumably soil influences should be reflected in characteristics of the rooting system, i.e., as adaptations. Thus one would expect adaptations to become evident in differences of gross morphology such as are found between species. Forest ecologists have long used such specific terms as shallow root, heart root, and tap root to denote major distinctions. They have claimed that a mixture of shallow- and deep-rooted species offers optimal utilization of soil capacity through reduced competition in the rooting sphere in the same way as the combination of fast-growing intolerant species with slower tolerant species is beneficial in crown space. There are, in fact, indications that similar differences in root structure exist among genotypes of the same population and that these differences contribute to competitive ability and competitive influence (see review by STERN, 1969).

Unfortunately, investigations of root systems are expensive and few have been recorded in the literature. Still, the study by BIEBELRIETHER (1964) indicates that such differences as exist among species (see KÖSTLER et al., 1968 for summary) may also be found among provenances of the same species. But an interpretation of the observed differences is not always easy because of the above-mentioned correlations among soil factors on the one hand and climatic factors on the other.

A unique role in this regard is played by the humus that is formed underneath forest stands. It has long been established that humus formation depends upon species and certain characteristics of the soil and climate. Thickness and condition of the humus are influenced in like manner. Soil structure and mineral content can be influenced by the forester to a lesser extent than by the farmer or gardener. As a result, forest ecologists have always paid considerable attention to the relationship of forest vegetation and soil development. Probably the stability of an ecosystem depends largely upon the interaction of vegetation and soil. The reciprocal dependencies in this regard primarily include the following: the nutrient cycle, soil genesis in relation to humus formation (degradation of the soil to podsol or other soil types), and the role of raw humus when assuming functions of nutrient and water storage. Thus, interactions of forest vegetation with soil include as participants not only climate, soil, and the root systems, but also properties of falling leaves, characteristics of the crown canopy (permeability to light and heat), and others. A detailed discussion is not possible here and we must refer to the literature.

Physiological characteristics of root systems have probably been a subject of research to the least extent, again because of the experimental difficulties encountered with the widespread root systems of forest trees. Investigations concerning mycorhizae are part of this subject, i.e., of fungi that live in symbiotic or parasitic relationship in or on the rootlets of forest trees. In some instances, as with *Alnus* species and legumes, mycorhizae perform the function of nitrogen fixation and storage. The genetic aspects of mycorhizae and of their hosts have not yet been investigated and these few remarks, therefore, must suffice at this time.

The capability to absorb and utilize mineral nutrients and trace elements also differs greatly among genotypes, as is known from breeding of farm crops. In our own experiments dealing with the marking of pollen by means of manganese, it was found that the natural manganese content of Scots pine in the same stand varies widely. Part of the reason appears to be unequal uptake of this trace element by the different genotypes. This, of course, is no surprise: wherever adaptive characters exist they exhibit genetic variation. Few rules of population genetics are as reliable as this one, and for this reason it seems easy to predict the results of some studies that will possibly be undertaken at some future date.

Since there is a scarcity of experimental results from studies of forest trees for reasons given above, it may be permissible to refer to findings on races on different soil types from studies of lesser plants. We shall choose initially the detailed and vivid investigations of British geneticists dealing with the evolution of grasses on mine dumps. High ion concentrations of heavy metals (lead, copper, zinc, nickel, manganese) that are normally toxic to plants characterized these sites.

JOWETT (1964) explored the lead tolerance of *Agrostis tenuis* on old mine dumps in England. He found a varying degree of tolerance at different sites. Distinctly different races with respect to lead tolerance were present within a small area. In spite of this, considerable genetic variation remained within the populations. One may conclude, therefore, that selection pressure was probably not very strong and variable (see IV.3.1. below). With increasing tolerance to lead these races acquired tolerance to low phosphorus and lime content; they also became morphologically more uniform and flowered earlier. *A. tenuis* is a perennial, capable of preserving proven genotypes through vegetative reproduction — without the process of recombination that is enforced by sexual reproduction. The same author (JOWETT, 1958) also discovered that lead tolerance in *Agrostis* species was coupled with heritable tolerance to copper and nickel.

MacNeilly (1968) found resistance against copper in *A. tenuis,* again on dumps of a small mine. In this case prevailing winds determined seed and pollen distribution. In spite of consistent migration in the same direction, transition from the nonresistant to resistant types in wind direction was limited to a zone of approximately one meter, i.e., it was very abrupt. In contrast, on the lee side, resistant types were found with gradually decreasing frequency, even in soil lacking copper. Thus, there was selection against resistant types but with much less intensity than on the mine dump area against the nonresistant types.

MacNeilly and Bradshaw (1968) confirmed these results from several copper mines in Wales and Ireland. A high heritability of copper tolerance was also determined. On soils with low copper levels selection was against tolerant types. Of special interest from the point of view of methodology is their deduction that results must be based on mature plants to obtain a clear picture.

These investigations reveal the complex nature of the development of edaphic races. This phenomenon involves selection intensity, periodicity of reproduction (through annual or perennial plants), migration rate, and correlated response to selection. Some familiarity with these and other factors is required before conclusions can be drawn. We shall later return to their influence and appraise them further.

Antonovics (1968 a, b) experimented with a population selected for tolerance and considered its reaction theoretically with respect to migration. He discovered that perennial forms develop tolerant populations more rapidly than annual forms (as was to be expected). Tolerant forms of *Agrostis tenuis* and *Anthoxantum odoratum* had reduced the probability of pollination by immigrating pollen by becoming more self-fertile. Genes for self-fertility would be expected to spread rapidly and linkage with genes for tolerance should be an advantage. The prerequisite in each case is strong selection pressure, but also dominance of the gene for self-fertility and recessiveness of the gene for tolerance (see Section IV.3.1 b).

In a theoretical study Jain and Bradshaw (1966) investigated the "seams" between the two soil types. One of their findings is that differences in generation interval also have a bearing. They include a new appraisal of the selection coefficient and of other variables. Of interest are their estimates of the selection coefficient of the genotypes on opposite sites: for tolerant forms on nontoxic sites these range from 0.09 to 0.99, 0.90 being very common; and for nontolerant forms on sites with a heavy concentration of heavy metal ions they range from 0.05 to 0.50.

Presumably, severe conditions and rigorous selection of this kind would foster the development of a crossing barrier as the most effective "adaptive strategy." Such tendencies have indeed been observed in parapatric (i.e., mosaic) distributions (Jain and Bradshaw, 1966) by MacNeilly and Antonovics (1968), Jowett (1964), and other authors.

An interpretation of such situations also requires an evaluation of the kind of genetic variation of tolerance as a character. As indicated earlier, perennials profit over annuals through their ability to reproduce without recombination. This is especially valid where tolerance is due to nonadditive gene effects. Dessureaux (1959) finds the genetic variance of manganese tolerance to be almost exclusively additive, leading to rapid progress in selection.

What about conditions where niche differences are not as pronounced, where differences in selection coefficients are relatively small? ELLENBERG (1958) has examined this problem with respect to soil influences in general, although with emphasis on species differences. KRUCKEBERG (1954) summarizes results from serpentine soils. These soils, although extremely different from soils of other geological origin still have a complex makeup in comparison with "normal" soils. Their peculiarity does not consist of a single decisive factor such as the content of ions of a heavy metal.

BRADSHAW (1960) and SNAYDON and BRADSHAW (1961) discovered differences in reaction of *Festuca ovina* sources from acid and from calcareous soils that are as great or greater than the differences described by ELLENBERG (1958) between species preferring calcium (calciphytes) and those that avoid it or can exist with small amounts (oxylophytes). However, in his *Trifolium repens* experiments with potted plants and field plots, SNAYDON (1961, 1962) did not discover any marked differences between sources from acid and calcareous soils. However, if these sources are allowed to compete on different soils, the source grown on its original soil is always superior to other sources. Similar results were obtained in experiments by SNAYDON and BRADSHAW (1962a), when testing certain sources of one species against those of other species, such as clones of the clover *T. repens* against sources of the grasses *Festuca* and *Agrostis*. Thus interspecific competition is a factor of selection that must be appraised in a similar way to intraspecific competition.

The same authors (SNAYDON and BRADSHAW, 1962b) have found in *T. repens* a correlation of 0.96 between phosphorus content of the location of origin and response to phosphorus content of the soil in experiments. Therefore, populations from soils with low phosphorus content are capable of greater phosphorus uptake and storage than populations from soils of high content.

These relationships are no surprise (MELCHERS, 1939) and many other examples can be found. KUMLER (1969), for instance, detected two extremely different edaphic races of *Senecio silvaticus*. The "dune race" grows larger with abundant nutrients and the "mountain race" is more sensitive to a change in those nutrients that are present in minimum amounts. PHARIS and FERRELL (1966) found distinct differences between seedlings from coastal and inland provenances of Douglas fir as measured by the date of mortality after exposure to drought, but Arizona provenances responded like coastal sources, as described earlier.

Thus, if we would try to answer VON SCHÖNBORN's (1967) question as to whether edaphic races of forest trees really exist, we would have to conclude, with him, that most experiments pertaining to this subject have suffered from inadequate methods. Furthermore, it is doubtful whether the question should be asked as to whether there are edaphic races in forest trees. Rather, the question should be whether or not there is any reason to think that forest trees should be assessed differently from other plant species. In forest trees, intra- and interspecific competition are important but provenance experiments including these factors are entirely lacking (STERN, 1969).

c) Adaptation to Competition

DARWIN indicated the significance of competitive ability in determining the fitness of individuals and populations, and HALDANE (1932) pointed out the consequent necessity to measure and interpret fitness under specified conditions of competition. This suggestion is particularly well understood by the forester, who cannot help but be impressed constantly by the presence of competition and its effects. A large part of his work consists of the regulation of competition by thinning to concentrate increment quantitatively and qualitatively on the best members of a stand. The decrease in the number of stems with increasing stand age as recorded in yield tables is a reflection of the intensity of competition in different age classes. As a result, the literature on the phenomenon of competition is rather voluminous.

The interest of the biological disciplines in competition centers on different aspects. Thus the physiologist will be primarily concerned with the effects of competition upon certain functions of an organism; the plant geographer in its role when determining the limits of distribution; the ecologist in its capacity to develop new niche dimensions or to change the optimal conditions of known niche dimensions; and the geneticist in its influence upon the selective value of genes, the fitness value of individuals, or the adaptive value of populations. It is therefore not surprising that the word "competition" has had a variety of meanings, especially because proper or improper parallels in human society are close at hand, such as competition between individuals, firms, and groups (COHEN, 1966). In this work, competition is defined as the sum of all mutual influences among trees growing in a stand. Thus the net result of all competition effects upon a given individual may be positive, negative, or neutral, depending upon the prevailing influence.

Mutual influences imply that every individual is capable of exerting competitive influences upon other individuals as well as withstanding their competition. The first complex of characters has been termed competitive influence, and the second competitive ability (see STERN, 1969 for review; the considerable literature on the subject cannot be cited and evaluated at this time). One would expect strong and positive correlations between both complexes, but this is not always so, as shown particularly by SAKAI (1961). His "genetic value" is approximately equivalent to our competitive ability. It should also be noted that competitive influence and competitive ability of a phenotype, etc., are not permanent characters, but characters valid only for certain situations — i.e., for competition with a given phenotype or groups of phenotypes. Furthermore, environmental conditions other than the competitors also play a role. Thus competitive ability and competitive influence can only be fully understood when viewed against the background of the environment where measurements were taken.

Competition in a more restricted sense, (contention for some limited resources) may occur in all developmental phases of trees and include all resources that individuals of the same or different organism strive to utilize. Such resources can be light (as an energy source), water, and mineral nutrients, but also (even in plants) an area component of the environment, since every individual requires a certain space for its full development. Consequently, competition must be detectable, if not in all, at least in most, niche dimensions, and hence should have a bearing on several dimensions of the niche of a population. This is discussed further in Chapter V. 2a.

The niche of a population includes the environment of all developmental stages. This is particularly important for organisms such as insects, which have discrete niches in the individual metamorphic stages. Here the niche as a whole can only be

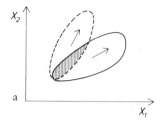 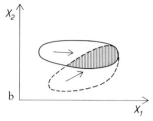

Fig. 5 a. Arrows within the niches indicate the shifting direction of niche occupancy with time. Overlapping occupancy and competition resulting from this is limited here to the earlier developmental stages. This would apply to two species, one of which would grow into the upper stratum of a multi-storied forest, the other into a lower stratum. b. The niches of the two species overlap in the later developmental stages. In spite of this, there may be no competition (see text)

represented by several nonoverlapping hyperspaces, each being typical for a developmental stage. With trees, developmental stages are less distinct, but comparisons of adaptations in different stages indicate clear differences at several niche dimensions. Thus needles of young seedlings of *Abies, Picea,* and other genera are adapted to much lower light intensities than needles of older trees. Niches of such trees taken over all developmental stages could probably be conceived as a continuum in niche space, certain parts of the continuum being typical of specific stages.

Figures 5 a and b represent hypothetical examples. The first case deals with two species having overlapping niches in early developmental stages. Therefore, competition is limited to these stages, perhaps the seedling and sapling stages. This may apply to two species in a multistoried forest with many species, where their seedlings compete equally well under certain light and moisture regimes, and then grow into different strata of the forest and compete no longer. Figure 5 b depicts the opposite case. Niches of the species overlap in the later stages. However, because the decision concerning the niche component "area" has already been made while the species occupied and developed on site, competition may not arise. Thus, in the first case, at overlapping niche dimensions, species adaptation should be very similar and differ at later stages. In contrast, in the second case the same or very similar optima in all or several dimensions lead to the expectation of similar adaptations in later stages. This is a case of character convergence, i.e., a development toward the same optima in the primary adaptations. RICHARDS (1969), in an example discussed earlier, describes this situation from a West African rain forest, where the tree species of the upper stratum are hardly distinguishable in leaf and crown characters although they belong to different species, and even genera. Such convergence in direction of the same optimum is to be expected when, as apparently in this case, niches of the species are the same in the last developmental stage. Thus competition is lacking and hence a need for niche differentiation does not exist.

Competition among individuals of the same species and among those of different species need not necessarily lead to the same or similar requirements in all competition-oriented niche dimensions. Since intraspecific and combined intra- and interspecific competition usually leads to the development of adaptive strategies, these will be discussed further in Chapter IV. It is sufficient to mention here only some of the many experiments related to this problem. For example, MOORE (1952) found that in competition experiments with one line each of *Drosophila melanogaster* and *D. simulans,* one of the lines was eliminated after several generations in nearly every replication. In one case, however, after initial inferiority, the normally weak population prospered and eventually replaced the normally strong population. Repetitions of the experiment, using the one surviving line of the inferior species and the initial line of the other species, demonstrated that a special line of the former had developed that was now superior in competing ability to the latter. This and many other experiments once again substantiate the validity of DARWIN's and HALDANE's assumptions concerning the importance of competition in the adaptation of populations. In higher plants, SNAYDON (1961, 1962) found that populations ("sources") from various soil types differed greatly with respect to competition and were always superior on their native soil. In other experiments (for example, AYALA, 1966, BARKER, 1968) it could be shown that with increasing population size of the initially superior species, and thus with more and more intraspecific competition, the competitive superiority of this species disappeared and was not recovered until population size fell to the point where interspecific competition was again much more frequent and predominant. Here equilibrium conditions in competition may develop, that may fluctuate in time or reach a steady state to guarantee the coexistence of both species, conditions such as are observed in the climax stage of mixed forests. In these cases one of the prerequisites is niche differentiation for certain stages of development. However, with site heterogeneity, "selection for mutual avoidance" (ASHTON, 1969, and others) may take place. "Selection for mutual exclusion" (RICHARDS, 1969) is another, preferable designation. In such cases competition limits distribution to certain sites where a species competes successfully, and leads to its disappearance from other sites that it could occupy but where it is replaced (TANSLEY, 1917, MOORE, 1959 for an example in *Eucalyptus*).

Much is known about the existence of competition and its importance regarding adaptation, but little is known about the adaptation itself. The best investigations concern the adaptation of seedlings. Here considerable differences exist that already were known to the older forestry scientists and led to the distinction between light-tolerant and light-intolerant species. Seedlings of *Abies alba* are capable of existing for decades in the shade of old trees where the light conditions would not even permit the full development of one-year-old seedling of other species. Furthermore, the tree species of different strata typically have divergent light requirements, just as they occupy different positions in succession. We also know that the root systems of different species (or even different genotypes of the same species, see LEE, 1960; and STERN, 1969 for summary and literature) allow a diversified utilization of soil nutrients and water. Yet it is difficult to obtain a reasonably complete picture of the physiological basis of competition-related characters among genotypes of the same species or of different species, especially in forest trees. Reasons for this difficulty are not only the complexity of competition-related characters resulting from many niche dimensions and the pertinent adaptations that need to be investigated, but also the necessity to account for changes of the niche during the different developmental stages, and last, of course, the purely technical difficulties with such experiments.

Corresponding conclusions must be drawn on the subject of positive influences among neighboring trees. Although mutual protection against wind at the forest's edge or in areas of extreme exposure is distinctly recognizable at the crown canopy, there are other interesting questions that remain unexplored, such as the positive mutual influence among individuals of different species or among different genotypes of the same species — with the possible exception of simple cases.

A last complex of mutual influences that should be mentioned at least is the so-called allellopathy. Birches in stands of *Robinia pseudoacacia* (GÖHRE, 1952; KOHLER, 1963) die possibly as a result of substances produced and released by *Robinia*. MERGEN (1959) found phytotoxic substances in the leaves of *Ailanthus* that may prevent plant growth underneath trees of this species. WEBB et al. (1967) discovered a high production of phytotoxin in the leaves of a nongregarious species of the Australian rain forest that would make it impossible to grow the species in pure stands. Phytotoxic substances have also been reported from a walnut, *Juglans regia*. More examples could be given, but instead the reader is referred to the literature. All such results indicate that some species are capable of improving their competitive position through "chemical warfare" by killing or retarding growth of their competitors.

d) Resistance to Diseases and Damaging Animals

VAN DER PLANK (1968) distinguishes "vertical" from "horizontal" resistance: "When a variety is more resistant to some races of a pathogen than to others the resistance is called 'vertical' or 'perpendicular.' When the resistance is evenly spread against all races of the pathogen it is called 'horizontal' or 'lateral.'" Evidently the meaning of these terms is somewhat analogous to "general combining ability" versus the "specific combining ability" of the plant breeder, or to HARPER's (1963) "ecological combining ability," which again can be "general" or "specific." We will be satisfied with these definitions: if a genotype is generally resistant to many genotypes of the pathogen, then this genotype is generally better adapted; populations that introduce specific resistance of certain genotypes against certain genotypes of the pathogen play an adaptive strategy.

The evolution of parasitism is coevolution, like the evolution of systems of competing species. To be successful, a parasite must be precisely adapted to a complex of physiological and morphological properties of the host, as investigations of specific resistance and virulence have indicated (VAN DER PLANK, 1968; WILLIAMS, 1964). Every new host genotype that is resistant, or more resistant to the races of parasites, possesses a selective advantage and increases in frequency. It also sets in motion a process of selection in the parasite: every new parasite genotype that is capable of surmounting the newly developed resistance barrier has a selective advantage and accumulates in the parasite population. Resistance of this kind can originate in every phase of the adaptation of the parasite to the metabolism, development, and morphology of the host. Individual genes causing divergent behavior or the host in one of these phases may be capable of upsetting the development of the parasite. On the other hand, a single gene of the parasite may be sufficient to eliminate the newly acquired resistance. In adapting itself to the host, the parasite population opposing vertical resistance may undergo stepwise development controlled by many genes: for every resistance gene in the host there may be a gene for virulence in the parasite. This is the basis of the gene-for-gene hypothesis advanced by FLOR (1954). As a matter of fact, it has been discovered that there are hundreds of genes for virulence and resistance in host-parasite systems.

Therefore, the relationship between host and parasite can be disarranged in many ways (as classified by WILLIAMSON, 1964):

(1) through escape, for example, when larvae of the European green oak leaf roller moth, *Tortrix viridans*, which depend upon young leaves, hatch on late-flushing oak trees;

(2) through the inability to reach the host tissue because of, for instance, the presence of toxic substances, a heavy cuticle, prevention of germination of spores or misdirected growth of the germ tube, or blockage of natural openings such as stomata and lenticells where the germ tube normally enters;

(3) through the parasite's failure to develop in the host tissue after entry as a result of biochemical inhibitors, anatomical specialities, or hypersensitivity causing death of the host tissue and with it the parasite around the area of infection. Examples for nearly all of these possibilities may also be found in forest trees (GERHOLD et al., 1966).

The development of immunity after infection, such as is common in animals subject to infectious diseases, is comparatively rare in plants. Genetic mechanisms for the development of antibodies against infections by microorganisms are considered true adaptations of the host. Here, too, adaptive strategies may often evolve, such as the biochemical polymorphisms of blood and serum groups. A true adaptation of the host would also be the development of tolerance, i.e., of insusceptibility against damage.

Resistance can result from morphological mechanisms that are developed especially for this purpose, or it may be the by-product of a genetically conditioned change of a host character that accidentally affects the adaptation of the parasite.

For the host population the development of a system of general resistance is always an advantage. As a result of genetic variability on both sides, these cases again include genetic variants of the host with weaker and stronger resistance, and of the parasite with lesser and greater virulence. The answer to the question as to whether in a specific case there is present high general resistance or an interim stage of temporary high specific resistance is entirely a matter of experience, as indicated by VAN DER PLANK (1968). For our forest trees growing in their natural ecosystem, one may assume that high general resistance prevails against the principal parasites and other damaging agents. This view is reinforced by observations on introduced diseases. For example, *Cronartium ribicola* occurs on the five-needled pines of Europe and Asia, where it does not cause appreciable damage. On the five-needled pines of the north American continent, however, it has become a very harmful disease. Even here less susceptible, preadapted genotypes are found that may constitute the basis of selection for high general resistance. Yet it must be realized that the high general resistance of European and Asiatic pines is also accompanied by signs of specific resistance (genetic variation in both virulence and resistance). The best strategy for the breeder wishing to accelerate the process of selection for high general resistance (if resistance can be obtained) might be to work with a broad genetic material that alone can achieve an equilibrium in the host–parasite system at a high level of general resistance (STERN, 1969).

Every host-parasite system is subject to environmental influences. For example, *C. ribicola* is not found in some areas of North America for climatic reasons. The resistance of Douglas fir against *Rhabdocline pseudotsugae* is determined in Europe equally by provenance (genotype) and environment, whereas the disease is harmless where Douglas fir is native. Even old host–parasite systems, like those of *Pinus silvestris-Lophodermium pinastri*, are subject to fluctuations as a result of annual changes in weather. Populations density of the host, its occurence in pure and mixed stands, is not without influence on the parasite. In some cases low population density of the host and a consequent low infection probability appear to be the only way out for the host. Thus epidemiological and other ecological factors must be taken into account when judging a host–parasite system.

This is particularly evident in the case of harmful forest insects. The dynamics of insect populations appear to be largely independent of genetic properties of the host, being determined by other factors such as frequency of enemies, the weather, etc. As a rule, however, adaptations of the host population exist even if only in the form of physiological mechanisms that provide sufficient tolerance to insect feeding. In addition, active mechanisms are probably found, such as resin flow to kill bark beetles. In many circumstances the populations of harmful animals in turn are controlled by enemies, diseases, or extremes of weather.

In all harmful animals, behavior patterns play a role that may be genetic in origin or acquired by learning. Studies in this field are only beginning. It becomes more and more evident, however, that forest ecosystems can only be understood when the role of animal species is fully appreciated. For plant species, animals are important not only as damaging agents but also as pollen vectors (therefore, for an important part of the genetic system), for seed dispersal, and for the breakdown of organic matter. Here and in the preceding section it is very clear that the coevolution of species in an ecosystem cannot be neglected. Furthermore the genetic variability of all the species involved and the resulting evolutionary possibilities and tendencies imply that in addition to adaptations, adaptive strategies must also be considered.

e) Special Adaptations

One of the factors of decisive importance for the survival of a tree population is often the capacity to persist in the presence of forest fires. In many areas the composition of vegetation is significantly influenced by forest fire frequency (see Section II.1a, GILBERT, 1959; SWEENEY, 1956, and others) because of differential resistance of plant communities to forest fires or capacity to regenerate.

Tree survival following fires is achieved in several ways. A heavy bark, which may be difficult to ignite, frequently offers sufficient protection against ground fires. The ability to regenerate the leaf area on stem and branches after crown fires is also important. Adaptations to fire exist in some species in the normally most susceptible juvenile stage. An example is the grass stage of *Pinus palustris* and of other pine species. During their earliest years, seedlings of these species first develop a strong, deep root system while growing little in their top portion. After formation of a robust root system, height growth sets in at a comparatively more rapid rate. During the grass stage, fires cause little damage to seedlings.

To secure regeneration of the population after forest fires, root suckers for vegetative propagation or special protective mechanisms for seed are particularly common devices. The latter may be perfected to the extent that germination of the seed may be possible only after a forest fire, as, for instance, in some pine species. Here resin bonds may keep cones closed so that a seed is not released without heat treatment. Disregarding exceptional cases where cones are opened and seeds released mechanically by squirrels or woodpeckers, such species actually cannot regenerate without forest fires.

The importance of these mechanisms for adaptation is emphasized by the observation that the same species possessing serotinous cones in areas where the incidence of forest fires is high bear normally opening cones in others. In fact, serotinous cones would prove to be almost fatal for the population if forest fires ceased for any length of time. The fitness value of individuals with serotinous cones then drops nearly to zero, whereas individuals with nonserotinous cones will have a value of nearly one. Such extreme differences in fitness between two genotypes induce the development of polymorphisms in areas where regeneration does not occur exclusively after fires, but both in the presence and absence of fires. Jack pine, *Pinus banksiana*, for example (see IV.3c), possesses populations with exclusively serotinous cones, populations without serotinous cones, and also populations that are polymorphic for this character. *P. clausa* displays the same polymorphism in the population around Juniper Springs, but its western populations consist of trees that all possess cones that open normally (T. O. PERRY, personal communication). LEDIG and FRYER (1971) find 100% serotiny of the *P. rigida* individuals in the Pine Plains of New Jersey, and decreasing percentages with increasing distance from this part of the species' range, which appears to be particularly subject to forest fires. LUTZ (1960) could show that in areas with frequent forest fires, species of the genus *Picea* also have the capacity to develop cones similar to those of the fire species in the genus *Pinus*. *Picea mariana* in Alaska and other regions is an example of this. Of course, not all polymorphisms of cones are related to forest fires, such as the cline in cone morphology described by SCHOENIKE et al. (1959) or the much-cited polymorphism of the cones of *Picea abies*, which will be discussed later.

Adaptations to forest fires should offer particularly interesting possibilities for ecological-genetic research with forest trees in the next few years. This subject is discussed further in Chapter VI.

It has already been stated that wood structure should reflect adaptation to different environments both qualitatively and quantitatively. Wood properties are subject to natural selection, directly through their functions (support, conduction, storage, and gas exchange), and indirectly through correlation with other characters, e.g., phenological characters (LANGNERS and STERN, 1966; DIETRICHSON, 1963). Experiments of forest tree breeders indicate that some of the characters most important for the functions of wood possess great variability and therefore could be changed rapidly and drastically by selection (see HATTEMER, 1963 for review and additional literature). Is seems to be clear that wood functions would also be changed, although the direction of change is hardly predictable at this time. All things being equal, selection for a high proportion of latewood in Norway spruce, for example, would reduce the proportion of earlywood fiber with large cavities specialized for conduction of water and mineral nutrients. The age trend in arrangement of wood characters, known since Professor HARTIG's studies, may have a similar explanation: an increasing proportion of the supporting tissue with greater tree age and crown size secures more protection against wind, snow, and ice breakage, not only because of a higher proportion of these tissues in the tree stem but also because of their more favorable arrangement with respect to statics.

Regrettably, most investigations of wood biologists in the past have favored the descriptive approach and neglected the genetic and evolutionary questions. Here, too, many problems exist, the exploration of which is interesting from both theoretical and practical viewpoints and could contribute to a better understanding of the causes of genetic diversity of wood properties (BANCROFT, 1930; DUFFIELD, 1968, and others).

In higher organisms like trees, the whole course of development must be adapted to the changing requirements of the environment. Therefore, WADDINGTON (1957) contrasted "homeorhesis" of total development with "homeostasis" in distinct developmental phases. In the same context, THODAY (1953) distinguishes "phenotypic flexibility" from "developmental homeostasis." Developmental homeostasis is defined as the irreversible modification of organs or physiological processes, the specific formation of which permits optimal adaptation in a specific stage of development. Examples are juvenile forms of leaves and other organs in some tree species, the appearance of reproductive organs late in life, etc. The phenomena of cyclophysis and topophysis, familiar to horticulturists and tree breeders, are also consequences of developmental homeostasis. The terms refer to the special behavior of plants grown from cuttings taken from trees of different ages, taken from different parts of the crown, or taken from branches of different order, the plants still reflecting through their morphological and physiological characters the special nature of the part of the donor tree. However, it is not certain that effects of cyclophysis and topophysis are irreversible. In some cases, as in cuttings of lateral branches of *Araucaria*, the effect of topophysis appears to be maintained permanently, but in other cases it diminishes gradually in the course of development of the plant. Yet, in the final analysis the criterion of irreversibility is also not decisive for the interpretation or demarcation of developmental homeostasis. CROWE (1954) summarizes the consequences of life history phenomena at the population level and SCHAFFALITZKY DE MUCKADELL (1959) presents a similar review for the purposes of silviculture.

Phenotypic flexibility is defined as the reactions of the developing organism to special qualitites of the environment in certain developmental stages, reactions which, in turn, may influence subsequent stages. An important part of phenotypic flexibility is tolerance of extremes in environmental variables.

A particularly instructive case of this kind is described by BORMANN (1965) for *Pinus strobus*. If a tree is checked in growth and descends into the "suppressed" class, xylem formation decreases first. In the last stage, xylem formation at the stem base ceases entirely. Finally, even production of phloem is terminated. In spite of this, transport of materials is still possible. In some instances suppressed trees may be nurtured by their neighbors by means of root grafts, as indicated by LAESSLE (1965) for *P. clausa*. The process described by BORMANN may extend over a long time period. He found that suppressed trees still managed 50% of total height growth in the second half of their 60-year-long life period, but only 19% of diameter growth. There is no doubt that phenotypic flexibility of this kind is of great adaptive value; it permits trees to survive for a long period when suppressed, and to await a more favorable opportunity for development.

The importance of phenotypic flexibility for adaptation, or, more generally, for the evolution of populations, has been outlined by BRADSHAW (1965). In populations adapted to a heterogeneous environment, more phenotypic flexibility is to be expected right from the beginning than in highly specialized organisms. For example, in forest trees, weedy (pioneer) species should grow and regenerate over a larger range of environments than species found near the climax stage. This expectation is by and large supported by experiments.

Obviously, a population or species has at its disposal other means to cope with a heterogeneous environment, especially its "genetic flexibility." This is defined as its capability to react to selection, to adapt to new environments by changing its genetic composition. LERNER (1954) considers especially the case where the diversity of genetic variants is maintained, in spite of selection. He finds a number of possibilities to achieve this and summarizes them as mechanisms of genetic homeostasis of populations. Of course, such mechanisms have a bearing on adaptation as well, but since they are based upon distribution and conservation of genetic variability, they may be considered as adaptive strategies and treated separately in Chapter IV.

At this stage it has become clear that the distinction between adaptation and adaptive strategies is more or less a question of expediency. A clear distinction probably cannot be drawn, although in an experiment or in a discussion more emphasis can be given to the one or the other aspect of adaptation: adaptation to a relatively constant niche or to a heterogeneous environment. It may also have become obvious that the net adaptive value of a population is determined by many components, which may be independent of each other or correlated (See HASKELL, 1961 for examples of character correlations of seedlings and adult plants). All of these components become effective during shorter or longer development phases of the organism. Only a few have been selected for this discussion, namely, components reflecting simple conditions that we meet again and again in forestry literature.

2. Reproductive Cycle

Adaptations of the vegetative cycle and its phases have long been the primary subjects of forest ecological research. The regeneration of natural and managed forests, which includes seed dispersal, conditions for germination, and establishment of seedlings, is part of this field of inquiry. The processes preceding these stages, such as flowering, fruiting, and their adaptations, are not as well known. Most of the studies were carried out in the Scandinavian countries (Sweden, Finland, Norway) where their northern forests have long been under intensive management. Here the conditions for flowering and fruiting are extremely unfavorable.

In contrast to these research efforts in the north, very little has been done in the south and our knowledge of the flowering and fruiting of tree populations in the subtropics and tropics is very sketchy.

Flowering, fruiting, and seed dispersal of tree populations are important for an understanding of genetic systems of populations. Therefore, in recent times these phenomena have been primarily the concern of forest geneticists. Studies included problems that, although of no special interest to classical forest ecology, appeared in a new light with increasing knowledge of the biology of forest tree populations and forest ecosystems. This is valid, for example, for the distribution of seed and pollen as factors determining population structure, for the rate of cross-pollination, the index of recombination, etc. The work of forest biologists concerned with these problems is becoming more and more affected by the progress in general genetics, and particularly in population genetics. This work in forest biology not only has a theoretical aspect, but can be utilized for practical forestry.

a) Adaptations to Climate

Physiological processes related to flowering and fruiting are of necessity adapted to the regional climate where evolution occurred. This adaption is often indirect, as in species of higher plants in subarctic regions where pollination by insects is absent and consequently the mating and breeding structure of populations differs. However, under the heading "Adaptations to Climate," we shall discuss only some adaptations to seasonal cycles. Here again, conditions in climatically extreme areas offer the best examples.

The dates of flower appearance, blooming, and seed maturity indicate important stages in the phenology of a plant species. Although they differ for various species at the same locality, these dates are considered adaptations to the climate of the locality where the population occurs and adjustments to the special conditions of that locality. Yet, even the time period when meiosis, or a certain stage of meiosis, takes place represents a phenological stage that may determine in part the success of adaptation of the generative cycle. Thus the complete development of the generative organs and of the seed must be adapted to the seasonal course of weather, by being programmed suitably. Since the male (staminate) and female (ovulate) flowers of some species are separate, or even appear on different individual trees, the developmental cycles of male and female sexual organs must be synchronized, at least in the final stage. Furthermore, in cross-pollinating species all individuals should flower simultaneously, or the flowering periods should overlap to such an extent that chances are maximized for the pollination of each female flower by pollen of a different plant.

Consequently, ANDERSSON (1965) has examined the influence of climate in three stages: the year of bud formation, the year of flowering, and the year of seed ripening. Naturally, this applies only to species whose seeds ripen one year after flowering.

In subarctic and boreal climates not all species flower and bear seed every year. TIRÉN (1935) has investigated this, particularly with respect to Norway spruce. He assumes that a physiological threshold value exists that, when exceeded, switches development of primordial tissue toward a flower bud. He further assumes that conditions for the formation of male and female flower buds are not necessarily the same. The physiological threshold values should be codetermined by the environment of the year of flower bud formation, i.e., in general the year before flowering.

Table 1. Distribution of female flowers in six full-sib families of *Betula pendula* resulting from a diallel cross of four early flowering trees, rated in 1961 (upper line) and 1962 (lower line). Flowering intensity is given in six classes: class "0" denotes the percentage of trees without female catkins, class "1–20" the percentage of trees having 1–20% short shoots with female catkins, and so on. (STERN, 1963)

Cross	Intensity class (%)					
	0	1–20	21–40	41–60	61–80	81–100
	% of trees with female catkins					
66 × 67	5.2	16.5	30.4	18.3	19.1	10.4
	0	0.9	2.6	20.0	47.0	29.6
66 × 68	5.2	20.8	10.4	25.0	26.0	12.5
	0	0	0	17.7	40.6	41.7
66 × 69	7.3	19.3	28.4	22.9	10.1	11.9
	0	0.9	0	1.8	22.0	75.2
67 × 68	9.8	20.5	31.2	25.0	8.0	5.4
	5.4	4.5	14.3	37.5	27.7	10.7
67 × 69	5.0	21.7	32.6	21.7	14.2	5.0
	1.7	2.5	5.8	26.7	40.8	22.5
68 × 69	0.9	6.1	16.7	24.6	24.6	27.2
	0.9	2.6	2.6	14.9	35.1	43.9

TIRÉN's expectations, based on observations in Norway spruce, also appear to be relevant for genetic aspects of flower formation. STERN (1963) found that in families of *Betula pendula* resulting from crosses of selected trees, the tendency to form male or female flowers was inherited fairly independently (see Table 1). Some families formed almost exclusively male or female flowers, others both types. When counting the proportion of flower buds on each tree — for female flowers the proportion of short shoots with flower buds, for male flowers the proportion of the terminal buds on long shoots with flowers — distributions were found for individual families that corresponded to those to be expected from TIRÉN's concept (distributions truncated at the critical threshold value). Selections for early flowering were obtained from the same birch material. It became apparent that after only a few generations, extreme types were obtained that developed flower buds only and therefore were unable to survive (STERN, 1961). Another of TIRÉN's observations — that in some years female flowering and in other years male flowering was favored — could also be confirmed.

The three factors probably considered most important for flowering in the literature are flower production in the year preceding bud formation, summer temperature and summer drought during the year of bud formation, or a combination of all three. There are, of course, possibilities of influencing flowering experimentally by other means, but these are of less interest here.

TIRÉN (1935) presents long-term statistics of cone production in Norway spruce stands in Sweden (Fig. 6) indicating a fairly regular cycle. Superimposed on this cycle are special effects, presumably mainly those caused by summer temperatures. Several authors have noted that flower and cone production should curtail wood increment but DAUBENMIRE (1960) could not confirm this.

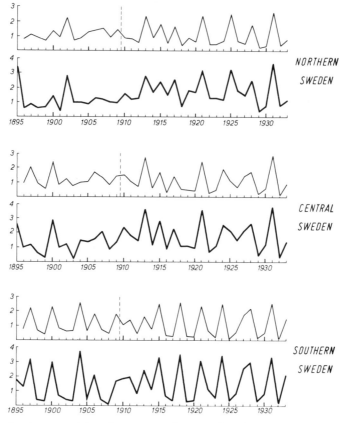

Fig. 6. Flowering (upper scale) and cone crop (lower scale) of Norway spruce in northern, central, and southern Sweden. (After TIREN, 1935). Flowering was estimated up to 1909, and assessed more precisely in experimental stands from 1910 on

Soon after the first stages of development toward the flower bud, meiosis takes place; in the male flowers of *Betula pendula* this occurs as early as June. Meiosis represents one of the significant stages in the formation of generative buds and must also be fitted to the course of seasonal weather. In Norway spruce at Gällivare (northern Sweden), ANDERSSON (1965) found disruptions in meiosis of pollen mother cells at temperatures of $-4°$ C, leading to the formation of abnormal, nonfunctional pollen grains. In this way he explained the defective nature of pollen often found in the northern part of the species' range. The short growing season, to which Norway spruce is adapted in its northern range, forces the species to initiate flower formation early in spring and to risk loss of a part, or all, of the gametes (both male and female) through late frosts.

The short growing season in these regions also leads to incomplete seed maturity because of the early onset of winter. Thus the low quality of Norway spruce seed from northern regions (and high elevations in mountains, where the same conditions prevail) is principally brought about in two ways: a low quantity of functional pollen leading to a high percentage of empty seeds, and poor seed maturity resulting in a low percentage of germination. The following conclusions of ANDERSSON (1965) are of interest to forest ecologists and geneticists from both a practical and a theoretical viewpoint:

"In Sweden, hitherto, when selecting plus trees ... one has primarily taken into consideration characters such as growth rate, stem form, branching habit..., wood quality..., and as far as possible, the resistance and hardiness during the tree's vegetative phase. On the other hand, the generative fitness of plus trees to the climatic conditions at their habitats has not been systematically examined during either the selection of plus trees or the composition of seed orchards. This should not be necessary in regard to seed yield per cone and seed germination capacity in all climatic zones in Sweden. The generative adaptation of the trees, however, plays an important role in those high level areas in Northern Sweden (particularly about 300 m and above), which have biological conditions for natural regeneration. Therefore, if it is desired that the forest plantations, raised from orchard seed, should have a better ability to generate naturally than that of the existing natural stand, then it is very important from the point of view of reproduction, to take into consideration, among other things, the reproductive fitness of the trees when selecting plus trees for the composition of seed orchards."

In his investigations the author in fact discovered large genetic variation in seed maturity. Little is known concerning individual genetic variation of damage to pollen mother cells, yet the causes of such damage have been considered (ANDERSSON, 1969). This author moved scions of Norway spruce into different elevational zones and studied the degree of frost damage as indicated by the percentage of damaged pollen mother cells. Again, an increase in damage with lower temperatures was noted.

ANDERSSON et al. (1969) reviewed the literature on meiotic irregularities in conifers. Irregularities occured in 15 species of six genera after exposure to either high or low temperatures. There was agreement that such irregularities could be expected in certain stages of meiosis, which, accordingly, are especially important for adaptation.

Concerning this point, interesting results are available from introduction trials with species of *Larix* (EKBERG and ERIKSSON, 1967; ERIKSSON et al., 1967; ERIKSSON, 1968a, b). These authors studied meiosis in pollen (microspore) mother cells of *Larix decidua*, *L. leptolepis*, and *L. sibirica*. The result for the experimental area at Stockholm is shown in Fig. 7.

The following differences are observed among the three species. In *L. sibirica* meiosis is initiated earliest; the diplotene, diakinesis–telophase II, and first microspores are observed first. In *L. decidua* the first stages of the pachytene are observed somewhat later, as are the diplotene, the diakinesis–telophase II, and the first microspores; the time period over which diplotene, diakinesis–telophase II, and first microspores are observed is shorter. In *L. leptolepis* the pachytene stage begins at an intermediate period and is limited to a much shorter time interval than in the two other species; the first stages of the diplotene appear late and the last ones are again observed later than in the two other species; diakinesis–telophase II occur approximately at the same time as in *L. decidua*; the first microspores are observed at an intermediate period.

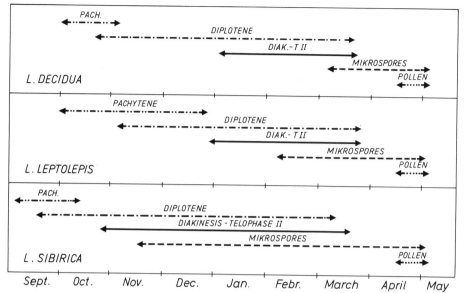

Fig. 7. Seasonal course of meiosis in the male strobili of three species of *Larix* near Stockholm. (From ERIKSSON et al., 1967)

Of course, the experimental material serving as the basis for the results of these authors was limited. In spite of this, they found one clone that demonstrated the importance of individual genetic variation for the adaptation in these phases of the generative cycle (compare tables given by ERIKSSON, 1968a). Thus the relationships shown in Fig. 7 merely give a rather general picture. Further findings are presented in the additional papers of the authors cited above, such as comparisons of results obtained over several years at different localities in Sweden, including comparisons of climate, demonstrating a distinct temperature dependence of the percentage of sterile pollen. Damage patterns of sterile pollen were rather uniform (Fig. 8). All active phases of meiosis are susceptible to low temperatures. According to ERIKSSON (1968a), the prerequisites for regular pollen formation are attainment of the diplotene stage before the first fall frosts, stability of the resting stage (diplotene) against temperature fluctuations between $-5°$ C and $+5°$ C, and rapid development of the tetrad stage after passing through the diplotene. The large percentages of sterile pollen in all three species at unsuitable plantation sites (up to 100%) indicate the importance of this kind of adaptation in the general fitness of a population, e.g., for the introduction of exotics. In another example, TIMOFEEF (1965, oral communication) observed differences in the flushing date in an introduced provenance of *L. sibirica* near Moscow of six days after the first generation and four days after the second generation. This indicates the possible importance of correlations among phenological characters; a direct selection for date of flushing is unlikely.

The preceding examples dealt with exotics that, although adjusted in their vegetative phase to foreign climates at the new site, were not adapted generatively. Thus, it is somewhat surprising that ANDERSSON (1965) labeled the native Norway spruce of northern Sweden as being rather poorly adapted in its generative cycle, as it apparently suffered under the same difficulties as the species of *Larix* mentioned above, with regard to meiosis as well as to seed maturity. He considers it possible to improve the degree of adaptation through selection; i.e., he assumes that optimal adaptability has not been reached in the native population. Reasons for this could be the relatively small number of generations passed since immigration into northern Europe or steady migration from southern regions, especially through pollen flight. Similar difficulties in adaptation of the generative cycle of northern Scandinavian forest tree populations have been recorded by SARVAS (1966). Presumably the species concerned have reached the limits of their genetic flexibility in this area, but this is contradicted by ANDERSSON'S (1965) deduction that genetic variation exists in the characters of decisive importance for adaptation in the generative cycle. Yet we are far from a balanced appreciation of these relationships since relevant investigations are in their early stages.

We conclude, therefore, that the time before flowering is not only decisive for the formation of flower buds but also, at least in some cases, for the success of the developed flower buds. Naturally there are possibilities of failure even after com-

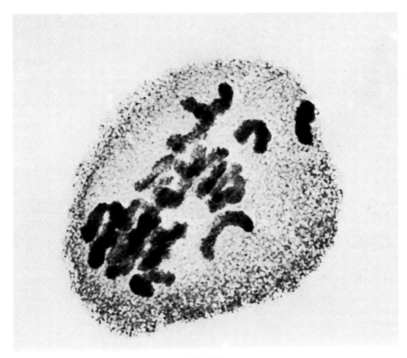

Fig. 8a

Fig. 8 a–d. Irregularities in pollen formation of ill-adapted coniferous trees (courtesy Dr. GÖSTA ERIKSSON, Stockholm). a Univalent formation in *Picea abies* caused by high temperature. Pollen mother cell from a grafted plant in a plastic greenhouse, April 22, 1969. b 'Stickiness' in *Larix sibirica*, January 25, 1966. c Degeneration of chromosomes in *Larix leptolepis*, February 27, 1967. d Formation of micronuclei as a consequence of univalent formation in a previous stage, *Larix sibirica*, November 28, 1966

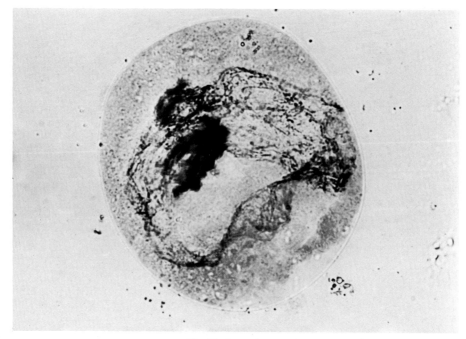

Fig. 8b. Legend see p. 39

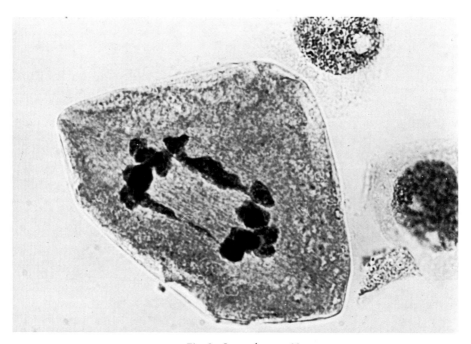

Fig. 8c. Legend see p. 39

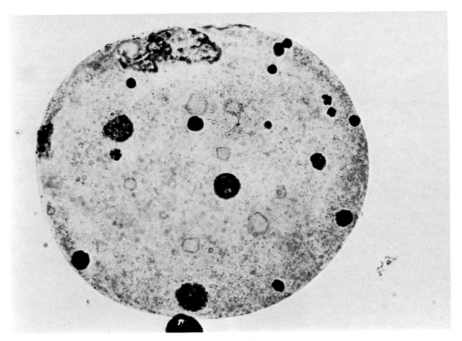

Fig. 8d. Legend see p. 39

pletion of meiosis. SARVAS (1962) has made long-term studies of the development of male and female strobili of *Pinus silvestris*. Development must be synchronized so that receptive female strobili are available when male flowers release pollen. Failure of synchronization would be catastrophic, particularly in wind pollinators where flowering time is relatively short (WHITEHEAD, 1969). Sarvas found that the flowering abilities of male and female strobili are reached at the same time, even in years when the course of temperature (presumably the primary influence upon both male and female development) appears to favor unequal development in the preceding stages. Both male and female strobili of Scots pine are susceptible to frost in the more advanced developmental stages. These stages are reached at such late dates that frost damage is rare but is still occasionally possible (DENGLER, 1940; SARVAS, 1960, 1968). Thus the date of flower maturity must be adapted to the rhythm of the annual growing season at the natural site of a given species.

   That differences in flowering date actually exist is shown by many experiments, although these differences are not always as pronounced as among hazelnut varieties of different provenance (STRITZKE, 1962). In commercial plantations, skill in combining varieties determines productive capacity (*Corylus avellana* is self-incompatible; it is one of the few species that flower in February, long before leafing-out). Here the different varieties (clones) originated in different areas, and differences in mean flowering dates of populations from different geographic sources have been observed in many experiments. Yet some authors also find genetic variation of the different developmental stages of male and female flowers within geographic sources. According to PANIN (1960), the flowering dates of early and late flushing

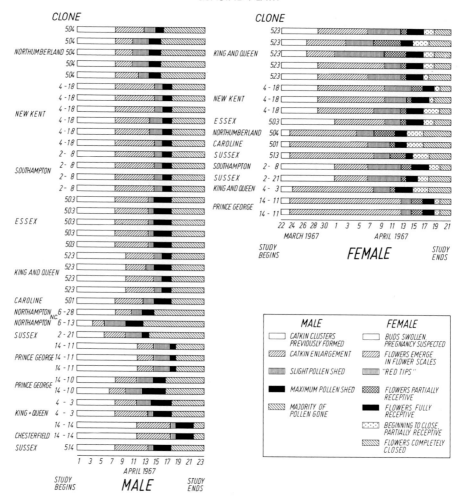

Fig. 9. Phenological observations in two seed orchards of *Pinus taeda* from the coastal plain (above) and the Piedmont (p. 43) of North Carolina, and in a seed orchard of *Pinus echinata* (p. 44). (From WASSER, 1967)

Norway spruce differ by approximately three days (a surprisingly small difference when we consider that differences in flushing are perhaps better measured in weeks rather than in days). SCHMIDT (1970), for example, found genetic variation of the phenological stages of both male and female strobili in Scots pine. In spite of notable variation among the flowers of the same tree, particularly the different parts of the crown, the heritability of most phenological characters is considerable. WASSER (1967) made similar observations in three seed orchards of southern pines located near Richmond, Va. The species involved were *Pinus echinata*, *P. taeda* from the Piedmont, and *P. taeda* from the Coastal Plain. Results are reproduced in Fig. 9. Most clones are represented by more than one ramet. If one compares, for

PIEDMONT

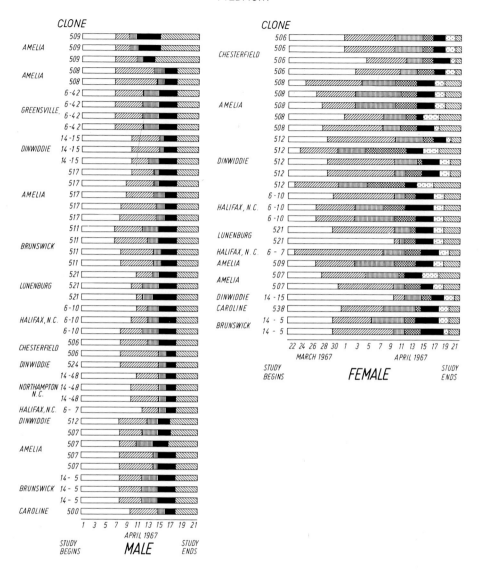

example, the data for male flowers of clone 16 of *P. echinata* with those for the female flowers of clone 20, it is apparent that clone 16 has shed nearly all of its pollen on May 2 whereas the female strobili of clone 20 are just becoming fully receptive. For *P. taeda* from the Piedmont, clone 509 flowers too early to pollinate the female strobili of clone 506.

Probably the largest investigation of this kind was undertaken by POLK (1966). He found, for instance, that among 70 seed sources of Scots pine, grown at the same location, the mean period of pollen flight differed by as much as three weeks, which

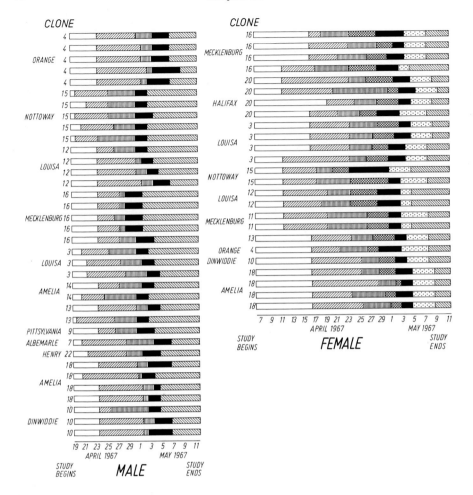

amounts to an absolute isolation of these sources. In Douglas fir the mean flowering dates of clones within the same seed source may still range over ten days.

On the other hand, STERN (1972) found that in mature stands of Scots pine, trees
labeled by manganese spread their pollen equally throughout their whole pollen-
flight period. Therefore, he assumed that in older trees flower position in the crown
and other causes tend to eliminate the distinct differences in flowering date observed in young trees.

Pollen that has arrived on the stigma, or pollen chamber in the case of conifers,
usually germinates in a short time. In species forming fully developed seeds in the
fall or winter following pollination, fertilization takes place only a few days after
germination of the pollen grain. The division of the megaspore mother cells in
Norway spruce takes place shortly before or during the receptivity of the female
strobili, and early archegonial development is already evident two weeks after
pollination. The time interval from pollination to fertilization is only four weeks.
The time then required for seed development is two to three months. On the other

Table 2. Proportion of Scots pine pollen distributed daily throughout the flowering period, by tracing pollen marked by manganese within a 12 meter radius around a pollen source. Values marked by x were recorded on days with minimum pollen flight (From STERN, 1972)

Day	1 AHLTEN 1 (1969)	2 AHLTEN 1 (1970)	3 AHLTEN 2 (1970)	4 OERREL 1 (1970)	5 OERREL 2 (1970)
1	0.31	0.56	0.39	0.26	0.40
2	0.44	0.65	0.29	0.35	0.34
3	0.35	0.55	0.38	0.35	0.34
4	0.10×	0.56	0.34	0.31	0.31
5	0.11×	0.57	0.34	0.29	0.35
6	0.38	0.51	0.32	0.31	0.32
7	0.47	0.61	0.29	0.47×	0.23×
8	0.14	0.42×	0.39×		
9	0.35				
10	0.57				
11	0.38				
12	0.55×				
13	0.48× ×				

hand, in Scots pine with a time span of 1 ½ years between pollination and seed maturity, division of the megaspore mother cell takes place three or more weeks after flowering, and it then needs a further six weeks for full development, after which it remains in a resting stage until the following spring. Thus the full development of the two gametophytes is achieved one year after pollination, often even after 13 months. As a result, not much time remains for seed development. A short growing season is therefore the limiting factor for seed maturity — an essential component of generative fitness as ANDERSSON (1965) emphasizes. Before techniques of X-ray examination had become available it was not possible to determine causes of differences in seed quality of Norway spruce and Scots pine from year to year. The degree of maturity of the embryo and (primary) endosperm (macrogametophyte) is the decisive factor (SIMAK and GUSTAFSON, 1959, and others).

We have so far considered wind-pollinated species that require adaptation to climate for two reasons: (1) wind-pollinated species must flower within a relatively short time span to make cross-pollination effective through production of a sufficiently large volume of pollen; and (2), their vegetative cycles must be adapted to the seasonal rhythm of weather. The latter reason is also valid for insect-pollinated tree species of the temperate and boreal zones as well as the tropical and subtropical zones with dry and wet periods. However, other factors are also involved here: the adaptation of the generative cycle ot tree species that are pollinated by animals must be seen in relation to the adaptation of the pollinating animals. In most cases it is a process of coadaptation of two (or of several) organisms, the life functions of which are influenced by climate to some extent, but where other factors also play a role. Two interesting examples for pollination under extreme conditions are given by HAGERUP (1951) for the Faroe Islands and POTTER and ROWLEY (1960) for the San Augustin Plains in New Mexico.

In the humid tropics and subtropics without marked seasonal fluctuations of weather there is no necessity for adaptations of the preceding kind. In spite of this, mechanisms are found here that provide for simultaneous flowering of the same species, or at least in overlapping periods, thereby ensuring possibilities for cross-pollination. On the other hand, the flowering period may extend over several weeks and months, and overlap the period when mature seeds are available. It was assumed that wind-pollinated species are rare in the tropics and subtropics, but this is not necessarily so according to ASHTON's (1965) investigations.

Unfortunately, studies of flowering and fruiting of the tree species in the humid tropics and subtropics are only beginning. Nevertheless, results from such studies are needed to complement those obtained in decades of research in temperate and boreal zones, and to allow a better understanding of the evolution of tropical tree species and ecosystems, as well as for the planning of economic measures in these areas (BAWA and STETTLER, 1969).

b) The Mating System

Biologists (and plant breeders) have for many decades distinguished cross-pollinated and self-pollinated species. In fact there are "obligate" self-pollinators whose flower structure does not permit cross-pollination (cleistogamous species), and "obligate" cross-pollinators that exclude selfing by special genetic mechanisms. Today the opinion prevails that these extreme cases are not common or do not exist. Rather mating system of a population should be understood as an adaptation, an optimized subsystem of the genetic system as a whole (see Section II.1). In sexual reproduction, chromosome and gene recombination permit organisms (except those that are close inbreeders) of each generation to offer a new multiplicity of genetic variants, if the unpredictable nature of the expected environment favors this solution. In populations with forced recombination (i.e., those that are predominantly cross-pollinating) it is to be assumed from the start that they have to contend with the insecurity of a heterogeneous environment. Populations with strongly limited recombination (where selfing predominates, apomixis and vegetative reproduction are common) may be fitted into niches with constant values in their important dimensions, may be expectionally tolerant, or may be prevented for some reason to realize their optimal mating system.

Plant species in the polar regions, for example, can hardly distribute pollen through animals. Possibilities for effective cross-pollination through wind are also limited as a result of the normally low population density and pollen production (see FAEGRI and VAN DER PIJL, 1966, and KUGLER 1970 for summaries of the imposing pollen production of wind-pollinated outcrossing species). This leads to the necessity of compensating for the small pollen quantity, i.e., using it economically. Evolution forces the development of flowers that favor selfing, or even apomixis, in spite of the fact that the extreme fluctuations of weather, especially in these regions, tend to favor the development of mechanisms that would provide much genetic variation in every generation (see DUNBAR, 1968, for detailed description). Therefore, the actual mating system of a population represents a compromise determined by evolutionary history and evolutionary potential of the population as well as requirements of the present environment. The system must approach the optimum (which, in the last analysis, is always hypothetical), at least to the extent that it allows permanent survival of the population. A similar compulsion for economic utilization of its own pollen may exist in very different environments, for instance, in the tropical rain forest where the multiplicity of species leads to low population density and a scarcity of foreign pollen.

We shall examine at least some relatively well-known mating systems of forest trees with respect to their adaptive characteristics. *Theobroma cacao* is an equatorial species with its distribution center in lower elevations of the eastern slopes of the Andes Mountains growing in a lower stratum. Here the species shows broad genetic variation and is self-incompatible (COPE, 1962a, b). With increasing distance from the center, the proportion of self-incompatible individuals diminishes, as does the genetic variability of populations. *Theobroma* is pollinated by midges. Cross-pollination is then possible only when trees in a stand occur in groups, as is the case in the central area. Thus the lower population density toward the edge of the range may have caused the observed increase in the proportion of self-fertile trees. It is also noteworthy that populations introduced to other areas are also self-fertile. In the case of *Hevea brasiliensis*, another species of the tropical rain forest that has been much investigated, apparently similar conditions prevail. Most of the known clones are self-fertile but self-incompatible types also exist. This is another species of the lower tree story and rare in many parts of its range, making self-fertility a necessity for generative reproduction. Midges serve as pollen vector and restrict pollen distribution to short distances (PURSEGLOVE, 1965).

In *Juglans regia*, SCHANDERL (1964) discovered that the frequency of cross-pollination is solely dependent upon the proportion of foreign pollen in the "pollen cloud" (*J. regia* is wind-pollinated). Self-pollination is equally successful, and if pollen is entirely absent, seed can be produced apomictically. A very similar situation exists in mountain ash, *Sorbus aucuparia*; the general pattern prevailing in the *Rosaceae* (not only the trees of this family) is one with much flexibility in its mating system (SCHANDERL, 1932, 1937, 1962; RUDLOFF and SCHANDERL, 1937; BRANSCHEIDT and PHILIPPI, 1940, and others; FAEGRI and VAN DER PIJL, 1966 for summary).

In some species like the European ash, *Fraxinus excelsior,* there are not only separate male and female but also trees with flowers of both sexes: ash is "trioecious." This has been interpreted as a transitory stage in the evolution of a species (although it may also be considered an adaptive strategy — the play with three instead of with two morphs; the dendrological literature presents further examples). But this is not necessarily so; cases are known where trioecy, or, more generally, heteroecy, provides for an optimal mating system. STAUDT (1967, 1968) has made detailed studies of this problem in *Fragaria*. He discovered that heteroecy must have arisen together with polyploidy, which is common in this genus. All diploid species are hermaphroditic and either self-fertile or self-sterile. In contrast, the natural populations of polyploid species are always heteroecious; they consist of either male and female, or of female and hermaphroditic (gynodioecious), or of male, female, and hermaphroditic (trioecious) individuals. The author investigated inheritance of trioecious morphs in several populations of *F. orientalis*, some of which originated in widely separated areas. He found that the "sexes" depend upon three alleles of a locus, the female allele being dominant over the two remaining and the male being dominant over the hermaphroditic one.

In this connection we are reminded of the well-known investigations of JONES (1934), which indicate how the mating system can be changed in a relatively quick and simple manner (in his case with maize), from monoecy to dioecy. The allele *sk* discovered by him suppresses formation of the style whereas the allele *ba* suppresses development of the female flower as a whole, and the alleles ts_2 and ts_3 transform male into female flowers. These four genes alone make it possible to change *Zea mays* into a dioecious species. In the case of *Betula pendula* (STERN, unpublished), a single generation of selecting mainly male or female plants, respectively, was sufficient to produce at least pure females. Here, too, the creation of dioecy would only require several generations. WESTERGAARD (1958) points out examples of the reserve: *Cannabis* and *Vitis* are both dioecious; for reasons of a lack of fertility of male plants of *Vitis* and low yield of male *Cannabis*, both species were transformed by man into hermaphrodites.

The experimental possibilities of inducing self-incompatibility or, more generally, crossing incompatibility in plant populations are generally known or belong to the subject of adaptive strategies and therefore need not be considered at this time. It was the purpose of this section merely to point out that even the mating system of a population represents an adaptation, which, as a component of the genetic system (see Chapter III), carries weight in determining the adaptive value of a population. BAWA and STETTLER (1969) are certainly correct when they emphasize that, especially in tree species of the tropics and subtropics, much needs to be done in this field to evaluate their position in the ecosystem and anticipate their behavior in a breeding program. For a detailed account the reader is refered to STEBBINS (1950).

c) Pollination

All higher plants, including the tree species, are capable of sexual reproduction. Many reproduce themselves only generatively. To fully utilize the evolutionary possibilities open to a population from sexual reproduction, at least a certain percentage of the progeny should originate from cross-pollination. This requires the transfer of pollen of one tree to the female flowers of other trees, involving in many cases guided transport over greater distances. This problem is solved in the simplest manner by the wind pollinators (anemophilous plants), but in a way that is also least economical and by no means applicable in all situations.

WHITEHEAD (1969) enumerates the requirements of a well-functioning wind pollinating system. Let us begin with characteristics of the pollen. To be capable of transport by wind, the pollen of wind pollinators should have a low rate of fall. DYAKOWSKA and ZURZYCKI (1959), EISENHUT (1961), and POHL (1937) (additional figures given by KUGLER 1970) give a rate of fall of 5.6–8.7 cm/sec (centimeters per second) for *Picea abies* and 2.5–4.4 cm/sec for *Pinus sylvestris*. Values for the pollen of *Alnus glutinosa*, *Betula pendula*, *Juniperus communis*, and *Corylus avellana* occupy approximately the same range as *P. sylvestris*, whereas that of beech (*Fagus sylvatica*) is similar to Norway spruce. Slight differences in the values given by individual authors may be explained partly by dissimilarity in method but also by variation in the pollen itself. ANDERSON (1947) and MARCET (1951) have demonstrated the presence of differences in pollen size (and corresponding rate of fall) of *Picea abies* and *Pinus sylvestris*, respectively, which was related to both environment and heredity. ANDERSSON (1954) has made a more detailed investigation of genetic and environmental variation in the pollen of the two species. According to MARCET'S (1951) calculations, pollen of an East Prussian seed source of Scots pine should fly approximately 1.5 times the distance of other sources. Of course, there is variation in other pollen properties, such as specific gravity, contents of specific materials, growth of the pollen tube, etc. This should always be taken into account when considering data found in the literature. On the other hand, differences in rate of fall whatever their origin should not be taken too seriously. SILEN (1962), for instance, found large pollen quantities of *Pseudotsuga menziesii* at great distances from the next pollen source, although pollen of this species is considered to be relatively incapable of distant flight. Accordingly, KOSKI (1970) emphasizes the difficulties of finding a purely theoretical solution to the problem of pollen transport by wind, using measurable characteristics of the pollen on the one hand and the type and rate of air movement on the other. Still, pollen of wind-pollinated species needs special adaptations for dispersal by wind over great distances. In some cases special devices exist for this purpose, such as air bladders. However, pollen of insect-pollinated (entomophilous) species is also often transported by wind.

The basic calculations still valid for the influence of air movement are those developed by SCHMIDT (1925), although variations and expansions have been provided since that time. A summary has been given by GREGORY (1961) and applications in the realm of forestry research by KOSKI (1970). In every case the conclusion is that, given normal conditions (sufficient wind speed and turbulence), large quantities of pollen may travel great distances (ROEMER, 1931; FIRBAS and REMPE, 1936; REMPE, 1937, and others). Furthermore, pollen in the atmosphere maintains its viability for sufficiently long time periods (WERFT, 1951).

From a forestry viewpoint, interest was initially centered on the relationship between natural regeneration and seed production, which, in turn, depends upon availability of an adequate quantity of pollen. As a result, forestry scientists experimented with pollen dispersal at an early date (e.g., HESSELMANN, 1919). Yet probably the most detailed investigations have been carried out by SARVAS with *Pinus sylvestris* and *Picea abies* recently (SARVAS, 1952, 1955, 1957, 1962, 1968). In addition to flower biology (see Section 2a of this chapter), he has paid particular attention to the measurement of pollen quantity and its relation to seed production.

Botanists have indicated repeatedly that the pollen production of wind pollinators, especially of conifers, must amount to vast quantities in some years (e.g., POHL, 1937b; FIRBAS and SAGROMSKY, 1947). The so-called "sulfur rain," i.e., the accumulation of pollen along the shores of lakes and streams, which indeed requires large pollen quantities, has been expounded for a long time. Estimates are also available of the number of pollen grains produced for each functional ovary (e.g., 6,734 in *Betula pendula*, see POHL , 1937a), and of pollen density at certain distances from the pollen source (e.g., see POHL, 1933 for *Quercus petraea*), but it is difficult to relate such numbers to the expected probability of pollination of a female flower. This probability is, of course, identical to the probability with which a pollen grain may land on an area of the size of a stigma, or the size of the sticky drop on the pollen chamber in the case of conifers. In most cases this pollen grain will be from a different tree since most forest trees are self-sterile. SCHMIDT (1970) and STERN (1972) found that 0–82% of the pollen in crowns of Scots pine came from the same tree. This percentage will vary of course from tree to tree; it will depend upon the arrival rate of foreign pollen and the production of the tree's own pollen. A simple count of pollen in the pollen chamber is not enough to obtain a clear picture, and it is even less adequate to estimate this probability from the pollen catch. SARVAS (1968) has discussed the problems of such experiments in detail. The most reliable method is a comparison of pollen density and percentage of germinable seed, measured in a useful and standardized manner. Such a comparison is given in Figure 10.

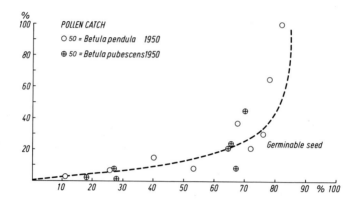

Fig. 10. Correlation of percent germinating seed and pollen density. (After SARVAS, 1955)

However, SARVAS (1955), who obtained these results, stresses that such comparisons are not equally reliable for all species. There are confounding factors that are of unequal importance and could be more weighty, for example, in Norway spruce than in the birches.

In years with plenty of male flowers the germination percentage of undamaged seed may reach 90% or more, indicating that sufficient pollen was available. In years with less pollen a larger percentage of empty seed must be expected.

A large number of investigations dealing with the pollen flight of wind-pollinated species is recorded in the literature but most of these had different, often very specialized, goals. The significance of pollen flight as the (probable) principal method of gene transfer into other populations, and pollen distribution within a stand and its effects upon the genetic structure of populations, will be considered later. We shall at least mention here the investigations concerning expected pollen flight into seed orchards and the suitability of measures to control it (PERSSON, 1955 and others, summary given by KOSKI, 1970), pollen filtering in forest stands (DENGLER, 1955a), and some other problems (DENGLER, 1955b; DENGLER and SCAMONI, 1944) that will not be considered.

Unfortunately, most of the results that have supported the preceding and will support the following discussion were obtained from species of the temperate and boreal zones. We know nothing about conditions in other climatic zones, and it is fairly certain that important differences exist. Therefore, it is quite possible that the story presented here may require revision in the foreseeable future.

At least a few remarks should be added concerning the adaptation in structure and function of the male and female flowers and inflorescent types of the wind-pollinated species. It is rather obvious that for a female flower the probability of receiving a pollen grain at the right time and place on its stigma, or bracts of conifers, does not depend only upon pollen density and wind speed (which determines the probability with which a pollen grain will fly at a given density of pollen per m^3 of air). So much has been published on this subject over many years that only references to two recent monographs will be given: FAEGRI and VAN DER PIJL (1966), and KUGLER (1970).

The upright position of the female inflorescence, preferred among the wind pollinators by both angiosperms and gymnosperms, appears to increase the probability for receiving windborne pollen. The position of female inflorescences of trees in peripheral parts of the crown may also augment the probability of pollination, and the concentration of inflorescences in the upper part of conifer crowns may have the same effect. The architecture of the female inflorescence (protruding or massed pistils of the angiosperms among the wind pollinators, purposeful structure of the female strobili in the gymnosperms) appears to be an equally successful adaptation. Protogyny is often interpreted as a means of increasing the rate of cross-pollination, but if one considers the genotypic variation of all phenological stages, including flower development, it could be seen as an arrangement to have flowers ready when pollen arrives. This again increases the chance of receiving windborne pollen. Protandrous flowers have been interpreted in a similar way, but POHL (1935) describes a case of "useless" protandry. Finally, the adaptation of the anthers in the male flowers to weather by opening only when relative humidity is low again increases the probability for movement over great distances.

The consequences of wind pollination for adaptation of the generative apparatus are therefore clear. However, the arrangement of the male and female inflorescences in the crown of anemophilous tree species also has consequences for the vegetative phase. Branching and crown form, and hence fitness for competition in crown space, are influenced by the selective advantages that certain arrangements of the generative phase offer. This will be even clearer after the discussion of pollination by animal species. We will point out at this time, however, that certain adaptations that at first appear to be related to the vegetative phase alone, can only be fully understood when considering the necessity of adaptation to the generative phase as well. The forester is inclined to neglect the generative phase entirely unless it is of decisive importance, as in tree populations of the northern tree limit that have been primarily discussed in this account.

Flower ecologists have distinguished the wind pollinators, i.e., the anemophilous or anemogamous plants, from the zoidogamous plants, i.e., those where pollen transport to the female flower is performed by animals (to the female part of hermaphroditic flowers or in self-sterile plants from plant to plant). Yet in many cases it is impossible to describe a species as purely wind-pollinated or purely zoidogamous. EISENHUT (1959), for example, found in *Tilia* that pollination by wind is far more frequent than by insects, although the genus was considered as purely

insect-pollinated at that time (see also GRANT, 1953; PORSCH 1956). For the gymnosperms such cases are known from the *Cycadales* and *Gnetales* (see DOGRA, 1964 for
a monograph of pollination mechanisms in gymnosperms). In some species of
Cycadales, the male strobili possess excretions that attract bees and other insects,
but rarely have these insects been found on the female strobili and hence they may
contribute little to fertilization. In the genus *Eucephalartos* of the same order,
certain beetles eat pollen from the male strobili and deposit their eggs into the
female strobili (RATTRAY, 1913). The role of these beetles in fertilization remains
obscure, but it may be a positive one, consisting perhaps of a directed long-distance
transport under conditions of low population density. Yet species with exclusive
zoidogamy and those where few animals are specialized as pollen vectors appear to
be limited to the angiosperms.

There are very different reasons for animals to visit flowers (in treating this
subject we will follow KUGLER, 1970). The principal reasons that insects are attracted to flowers are: the search for food (pollen, nectar, other especially produced
food materials, certain tissues of the flower and its parts) by food-gathering females, for flowers resembling the female where copulation can be imitated, for nes-
ting materials, and for places for the deposition of eggs; and instincts related to the
defence of territory. The attractions are the form, color, smell, and taste of flowers.
As a result of the multiplicity of pollinating animals, there must be a high degree of
adaptation on the part of the plant.

Pollen vectors among animals are apes, lemurs, Indian bears, and squirrels and
other rodents; also climbing marsupials *(Pharangeridae*, e.g., in eucalypts of Australia), flying marsupials *(Petaurus)*, mouse marsupials *(Dromica)*, and bats.
Among birds should be mentioned especially the hummingbirds *(Trochilidae, Nec-
tariniidae)*, spectacle birds *(Zosteropidae)*, flower pickers *(Dicaeidae)*, honey eaters
(Meliphagidae), loris *(Trichglossidae)*, and others such as *Pardalotidae, Drepanidi-
dae* and *Caerbidae*. Nearly all species of these genera (in some cases all species, and
often there are 100 or more species in a genus) depend exclusively or at certain
seasons of the year upon flowers as a food source. This may involve long-distance
travel to reach areas where flowers are available (migratory birds). There is a
marvelous variety of specialization in beak and tongue as adaptations to certain
flower structures and types. Flower-sucking species also occur in some other genera, as in the thrushes *(Turdidae)*, gnat catchers and kinglets *(Sylviidae)*, orioles
(Oriolidae), weaver finches *(Ploceidae)*, and others.

Among the insects, flower-pollinating species are among the Hymenoptera
(Symphyta, Terebrantes, Aculeata), Diptera *(Syrphidae, Conopidae, Bombylidae,
Empidae, Tachinidae, Muscidae, Tabanidae, Nemestridae, Cordyluridae, Stratio-
myidae, Bibionidae, Psychodidae, Chironomidae, Simulidae)*, Lepidoptera, Coleop-
tera and some others. These species may feed exclusively or occasionally on nectar,
pollen, and other materials or visit the flowers for other reasons (see above). The
plant species visited by these insects may be adjusted to few visitors or offer many
of them a chance for pollination.

The relationships between bees and the plants they visit have been most thoroughly investigated. On the part of the visitors, the adaptations of importance
include anatomical properties, physiological reactions, and behavior patterns. K.
VON FRISCH and his school have presented a multiplicity of results revealing adapta-

tions of an ingenuity on both parts that is nothing short of miraculous (VON FRISCH, 1953). Less well known is the performance of the other flower-pollinating species, such as those among the hummingbirds. According to GRANT (1966, 1967a, b) and GRANT (1968), the male bird of *Calypte anna* needs to visit more than 1000 flowers of *Fuchsia* daily, and utilize them fully, to meet only his average energy requirements. Within an observation period of very few hours, males of *C. costae* carried out more than 40 flights to search for food and visited more than 1300 flowers. Therefore, their cross-pollination performance may be substantial and equivalent to that of bees, which are better known in this respect (GONTARSKI, 1935 and others).

Specialization of some bat species that visit flowers would seem to indicate that their efficiency is equal to that of the obligate flower visitors among the insects and birds. Furthermore, if one takes into account that the highly specialized flower structure of the species visited excludes visitors other than those that are specialized themselves, the importance of the pollinating animals for the mating system and the evolution of these plant species becomes clear; conversely, the evolutionary importance of the adaptation of the animals just to these plant species is equally realized. VAN DER PIJL (1958, 1960/61) has outlined the ecological aspects of this situation and its consequences for the evolution of flowers in many examples. In some cases, closely related, sympatric species flowering at the same time are genetically isolated as a result of a different flower structure (see GRANT and GRANT, 1964 for one of the many examples). Other functions of the flower important in this connection, for example, consist of the production of specific types of food. If the pollinating species feeds exclusively on pollen and/or nectar, then these must contain all the necessary nutritional substance. Many investigations deal with this problem (e.g., PERCIVAL, 1961; LÜTTGE, 1961), and support this assumption. The color of the flowers visited differs, depending upon the section of the color scale preferred by the pollinating animal species. The preferred odors of flowers are also specific. Thus there is in fact a close dependence in the evolution of both partners, a coevolution of two species of organisms, as has been emphasized by many authors (see, for instance, UPHOF, 1942; FROST, 1965; EHRLICH and RAVEN, 1965, and others).

In this connection the forester will be interested in the fact that tree characters may be influenced that are directly related to wood production — aside from the subject of specialized flower structure and function through coevolution of tree species and pollinating organisms. In tropical and subtropical rain forests there is an extended period of flowering and fruiting in comparison with temperate and boreal forests. There is also the necessity for a certain sequence in flowering and fruiting of higher plants, including trees, to guarantee a regular supply of pollen and nectar (the need for some species to migrate has already been mentioned) so that specialized pollinators can be maintained. Furthermore, the position of the flowers in the crown could be adjusted to the requirements of the pollinators. Bats, for example, locate flowers by means of an echo sound system; birds search for them visually; insects search visually or use their sense of smell. Thus flowers visited by bats must definitely be situated in the outer and upper portion of the crown; for birds this may be helpful; some insects would also find them in other parts of the crown. As a result, there are consequences here for the structure of the crown and also for branching and stem form.

A related problem has been given much attention by earlier authors in forest genetics. They had asked questions regarding the influence of sex upon economic characters in dioecious tree species. This influence can have economic importance, as hemp and asparagus indicate in plant breeding. However, the assumed influences in poplar, i.e., a kind of sex dimorphism, could not be confirmed. In a representative material of *Populus tremuloides* with a sex ratio of 1:1, EINSPAHR (1960) found no correlation of sex to other characters. Most investigations of earlier authors do not indicate how representative sampling had been achieved.

With this observation we conclude the discussion of pollination. We have attempted only to indicate the most important adaptations relevant to this subject, and to call attention to their significance for the evolution of tree species. We shall see in the following that the method of pollination influences the mating system of a population in a significant manner.

d) Seed Dispersal (Presence)

The possibilities of higher plants to distribute their seed, and especially the ingenious devices for this purpose found on fruits and seeds, early directed the attention of botanists to the problem of seed dispersal. A number of monographs have been devoted to this special adaptation, and even the older works among them still make worthwhile reading (HILDEBRAND, 1873; RIDLEY, 1930; MATHENY, 1931; MÜLLER, 1955; VAN DER PIJL, 1969, and others). However, seed dispersal is only part of the plant's endeavor to be present wherever the opportunity arises to establish an individual or a population; all other adaptations that allow a population to be present belong into this group. These include, for example, physiological mechanisms such as pronounced seed dormanay or extreme shade tolerance of seedlings, each achieving in a different way the readiness of a large "waiting" population.

In this interpretation we are following VAN DER PIJL (1969). His monograph on dispersal devices of higher plants covers not only the different mechanisms, but also their ecological and genetic significance. The value of this approach in enhancing the understanding of the biology of ecosystems has been proven in many ways (see, for example, ZOHARY, 1937). For this reason we classify the dispersal mechanisms only according to points of view that are relevant to the genetic architecture of populations (see DANSEREAU and LEMMS, 1957). Although the evolution of the different adaptations is interesting in this connection, it will not be discussed. Seed or other particles that are dispersed (bulbs, parts of the plant, etc. — VAN DER PIJL, 1969 uses the general term "propagules") can be scattered over some distance by means of devices possessed by the plant itself (autochory). However, the usual vehicles of dispersal are wind (anemochory), water (hydrochory), and animals (zoochory). Special adaptations are represented by the ballistic devices of the autochors, the various flight mechanisms around seeds and fruits of anemochors, the swim and protective gear of the hydrochors, and the often very specialized and complex apparatus of the zoochors. The multiplicity of adaptations, especially in the last group, results of course from the very differnt possibilities of distribution offered by the great diversity of animal vectors (earthworms, insects, fishes, reptiles, birds, mammals). The seeds to be dispersed may be attached externally, pass through the digestive tract, or simply spread in consequence of any animal activity.

Seed is not deposited by trial and error only, e.g., carried away by wind to land at any unpredictable location. Flight distance, for example, is determined by rate of fall and the kind and rate of air movement. DINGLER (1889) found a mean rate of fall of 0.57 m/sec for Norway spruce and of 0.43–0.83 m/sec for Scots pine depending upon length of seed wings. HESSELMANN (1934) gives additional data. Thus the distribution of seed around the seed source may resemble the pattern of anemophilous pollen, which decreases in density with increasing distance from the source. In consequence, a relatively large proportion of the seed is deposited near the site that has proven its suitability for the species, and a decreasing proportion with increasing distance from that site. A part of the seed — i.e., a variable proportion — lands in a position with good chances of success; the remaining seed flying greater distances is distributed according to the trial-and-error principle. This principle may be varied widely, perhaps through the flying ability of the seed. Similar conditions may prevail in species with distribution of seeds (or of other propagules) by water — a means not rare in trees. Seed of *Alnus glutinosa*, for example, may float for a considerable time on water without suffering any damage and may be transported by streams over great distances. Vegetative parts of a plant may also be carried by water and may develop into new plants when carried ashore. This is supposedly not rare in willows and poplars. Most impressive, however, are the adaptations of tropical and subtropical species to water transport (SCHIMPER, 1891; GUPPY, 1906; RIDLEY, 1930; STOPP, 1956). The mangrove formation has attracted special interest for a long time, but this is wholly adapted to development in water. Seeds of some species may remain viable even in degrading seawater for several years. Occasionally living seeds of such species are found as far away as European seashores. In this connection, the results of studies concerning natural regeneration of flora of islands after catastrophes should be mentioned, particularly the resettlement of Krakatau Island following the disastrous volcanic eruption (DOCTERS VAN LEEUWEN, 1936). Not only the often-mentioned coconut tree, but also a number of other tree species, found their way to Krakatau in a surprisingly short time.

Similar conditions may exist when seed is transported by animals: one part of the seed is deposited in an advantageous position, the other distributed through the trial-and-error method; again the seed transported far may have the poorest chance to establish a plant. Yet there are cases where adaptation has been developed to the extent that the distributing animal species takes seed to an advantageous position, for instance, when birds "aim" to place seeds of the mistletoe plant onto tree branches. In many cases behavior patterns of the distributing animal contribute to the deposition of the seed at a favorable place, although not in such a spectacular way as in the case of the mistletoe. VAN DER PIJL (1966) emphasized the coevolution of higher plants and their distributors in his monograph on the ecological aspects of evolution of fruits (see also MEEUSE, 1958, 1966; TAKHTAJAN, 1959 and others).

In earlier periods, botanists and foresters primarily investigated distant transport. They were concerned mainly with questions of colonization and recolonization. FIRBAS (1935) discusses many examples taken from the forest trees of central Europe. The great reproductive capacity of higher plants was emphasized, which is particularly essential for colonization and recolonization (SALISBURY, 1942; SOBOLEFF and FOMITZEFF, 1908, and others). Difficulties of experimentation, of course, lie in the complex behavior patterns of the higher animals. Yet more recent investigations indicate that they, too, are highly efficient dispersal vectors. For example, SCHUSTER (1950) found that a single European jay carried several thousand acorns in one season, some of them over a distance of several kilometers. Thus long-distance transport is assured for a tree species, which is probably tropical in origin, and possesses unusually large seeds in comparison with other species of northern latitudes (MACATEE, 1947; VAN DER PIJL, 1957).

Colonization and recolonization of areas by tree species concern the forester for economic reasons. Colonies of a species at the limits of its range, often threatened by extinction and capable of existing only in the most favorable localities, have captured the interest of foresters again and again. One of the oldest studies here is that of SERNANDER (1901), who often found single trees of Scots pine at the distributional limit of the species. Many authors in fact have found that the seed of Norway spruce and Scots pine may slide great distances on ice or crusted snow, although their seeds are equipped for flight only over short distances. The first investigator to mention this method of seed transport appears to be HOLMBOE (1898); presumably the most reliable quantitative estimates of its importance were given by HEIKIN-HEIMO (1932). Of course a simple count of seeds is not sufficient; many of the seeds carried far may be empty.

HESSELMANN (1934) has attempted to develop a quantitative notion of the effectivity of seed transport in Norway spruce and Scots pine (excluding the case of transport over great distances mentioned above, which depends on so many conditions that its importance can hardly be expressed in probabilities). Other earlier authors (e.g., ENEROTH, 1929; HEDEMANN-GADE, 1929, and others) have demonstrated the relationship between seed fall per m^2 and seedling establishment with distance from the stand boundary. HESSELMANN starts with the seed production of a forest stand, which may vary greatly for the two species, depending upon region and seed year. HEIKINHEIMO (1932) found in Finland 1–1.5 million seeds or 3–4.5 kg (kilogram) in moderate seed years and 7–10.4 million or 36–38.4 kg/ha (hectare) in good seed years; HESSELMANN in Norrland found a lower maximum with approximately 2 million = 5.2 kg/ha in moderate and 2.7 million = 8.2 kg/ha in good years, which corresponds fairly closely to the figures of MORK 1933 for Norway; the largest seed production, 158 kg/ha, was reported by SOBOLEFF and FOMITZEFF from Courland. HESSELMANN's comparisons includes the production of closed stands and of stands opened for increased seed production.

The theoretical flight distance of seed of some species was determined by HESSELMANN on the basis of SCHMIDT'S (1925) formula, upon which the theory of pollen flight distance was also built. Considerable interspecific differences were found, e.g., the mean rate of fall of birch is 25 cm/sec, that of ash 200 cm/sec. The mean distance of seed dispersal of the same species amounts to 1600 and 25 m, respectively, given a rate of air movement of 6 m/sec and turbulence degree of 20. Yet considerable flight differences exist within species, depending upon the performance of seeds with various wing lengths, as graphically shown for Scots pine in Fig. 11.

The actual values may be greater since tree heights were not included in the calculation.

The chances of the seed to establish a tree also depend upon the time of release. Different species release seed in different seasons and the duration of release varies

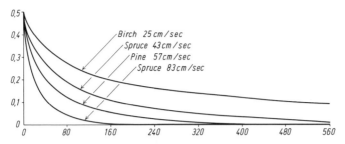

Fig. 11. Dispersal of seeds of different tree species in Scandinavia. The two seed types in Spruce differ in size of seed wings. (After HESSELMANN, 1934)

considerably. Seed of birch flies from the time of seed maturity in July until the following spring, but aspen seed only during a short period in summer. The seeds of both species are contingent upon favorable conditions for germination since their endosperm is small. Seed of Scots pine in Norrland flies from May to July with concentration in June (HESSELMANN, 1934), but in Finland from April to July with concentration in May (HEIKINHEIMO, 1932).

SNOW (1965) assumes that fruiting of the 19 species of the genus *Miconia* in Trinidad is distributed over different periods of the whole year because of resulting advantages in interspecific competition. The seeds are distributed by birds. The time period of seed maturity is determined here by other than the usual factors.

Species with light seeds that frequently travel far constitute the first arrivals in succession. Their frequency diminishes to a fraction of the original number as the climax is approached. Abundant seed production of the remaining individuals and extensive distribution are the means to occupy the areas available following catastrophic events and clear cutting. Yet few general statements can be made. On the one hand, there are species typical of climax forests in some parts of the range, such as Norway spruce in some central European plant communities, but pioneers in other parts of the range, as MELECHOW (1934) has shown for the same species in northern Russia. It would be interesting to determine how this is reflected in the related adaptations of Norway spruce populations. On the other hand, the different species have different means at their disposal to cope with similar situations. Birch and aspen exemplify species with light seed that is disseminated by wind. The seed of birch is available from July to April–May and there is a good chance that favorable conditions exist for germination and establishment of seedlings sometime during that period. Aspen seed, in contrast, is scattered only during a brief period and remains viable for a short time. But as soon as seedlings have become established, aspen species have the opportunity to form stands by means of root suckers. For example, BARNES (1966) generally found North American aspen species with clone groups averaging 300 m^2 in area. He postulates that these species, which are first in succession, maintain themselves for a considerable time because of the occurrence of tree groups made up by members of the same clone.

To realistically appreciate the role of a given adaptive phase in species distribution requirements, the whole breeding system should be investigated (see FRYXELL, 1957 for summary). Thus in some cases vegetative reproduction must be considered, and the preceding example of aspen is only one of the possible variants. MÖBIUS (1940) and WEBER (1967) reviewed the diverse possibilities open to higher plants. Apomixis may have the same functions or effects in a breeding system as vegetative reproduction (NYGREN, 1967).

Reproduction and distribution of tree species are determined not only by seed distribution but also by its presence at the right place and time. For this reason the seeds of many species possess special adaptations that allow them to germinate just at the right time. These adaptations may permit them to wait for the suitable time, often for years, occasionally for decades. An example previously discussed was the special protective mechanism of seed transported by water. These seeds are capable of remaining in water for a long time before germinating. The importance of this strategy in forest trees is reflected in statistics given by RUDOLF (1961). Of approximately 400 tree species examined, only 33% were without dormancy; 7% were endowed with seed coats impermeable to water that would permit germination only after a long time or after scarification; 43% possessed special physiological mechanisms regulating dormancy; and the remaining 17% regulated dormancy by means of several mechanisms. Therefore the "strategy" of seed storage is widely employed by tree species.

A number of publications deal with the maximum life expectancy of the seeds of higher plants (ODUM, 1965; CROCKER, 1938 and others). This varies, of course, with the conditions of seed storage. Foresters are familiar with these conditions because of the frequent need to stock seed for a long time before sowing. BARTON (1961) has treated this subject monographically.

There are also experiments testing the life expectancy of seeds of many species when conditions are similar to those in nature. That seeds of some species possess special mechanisms to remain dormant in the soil for long periods and to germinate when opportune conditions for seedlings prevail has been known to forest ecologists for a long time from investigations on cut-over areas or after catastrophes. The rapid development of the flora on such open areas could not result from seed flown in after the event. These dormancy mechanisms are indispensable, especially for the specific flora of clear-cut areas that are the very first in succession. We are particularly interested here in experiments where seed was buried in the soil. DARLINGTON (1951) reports an experiment started by BEAL in 1879. Seed of some 20 species was kept in a sandy soil at a depth of 45 cm. Five years after initiation of the experiment, seed of 11 species was still viable; after 30 years there were still nine alive, and after 70 years three species could still be germinated. TOOLE and BROWN (1946) published the results of an experiment begun in 1902 by Duvel. Seed of 107 species was kept in replications and under different soil conditions. Seed of 71 species still germinated after one year, 68 species after ten years, 51 species after 20 years, 44 species ofter 30 years, and 36 species after 39 years. From these figures and those given by RUDOLF, the importance of dormancy strategies for the respective species should be apparent.

These strategies may play a smaller role in the humid forest of the tropics and subtropics than in the forests of the temperate and boreal zones — at least for the tree species. Seeds with large amounts of nutrients seem to predominate, making it possible for the seedling to pass through the first stages of development while sustained by these reserve materials. In addition, there is extreme tolerance of shade, so that a large number of seedlings are kept in readiness instead of a large number of seeds.

A special method of keeping seeds alive exists in the fire species with serotinous cones, which were discussed earlier. SARGENT (1880) indicated that seed of *Pinus contorta* could be kept viable in the serotinous cones longer than a decade. If seed dormancy is of great adaptive value, as we have assumed for many ecological conditions, then species with a large range and contrasting environments must exhibit variation in the expression of this character. This will be illustrated by only one example. MORLEY (1958) found considerable differences among local populations in the proportion of dormant seed. On the average, seeds from colder, drier regions required lower temperature and less humidity to germinate than seeds from more humid and warmer regions. Finally, it should be mentioned that some plant species may achieve self-regulation of competition through adjustment of seed germination. PALMBLAD (1968) discovered that of nine species investigated, four employed this method to control competition.

In many species of higher plants the composition of the next generation is determined by the population of dormant seeds and, therefore, this generation must be influenced to a significant degree. In particular, EPLING et al. (1960) have stressed this point and indicated some of the consequences using an example of polymorphisms in *Linanthus*. Yet there are also other properties of the systems regulating distribution and presence that materially influence genetic systems and population structure (see CARLQUIST, 1968 for genetic systems in island floras, EHRENDORF, 1964 for similar properties of colonizing species).

III. Genetic Systems

The distribution of the genetic information to the members of the next generation is accomplished by the various species of living organisms, often in very different ways. The properties of the apparatus of gene distribution (the chromosomes), determining evolution of the population, may differ widely, for instance, in their possibilities of storing manifold genetic variability and bringing it into play in every generation, in their capability of producing particularly valuable genotypes through vegetative reproduction, and in other ways. For this reason all the characteristics of an organism that are related to these possibilities, and other characteristics determining transmission and utilization of genetic information in evolution, are being treated jointly under "genetic systems."

1. Concept of Genetic Systems

DARLINGTON (1958), who developed the concept of genetic systems (in the first edition of his work, published in 1939), gave a summary of their evolution. A grasp of these systems is also basic to an understanding of the behavior of tree species in their natural habitats (see STEBBINS, 1950 for a general outline). We are primarily interested in those components of the genetic system that reflect the peculiarities of forest trees or that indicate the most distinct differences in the genetic system of forest trees when compared with other higher plants. Perhaps one could roughly state that the concept of the genetic system is the counterpart of the concept of the ecological niche. The latter presents a formal basis for the comparison of ecological situations of different populations, the former an equally formal model to explain the evolutionary response of the population to the requirements of certain niches.

RIEGER, MICHAELIS and GREEN (1968) defined or summarized the genetic system as "any of the species-specific ways of organization and transmission of the genetic material in proto- and eucaryotic organisms, which determine the balance between coherence and recombination of genes and control the amount and type of gene combinations. Evolution of genetic systems means the evolution of those mechanisms effecting and affecting genetic variability. Factors which characterize a genetic system include the mode of reproduction, the type of population dynamics (breeding size, sex ratio, degree of panmixia), the mode of chromosome organization (genetic information all in one linkage group or distributed to several such groups), the chromosome cycle (normal meiosis in both sexes, or abnormal in one in the case of eucaryotes), the recombination index, and the presence or absence of genetic and chromosome polymorphism. The genetic system and its components determine the capability of a population to undergo evolutionary changes. Any genetic system is under genetic control."

Thus the genetic control of the genetic system, i.e., control through its own components, facilitates the evolution of the system itself. The system in turn enables a population as a unit of evolution to bring new adaptations into play or to try new adaptive strategies.

In his definition of the genetic system, WHITE (1954) emphasized its significance for evolution especially:

"Under the general term genetic system we include the mode of reproduction of the species (bisexual, thelytokous, haploid, etc.), its population dynamics (population size, sex ratio, vagility, extent of panmixia or inbreeding, etc.), its chromosome cycle (meiosis normal in both sexes or anomalous in one or both), its recombination index, presence or absence of various forms of genetic or cytological polymorphisms in the natural population, and, in brief, all those characteristics which determine its hereditary behavior over periods of time sufficient for evolutionary change."

The components of the genetic system that determine the degree of genetic differences among the gametes that form the zygotes have been grouped together into the breeding system following a suggestion of DARLINGTON and MATHER (1949). Within the larger framework of the genetic system, the breeding system is, of course, influential in determining the evolutionary possibilities of the population. However, even ecological factors and adaptations control genetic variability, as do population structure and the possibilities of evolutionary advance — such as, for example, the effectivity of dispersal devices, overlap of generations, etc. — as has been indicated in previous sections. It may be expedient, therefore, to remove some components of the breeding system and denote them as the mating system, i.e., those that are responsible for the mating of *specific* parents or the fusion of *specific* gametes. Three principal types of mating systems may be distinguished somewhat arbitrarily as follows: (1) random mating (occasionally designated as mating in a panmictic population), where every individual mates with the same probability with every other individual of the opposite sex; (2) genotypic assortative mating, where the probability of mating is determined by the degree of relationship (negative genotypic assortative mating occurs in obligate cross-pollinators, positive genotypic assortative mating in self-pollinators); and (3) phenotypic assortative mating, positive and negative, where phenotypic characters are responsible for the deviation from random mating.

The components of the breeding system (including those that belong to the mating system in the narrower sense) determine the genetic structure of the populations; in this sense the populations are characterized by their manner of gene distribution to the individuals of local populations and over all subpopulations (demes in the terminology of GILMOUR and GREGOR, 1939).

Some authors use the term "reproductive system," but its separation from other divisions of the genetic system is not always clear. For this reason reference is made to only three studies: WALLACE (1963) investigated the importance of the manner of reproduction of a population for evolution and breeding; THODAY (1964) discussed the integration of the components of the reproductive system; and ORNDRUFF (1971) applied the term "reproductive system" as a synonym for breeding system.

It could be interesting to compare the special characteristics of the genetic systems of the higher plants with those of the higher animals. In the next section another comparison will be made, namely, that between the genetic system of forest trees and the other higher plants. It is questionable, however, whether such general

comparisons are very fruitful (STEBBINS, 1950, 1960, and others). More precise answers should be obtained when subjecting two or more populations with initially
equal or very similar genetic systems to environments making different demands
upon certain components of the genetic system; or, conversely, subjecting populations with different genetic systems to identical or very similar environments and
observing the effects upon individual components. Such questions are relevant
because it is the situation at the start that largely determines which components of
the genetic system react and how a population finds its optimal breeding system —
if at all. Questions of this kind are the object of many, if not most, experiments of
population genetics and also of ecological genetics.

2. Main Characteristics of Genetic Systems of Forest Tree Species

In numerous instances the relevant literature alludes to the particular situation
of forest trees as compared with other higher plants. This situation results partly
from characteristics of the environment, partly from characteristics of their adaptations — both supposedly leading to special configurations of particular components of the genetic system. GRANT (1958), for example, sees the principal role of the
genetic system in pursuing an optimal balance between constancy and variability.
Like other authors he assumes for forest trees a high recombination index, which is
determined by chromosome number and crossover frequency. This may be correct
for many species but not for the widely distributed conifers. SAX (1932, 1933) found
average chiasma frequencies of 2.3 to 2.5 per bivalent in pines and larches, and
KEDHARNATH and UPHADAYA (1967) 2.4 per bivalent in two pine species. In all of the
species studied, the chromosome number is $n = 12$; their environments and adaptations corresponded to "typical" forest trees in all important characteristics. In a
similar study, MEHRA and BAWA (1969) found no evidence of very high recombination frequency in tropical angiosperm trees. DARLINGTON (1958) expects generally
higher recombination frequency in long-lived species, which would include perennial herbs and grasses.

Longevity offers greater opportunity to "experiment" while producing much
seed. STEBBINS (1958) assumed that the optimal genetic system of long-lived organisms is characterized by cross-pollination and high recombination rates, particularly in stable habitats. It is expected that populations of such species have stored a
large amount of genetic variability and probably give rise to new phyletic lines. An
extreme change in the recombination rate through selection is doubtless possible.
NEI and IMAZUMI (1968) have presented a review of the problem and summarized
the literature (also compare BODMER and PARSONS, 1962, and KIMURA, 1956 for a
simplified model).

SCHMUCKER and STERN (1969) have presented the following summary of chromosome numbers in conifers (Table 3).

According to Table 3 polyploidy in the conifers is rare. "True" polyploidy, i.e.,
multiplication of a basic chromosome set, appears to have occurred only in *Sequoia* and *Juniperus*. Whether this involves allopolyploidy or autopolyploidy will
not be considered. Polyploidy is not present in *Pseudolarix*, as MERGEN and GUSTAFS-
SON (1964) have shown; the chromosome number of $2n = 44$ has arisen through

breakage of chromosomes, leading to the distribution of the genetic information from a smaller number of chromosomes with two arms to a larger number with one. Similarly, in one species of *Pseudotsuga* with the haploid number $n = 12$, the genetic information of a two-armed chromosome was distributed to two one-armed chromosomes giving $n = 13$. In both cases this will have the effect of a higher recombination frequency, but as in *Sequoia* and *Juniperus*, the ecological necessity for these changes is not easily recognized.

The most interesting situation prevails in the *Dacrydium* and *Podocarpus*, family Podocarpaceae, with 20 and 80 living species respectively. Chromosome numbers range from 18 to 30 and 20 to 38 respectively. Ecological-genetic investigations in these two conifer genera would be of special interest. These could explore the question as to whether species with greater chromosome numbers need higher recombination indices. That this is problematic is indicated by the 90 species of *Pinus* with a consistent chromosome number of $2n = 24$. These occupy very diver-

Table 3. Chromosome numbers in Gymnosperms

Family	Genus	Number of living species	Chromosome number (if reliable known)
Lebachiaceae	*Lebachia*	14	
	Ernestiodendron	1	
Pinaceae	*Abies*	40	24
	Keteleeria	3	24
	Pseudotsuga	7	26, 24
	Tsuga	14	24
	Picea	40	24
	Pseudolarix	1	44
	Larix	10	24
	Cedrus	4	24
	Pinus	90	24
Taxodiaceae	*Sequoia*	1	66
	Sequoiadendron	1	22
	Metasequoia	1	22
	Taxodium	3	22
	Glyptostrobus	1	22
	Cryptomeria	1	22
	Cunninghamia	2	22
	Sciadopitys	1	20
	Athrotaxis	3	22
	Taiwania	1	22
Cupressaceae	*Cupressus*	15	22
	Chamaecyparis	6	22
	Thuja	6	22
	Thujopsis	1	
	Libocedrus	9	22
	Pilgerodendron	1	
	Callitris	20	22
	Neocallitropsis	1	

Table 3 (continued)

Family	Genus	Number of living species	Chromosome number (if reliably known)
Cupressaceae	*Tetraclinis*	1	
	Actinostrobus	2	
	Widringtonia	5	
	Fitzroja	1	
	Diselman	1	
	Fokiena	3	
	Arceuthos	1	
	Juniperus	60	22, 24
Podocarpaceae	*Pherosphaera*	2	26
	Phyllocladus	6	18
	Saxegotheae	1	24
	Microcachrys	1	30
	Dacrydium	20	30, 24, 22, 20, 18
	Podocarpus	80	38, 36, 34, 26, 24, 22, 20
	Acmopyle	2	20
Cephalotaxaceae	*Cephalotaxus*	6	24
Araucariaceae	*Agathis*	20	26
	Araucaria	15	26

gent ecological niches; with some species being "cosmopolitan," and others strictly endemic. Some species are typical pioneers, others are members of climax communities. In spite of such variable conditions, the resulting demands upon the breeding system, and therefore the recombination index, are not reflected in different chromosome numbers. The response of *Pinus* to the environment, resulting from such diverse demands, appears to lie on a different plane. Even in Central America, the apparent center of genetic diversity of the genus where many writers expect the continued development of new species (e.g., HAGMAN, 1967), no differences in the chromosome numbers of species or "races" have been reported.

The situation in the eucalypts (PRYOR, 1959) appears to be similar to that in the pines of Central America. Chromosome numbers of the several hundred Australian species of eucalypts are relatively constant, although the species are found in very diverse ecological conditions. In contrast, the willows exhibit considerably different chromosome numbers among species, and even within species groups with unequal basic numbers. However, the genus *Salix* is conspicuous for its frequent interspecific hybridization, and its evolution must therefore be seen in a different light from that of *Pinus* and *Eucalyptus*. The authors would like to note here that the conception of the different evolution of conifers, or of the different premises for their evolution, which has been presented by botanists of the old school and has since been repeated in the literature, cannot be supported in an objective manner. There is no evidence for the theory that the existing "old" gymnosperms have exhausted their evolutionary potential, etc.; quite the contrary, they still occupy a vast area and secure positions in stable ecosystems or in successions.

In some forest tree genera, polyploid series exist that may be interpreted as different species or races of the same species. Consistent results are not yet available. An example is the genus *Betula* with chromosome numbers ranging from $2n = 28$ to $2n = 112$, i.e., diploid to octoploid "species" or "populations." Here, too, the necessity for a higher recombination index is not apparent from the ecological niches of the species or populations. For example, *B. alleghaniensis* is hexaploid. It is a species of the climax forest and belongs to a taxonomic section of which the members generally occupy climax or near-climax positions. Other species of this section, e.g., *B. maximowicziana*, are diploid. *B. papyrifera* is reported to occur in several chromosomal races; it is a member of Section *Albae*, which consists entirely of pioneer species. *B. pendula* of the same section is diploid throughout, although this species occupies an area of similar dimensions and heterogeneity as *B. papyrifera*.

The investigations into frequency of chromosomal polymorphisms are of interest here. It is now recognized that polymorphisms of chromosome structures can contribute greatly to the complex adaptive value of populations; inversions are particularly important in this respect. Since the classical study of DOBZHANSKY and EPLING (1944), chromosomal polymorphisms have been used as genetic markers again and again (instead of single genes that were identifiable only in rare cases). The authors found frequency clines of inversion polymorphisms in *Drosophila obscura* and related species, indicating adaptive advantages of certain types. The resulting voluminous literature on this subject cannot be considered, even in excerpts. In essence, however, it confirms the interpretation given by DOBZHANSKY and EPLING. DOBZHANSKY (1967) has given a summary from the point of view that is of interest here; BEARDMORE et al. (1960) show that populations heterozygous for chromosomal polymorphisms produce a larger biomass — which may be of interest to the forester; many examples are available for the frequency of adaptive advantages conferred upon a species through new inversion polymorphisms (e.g., SPERLICH, 1966). ANDERSSON et al. (1969) have summarized the related literature on conifers, which indicates that chromosome polymorphisms are surprisingly frequent in this group. Thus SAYLOR and SMITH (1966) found that in *Pinus radiata* many of the normal trees investigated by them exhibited irregularities at meiosis.

If the reader will recall, inversions constitute gene blocks, i.e., supergenes, the loci of which cannot be broken up by crossover. Accordingly, they are considered units of recombination, although different alleles may be acquired within the inverted segment after the inversion polymorphism is established. If we may be permitted a comparison with other species, we would say that inversion polymorphisms appear to be more frequent in some conifers (gymnosperms) than has ever been reported in any other species of the animal or plant kingdom. But what this comparison means in a relative sense is not known.

Certainly the investigations, especially of forest tree breeders, indicate an immense variability in demes of many common forest tree species, and the recombination indices and other characteristics are not in opposition to an expected large variation. Nevertheless, these investigations do not permit conclusions such as have been drawn by some authors. The difficulty is to find or develop material in forest tree species for comparison of certain elements of the genetic system. For these reasons the following discussion will be limited to some well-known aspects of the genetic systems of forest trees, particularly their breeding systems, and to a comparison of different species. With regard to a detailed discussion of chromosome evolution in higher plants, the reader is referred to STEBBINS (1960).

One of the first problems is self-sterility, which is very common in the plant kingdom, and is to be understood in angiosperms mostly as self-incompatibility (between stigma and pollen of the same plant). This has been investigated in many species. HAGMAN (1970) shows that the common self-incompatibility system operates in *Alnus glutinosa* and *A. incana*. Growth of the pollen tube is retarded. The same system is often responsible for species incompatibility as well, and probably among these two species also. In an earlier study, HAGMAN (1967) provided a review of the types of self-incompatibility and self-sterility of other tree species, and recently (HAGMAN, 1971) he presented a monograph of self-incompatibility in *Betula pendula* and *B. pubescens*. STERN (1963), working with *B. pendula,* and CRAM (1952), working with *Caragana arborescens*, found similar systems; in both cases controlled crosses demonstrated the presence of the well-known self-incompatibility *(S)* locus with many alleles. There are also sufficient examples indicating the disappearance of self-incompatibility where cross-pollination is at a disadvantage, as in *Theobroma cacao* (COPE, 1962) or *Hevea*. Consequences for the population concerned are expected to be the same as for other species of higher plants (STEBBINS, 1957; HESLOP-HARRISON, 1966). The introduction or reintroduction of self-compatibility can hardly lead to the complete exclusion of cross-fertilization, for even in typical self-pollinators where selfing is favored by flower structure, a certain proportion of cross-fertilization is maintained (IMAM and ALLARD, 1965).

The possibility of switching from one mating type to the other is probably open to all populations. It is known how quickly selection may produce self-incompatible lines in cross-fertilizing species (e.g., OLSSON, 1960), and how much variation exists in trees for self-sterility as a result of the differences in self-incompatibility. The large number of S alleles may result from high mutation rates, as EWENS (1964) suggested. BATEMAN (1947) estimated the number of S alleles as ranging from 115 to 442 in red clover; upper estimate limits have been given by FISHER (1947) and PAXMAN (1963). RAPER et al. (1958) found in *Schizophyllum* 339 A alleles and 64 B alleles with 5% confidence limits at 216 and 562, 53 and 79, respectively. That COPE (1958) positively identified only five S alleles in *Theobroma* is not necessarily an indication of differences in the special self-incompatibility system of this species. The frequency of the self-incompatibility phenomenon in the angiosperms (EAST, 1940 had knowledge of approximately 3000 self-incompatible angiosperm species), and the success in achieving self-incompatibility by means of the simple selection experiments mentioned above, indicate that even the long-lived tree species can change from one to the other mating type in a relatively short time (compare ARASU 1968 for a review in angiosperms).

A good example from *Eucalyptus tereticornis* supporting this view is given by VANKATESH (1971). This species is generally cross-pollinating, like other eucalypts (see ELDRIDGE, 1970 and references cited there), but the flower structure in one tree was different, enforcing cleistogamy.

LANGNER (1959), working with *Picea omorica*, and FOWLER (1965), working with *Pinus resinosa*, obtained results that may best be interpreted as a return to self-fertilization. *Picea omorica* is an endemic species in Serbia, existing in a very small area and with low population density, which entails a short supply of foreign pollen. *Pinus resinosa* regenerates after fires from the seed of single trees. Here, too, the small amount of available pollen may explain the situation.

The self-sterility system of the conifers (as distinguished in this terminology from the self-incompatibility system of the angiosperms and other plants) offers some interesting facets that are of no concern in the angiosperms. To begin with, it has been known since the studies of ORR-EWING (1957) on Douglas fir that the self-sterility of conifers results from embryo collapse after self-fertilization (for further literature see HAGMAN, 1967). Thus self-fertilization is not prevented as in the prezygotic incompatibility systems, but the embryos resulting from selfing are eliminated in the postzygotic phase. KOSKI (1971) suggested that embryo collapse after selfing is caused by embryonic lethal alleles, and he estimated their frequency after selfing experiments with Finnish *Picea abies* as 7 to 11 per individual. According to his view, therefore, the lethal equivalents become effective as early as the embryonic stage and are responsible for self-sterility. SORENSEN (1969) likewise submitted an estimate of lethal equivalents in a conifer *(Pseudotsuga menziesii)*, suggesting frequencies of the same order given by KOSKI for *Picea abies*.

Similar postzygotic sterility mechanisms are also known from other species, for example, the angiosperms, and from trees such as *Theobroma cacao* (CHEESMAN, 1927; COPE, 1958, 1962a, b). In this case, however, self-incompatibility appears to exist between the embryos resulting from selfing and the tissues that surround them. S alleles exist and one should designate this process as postzygotic self-incompatibility. COPE (1958), for instance, found in *Theobroma* five S alleles, surely the lowest limit. Thus as long as detailed investigations are not available it is still an open question as to whether KOSKI's (1971) interpretation is correct. There is certainly some justification for doubt, for 7 to 11 lethal equivalents per individual for the embryonic stage alone represent a disproportionally high genetic load for the population. To these would have to be added the lethal equivalents of the postembryonic stage; their number is also considerable, as is known from many inbreeding studies with conifers.

A second unique feature of the conifer mating system is the presence of the so-called pollen chamber. Initially the pollen is trapped by a drop of fluid in front of this chamber, and as soon as the fluid disappears in the course of evaporation or otherwise, pollen is drawn with the fluid into the pollen chamber. This chamber has a limited capacity ranging from two to three in Scots pine and from four to seven in Norway spruce according to SARVAS (1962, 1968). With increasing pollen supply and a steady rate of self-pollination, the probability for cross-fertilization should increase (polyembryony should lead to the survival of the individual embryo resulting from cross-fertilization). With self-pollination rates of 0.1, 0.2, 0.3, and 0.4, these proportions would simultaneously indicate the share of self-fertilization if every pollen chamber had the capacity of a single pollen grain. With increasing capacity of the pollen chamber and pollen density a rapid change takes place (see Tables 4a and b).

The figures in Tables 4a and b are nothing more than fairly rough estimates, but indicate that the effective rate of selfing depends upon the capacity of the pollen chamber and pollen density, both variable quantities, in addition to the supply of the tree's own pollen. KING et al. (1970) have made *Picea glauca* crosses using

Table 4a. Rates of self-fertilization when the pollen chamber has a capacity of two pollen grains and self-sterility is absent. Pollen density refers to the occupancy of positions in the pollen chamber

Pollen density	% of the tree's own pollen			
%	10	20	30	40
100	0.01	0.04	0.09	0.16
80	0.04	0.09	0.16	0.24
60	0.06	0.13	0.21	0.30
40	0.08	0.16	0.25	0.34
20	0.09	0.18	0.28	0.37

Table 4b. Rates of self-fertilization when the pollen chamber has a capacity of four pollen grains. The remaining assumptions are identical to those in Table 4a

Pollen density	% of the tree's own pollen			
%	10	20	30	40
100	0.00	0.00	0.01	0.03
80	0.01	0.02	0.04	0.07
60	0.02	0.05	0.09	0.15
40	0.04	0,10	0.16	0.23
20	0.07	0.15	0.23	0.32

various mixtures of the tree's own and foreign pollen, with the tree's own pollen contributions amounting to 0, 12, 50, 90, and 100%. Increasing proportions of the tree's pollen reduced the percentage of full seed and retarded growth of the epicotyl. Therefore, results like this also depend upon the pollen quantity offered by the individual male flower. In angiosperms the premises differ as a result of their comparatively, and normally larger, capacity of catch pollen, and their possibilities for prezygotic sorting of their own and foreign pollen. The proportion of foreign pollen in the pollen cloud surrounding the crown of coniferous trees has been estimated by several workers. At least some typical examples will be given. SARVAS (1962) estimated the percentage of their own pollen in the crowns of male-flowering Scots pine trees as being in the range of 22–37, averaging 26. These figures are valid for the larger Scots pine stands of Finland. In different conditions the percentage of own pollen may be larger; DOYLE and O'LEARY (1935, cited by SARVAS, 1967), provided figures indicating that in Scots pine the quantity of own pollen greatly exceeded that of foreign pollen. STERN (1972) found a self-pollination rate ranging from 0 to 82%. The values shown in Fig. 12, estimated in one stand of the pine region in the Lüneburg Heath and another small isolated pine stand near Hannover, are of the same order. These results from experiments conducted in the spring of 1970 are representative of many others of a similar kind.

The only ready convincing evaluation of the influence of male and female strobilus position upon self-pollination was made by FOWLER (1965). Male strobili

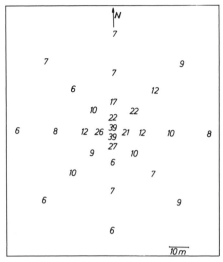

Fig. 12a

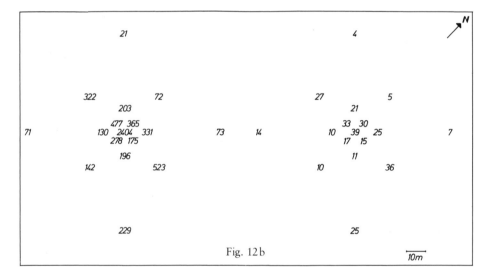

Fig. 12b

Fig. 12a–h. Pollen distribution around individual trees in stands of *Pinus sylvestris*. a Percent of pollen in the crown of a tree located in a stand near Hannover; counts taken at a distance of 6, 12, 25, and 50 meters from that tree. The pollen was labeled by manganese injection. (From FENDRIK, 1967). b Figures from one tree in the same stand given under (a). At left: number of labeled pollen grains. At right: their percentage in the total quantity of captured pollen. (From SCHMIDT, 1967). c Results from the same tree given under (b) but in the following year. d Results from another tree in the same stand. e and f Results from two trees in different stands of the Lüneburg Heath pine region (c–f from STERN, 1972). The Proportion of own pollen in the crowns of the labeled trees ranges here from 39 to 82% depending upon pollen production of the surrounding trees and the specific tree itself; the actual range ist 0–82% (or more) because many trees do not produce male strobili even in heavy-flowering years. g and h Pollen distribution around two Scots pine trees in Finland labeled by injection with P^{32}. (From KOSKI, 1967)

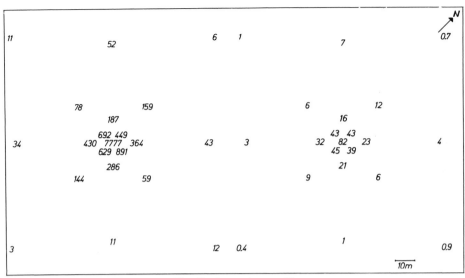

Fig. 12c. Legend see p. 69

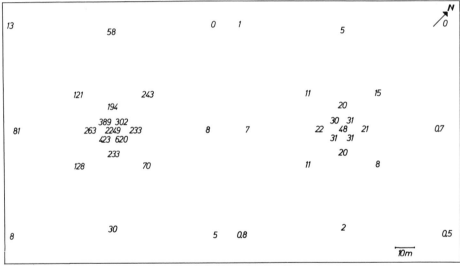

Fig. 12d. Legend see p. 69

are concentrated in the lower part of the crown, female strobili in the upper part.
FOWLER's results are presented in Fig. 13. They indicate that a long crown reduces
self-pollination drastically.

We have seen earlier (Chapter II) that in wind pollinators, male and female
flowering varies in response to both genetic and environmental factors, although a
need exists to limit the process to a short period. Since the main period of pollen
flight is often contracted into a few days, relatively small differences in mean dates
of male and female flowering of a tree would suffice to reduce self-pollination rate
drastically. SARVAS (1967) reports some related observations from *Pinus silvestris* in

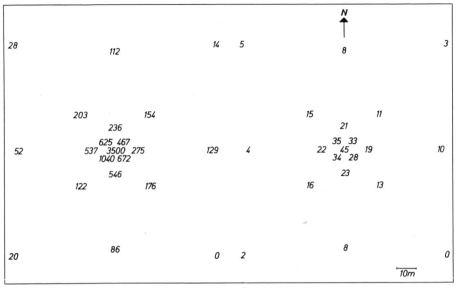

Fig. 12e. Legend see p. 69

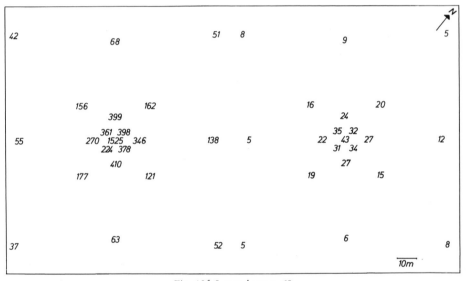

Fig. 12f. Legend see p. 69

Finland. The tree investigated by him is distinctly proterogynous, but the female flowers are usually not closed before pollen flight is terminated. (See Fig. 14)

The experimental results reported here to this point indicate: (a) that rate of selfing cannot be correctly estimated from studies of pollen flight; and (b) that the influence of many factors, which often change from year to year, cannot be estimated with any general degree of validity. In spite of many experiments with Scots pine and Norway spruce, the relationships have not been adequately researched even in these species, particularly the environmentally induced deviation around

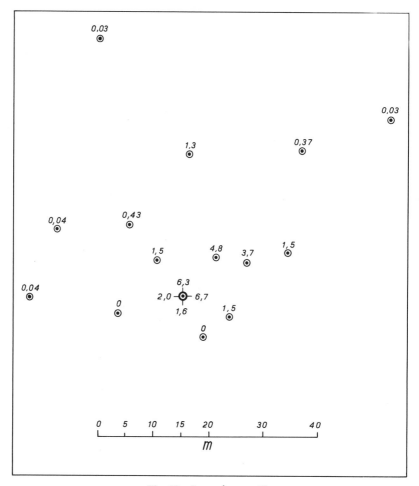

Fig. 12g. Legend see p. 69

mean values. Forest trees are often difficult objects for research of an experimental nature and this should be noted in regard to their mating systems. Detailed studies of the kind made on other plant species, which often included all relevant components of the breeding system, are still lacking (see VASEK, 1964 and later).

Experiments using marker genes should yield more reliable information on the actual self-fertilization rate than experiments on rate of self-pollination, which do not necessarily permit reliable conclusions for the reasons mentioned. Many such experiments have in fact been made. SQUILLACE and KRAUS (1963) worked in *Pinus elliottii* with rare chlorophyll mutants that resulted from selfing of trees in a stand, and estimated the rate of self-fertilization as 6%. KOSKI (1970), who based his studies on chlorophyll mutants and the data of EICHE (1955), calculated a self-fertilization rate in *Pinus sylvestris* of 9%. SARVAS (1962) concluded that in spite of an average self-pollination rate of 26%, the self-fertilization rate was only 7% (percentage of full seed resulting from self-fertilization). BANNISTER (1965), working

Fig. 12h. Legend see p. 69

with *P. radiata*, deduced a self-pollination rate of 10% and self-fertilization rate of 2%, and the results of Fowler (1965) from *P. resinosa* are similar. Barnes et al. (1961) demonstrated in experiments with mixed pollen of *P. monticola* that in spite of incompatibility a part of the embryos resulting from selfing survived the competition. But even these experiments, although coming closer to the goal of estimating the effective self-fertilization rate, are not entirely satisfactory. As many authors indicate, the given figures tell us very little about the actual role that seeds and seedlings from selfing play in nature in subsequent stages. One may assume that most of them will not reproduce because of the drastic inbreeding depressions observed in conifers — with the exception of the few species that have been mentioned. With this in mind, the mating systems of our conifers could be described in general as a system with obligate cross-fertilization. The question regarding inbreeding is then applicable only with respect to genetic drift (Chapter IV). The authors are in favor of this position.

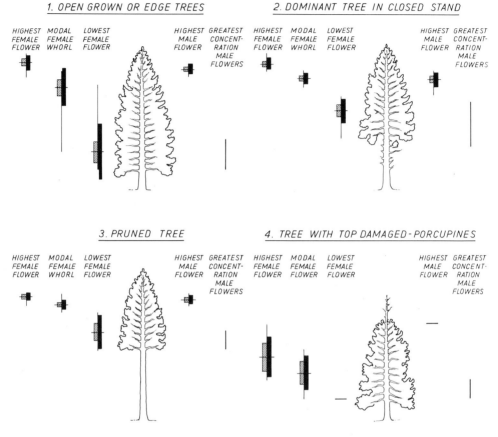

Fig. 13 a. Position of male and female flowers in the crowns of red pine trees of four crown types. The vertical line represents the sample range; the short horizontal line is the sample mean; the black rectangle to the right of the range line is the standard deviation above and below the mean; and the hatched rectangle to the left of the range line two standard errors above and below the mean. (From FOWLER, 1965)

The literature reflects some confusion regarding weighting of components in the mating system of Scots pine. It is therefore probably useful to classify these components once again in spite of the realization that such classifications are problematic. The main reason to do so is to determine the significance of self-fertilization for the genetic structure of Scots pine populations. This problem has been considered at several levels, both in theory and in experiments.

1. Prezygotic phases.
a) The proportion of own pollen.
   The probability of self-fertilization is determined first of all by the probability that a randomly selected pollen grain in the pollen cloud surrounding the crown of a tree is derived from the tree itself. This probability in Scots pine is approximately 26% (SARVAS, 1962), or may range from 0 to 82% (STERN, 1972).

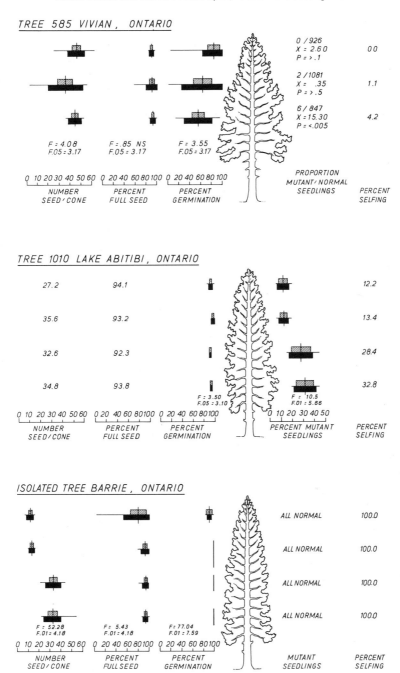

Fig. 13b. Natural selfing in *Pinus resinosa*. The horizontal lines represent the sample range; the short vertical lines the sample mean; the blackened rectangles below the range lines one standard deviation above and below the mean; and the hatched rectangles above the range lines two standard errors above and below the mean. (From FOWLER, 1965)

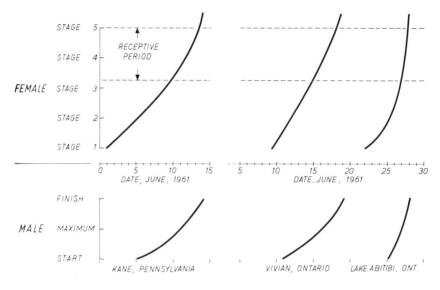

Fig. 13c. Time of pollen release and ovulate strobili receptivity of red pine trees at Kane, Pa.;
Vivian, Ont.; and Lake Abitibi, Ont. (From FOWLER, 1965)

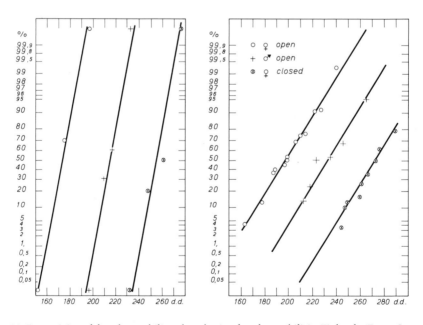

Fig. 14. Receptivity of female strobili and anthesis of male strobili in Finland. (From SARVAS,
1967)

b) The proportion of self-pollination.

This indicates the probability that any pollen grain in a randomly chosen pollen chamber originated on the same tree. The proportion of self-pollination and of own pollen are not identical because of intervention through proterogyny combined with individual differences in male and female flowering dates, and differences in the arrangement of male and female flowers in crown space. Flower arrangement may be of variable significance in different species of *Pinus* or different genera of the conifers. For example, trees in northern races of *Pinus silvestris* have long crowns with a distinct differentiation into zones with predominantly male and female flowers; in southern races umbrella-shaped crowns may lack such differentiation. In *Picea*, conical crowns generally lead to such differentiation. Overlapping male and female flowering periods and capacity of the pollen chamber must also be considered in this genus.

c) The proportion of self-fertilization.

This indicates how many eggs in an ovule are fertilized by the tree's own pollen. Mechanisms regulating competition between own and foreign pollen may become effective here. Everything known about this subject indicates that differences in the proportion of self-pollination and proportion of self-fertilization are immaterial or nonexistent (in contrast to the angiosperms).

2. Postzygotic phases.

a) The proportion of seeds with embryos following self-fertilization.

Embryo collapse following self-fertilization (due to incompatibility or lethal genes) reduces the proportion of seeds with embryos from self-fertilization compared with the initial rate of self-fertilization. Embryo collapse may also be related to competition between embryos of the same ovule.

b) The proportion of population members capable of reproduction following selfing, or the "effective self-fertilization rate." This is the proportion of individuals in the reproducing population that originated from self-fertilization. The phases 1a to 2a above include an increasing number of barriers that must be overcome by individuals from selfing before they can participate in the reproduction of the population. The last barrier, consisting of phase 2b, is also important and results from inbreeding depression of the plants that arises from selfing. Published experiments concerned with selfing of conifers indicate a large inbreeding depression in most species. Exceptions are the small number of species that occupy very special ecological niches, and only in this instance is the effective self-fertilization rate of any importance for population structure. In most cases it is considered immaterial.

This review indicates that our knowledge of the mating systems of those tree species that have been studied most is limited. Thus the scarcity of information on other species is not surprising. This is unfortunate, particularly with respect to the tree species of the tropics and subtropics. Studies of this group would likely provide an impetus to investigations in the temperate and boreal zone as well, where the specificity of the problems precludes an advance into more general areas. There are, of course, some exceptions, such as the investigations dealing with *Hevea* and *Theobroma*, but these genera may not be typical of this group.

IV. Adaptive Strategies

Every definition of the adaptive value of populations entails difficulties and uncertainties. The adaptive value cannot be derived easily from the coefficient of selection, but this is only one of the difficulties. Every function that may be useful to define the adaptive value is valid only for a particular usually narrow, situation. Meaningful results can only be expected where the definition reflects the actual process of adaptation correctly. The principal difficulties appear to lie in the fact that adaptation to different environments imposes different demands upon the population, which, in turn, may have several possible solutions at its disposal.

Most commonly used (although probably not useful for many situations) is still the "mean fitness," i.e., the mean value of fitness over all members of the population. It is not possible to present here the population-genetic theory of "mean fitness" and of related measures of adaptation in a complete and understandable way, and the reader is therefore referred to the literature, particularly CROW and KIMURA (1970), TURNER (1970), and LEVINS (1970).

In addition to a high mean fitness, a population with high adaptive value should also possess other qualities. Many attempts have been made to express these qualities in a precise way and to use them as separate measures of adaptation or components of a combined measure. To achieve this, researchers have drawn upon the theory of games (WADDINGTON, 1957; LEWONTIN, 1961; WARBURTON, 1967, and others) or upon cybernetics (SCHMALHAUSEN, 1949, 1960; PIMENTEL, 1965, and others). Such models usually lead to the designation of optima and the identification of population structures achieving relatively high adaptive values; they are less suitable to specify the population compositions and structures for maximum adaptation. The optima arise often from a simultaneous consideration of several, occasionally antagonistic, processes, resulting from special characteristics of the environment but also from characteristics of the genetic system, which in turn may also be "optimized." THODAY (1953) considered the capability for permanent survival in a habitat as a necessary component of a high adaptive value. MACARTHUR and WILSON (1967) stressed in addition the capability to reoccupy lost territory, which indeed is an indispensable component of the adaptive value of forest trees. In these cases, reasons for the high adaptive value of "defensive strategies" are uncertainties of the environment. Definitions of the adaptive value based on the premise of a stable environment lead here to erroneous expectations.

There are similar difficulties in other cases. The characteristics determining the adaptive value are often not evident. The characters controlling the growth of populations, the number of viable populations, and their period of existence are often not the same (LEVINS, 1970). There are also situations when selection achieves a stable population size commensurate with the resources. The writer (STERN, unpublished data) found that populations of *Drosophila melanogaster* kept for long periods in closed population cages grew relatively slowly although conditions were favorable for rapid population growth. Population growth of lines selected over many generations for slow development, the individuals of which required on average six days more than individuals from the original population for full development, was not much slower than that of the populations previously mentioned. The fastest growth was made by populations based on individuals from lines subjected to disruptive selection for duration of development. Next came the lines selected for rapid development and those selected for intermediate rate of development through stabilizing selection. These results and others in the literature indicate that the "mean fitness" represents an unequivocal measure of the adaptive value of a population only under certain conditions that perhaps could be more precisely defined.

The mean fitness has in fact been the measure of adaptive value that has been most commonly used since FISHER's (1930) fundamental study. A large number of writers, including, of course, FISHER himself, have examined its application and in most cases admitted its limited value. In addition to the studies already mentioned, the reader is referred especially to UNDERWOOD (1954), who attempted to develop different categories of adaptation; to BIRCH (1960), who investigated the difference between birth and death rate and its effect upon adaptive value, seeing it as a compromise between different selection principles; or to AYALA (1969), who described the dilemma that may result from the contrast between the adaptedness of populations and the fitness of their members. WALLACE (1952), to cite but one study, has discussed methods of estimating adaptive value in particular.

In spite of these limitations we shall subsequently regard the "mean fitness" as equivalent to adaptive value. This is justifiable here for convenience inasmuch as we attempt first of all to show how different the solution may be that is embodied in the "optimum population," depending upon the divergent conditions of the environment and characteristics of the genetic system. Given these presuppositions, it is probably correct to assume that natural populations differ in the direction of their optima (LEVINS, 1968). One must be aware of the danger of applying the following results uncritically to conditions found in nature. Rather, it is necessary in ecological genetics to check the conditions, past and present, that influenced the development of the populations under study. Unfortunately it is not possible to account for population-biological phenomena in only one way, by referring to a single, ruling cause.

1. Formal Concept of Adaptive Strategy

The most useful model for the derivation of an optimal adaptive strategy in different environments and properties of a genetic system was developed by LEVINS in the 1960's (summarized by LEVINS, 1968). This discussion will be limited to his approach and related to problems of forest ecosystems.

A model population that is stable, i.e., constant in time (LEVINS, 1962), is assumed. The environment exists in two forms, each favoring a different optimum phenotype (selection operates on the phenotype, and to make assumptions more realistic, we shall suppose one gene locus with one dominant and one recessive allele). Extension to a larger number of phenotypes and environments (ecological niches) is easily possible. An optimum population will be the one having the highest mean fitness, in both environments the highest "innate capacity to increase" as defined by ANDREWARTHA and BIRCH 1954, or in similar ways, all referring to the population with the largest number of offspring.

Corresponding to an n-dimensional egological niche, there will be an n-dimensional fitness space. Its dimensions are the niches for which the population in mind provides the especially adapted phenotypes, e.g., one or more phenotypes g_j with the frequencies q_j. With frequency independence of the fitness values w_{ij}, the fitness of these phenotypes in the ith niche is then

$$w_i = \sum_j q_j w_{ij}$$

where w_{ij} represents the fitness of the jth phenotype in the ith niche.

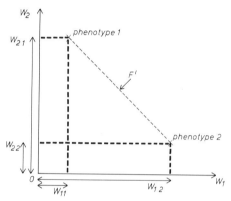

Fig. 15. The abscissa depicts fitness in the first and the ordinate fitness in the second niche. In this example, in the first niche fitness of the second phenotype exceeds that of the first, i.e., $w_{12} < w_{11}$. The reverse is true in the second niche where $w_{21} > w_{22}$

Given only two phenotypes and two niches, these two phenotypes can be depicted by two points in a two-dimensional fitness space as shown in Fig. 15. Let F be the set of fitness points in the n-dimensional fitness space. In a simplified situation, as in Fig. 15, there will be only two. These two points could now be thought of as the mean fitness of two populations consisting entirely of the one or the other phenotype. In addition to such "pure" populations, there will be also "mixed" populations made up of different proportions of the two phenotypes (q_j may range from 0 to 1). In conformity with the preceding equation, therefore, the mean fitness values of these mixed populations should be located along the line connecting $w_{12}; w_{22}$ and $w_{21}; w_{11}$. Thus the frequency of fitness points including fitness values of mixed populations is represented in fitness space by points lying on the straight line joining $w_{12}; w_{22}$ and $w_{21}; w_{11}$. This "extended fitness set" is designated by F'.

An example will illustrate this relationship. In populations of the European species *Quercus robur* there exists a polymorphism for date of flushing. Early flushing types are in the majority. Caterpillars of the oak leaf roller, *Tortrix viridana*, emerge from eggs at a date that is precisely synchronized with the flushing date of *Q. robur*. Caterpillars from eggs deposited on late flushing oak

Table 5. Mean fitness values $\overline{W}$ for the *Tortrix viridana* example as governed by niche frequency p and phenotype frequency q

		Niche frequency, p				
		0	0.25	0.50	0.75	1
		Values of $\overline{W}$				
Phenotype	1	0	0.25	0.50	0.75	1
frequency q	0.75	0.25	0.375	0.50	0.625	0.75
	0.50	0.50	0.50	0.50	0.50	0.50
	0.25	0.75	0.625	0.50	0.375	0.25
	0	1	0.75	0.50	0.25	0

trees die of starvation. The fitness value of the *T. viridana* population adjusted to the early-flushing oaks is 1 in a niche consisting exclusively of early-flushing oaks, and 0 in a niche with only late-flushing trees. If a type of *Tortrix* existed that was specialized for late-flushing trees, relationships would be exactly the reverse because the young caterpillars depend upon the tender leaves that they normally find near buds. Table 5 gives fitness values of "pure" and mixed populations of both phenotypes. It was also assumed that the two niches, early- and late-flushing oaks, occur with different frequencies and are indistinguishable to the egg-laying females of *Tortrix*. The fitness of the "pure" and mixed populations of the two phenotypes over both niches, which also have different frequencies, is then given by the equation:

$$\overline{W} = pq\,W_{11} + (1-p)\,q\,W_{21} + p\,(1-q)\,W_{12} + (1-p)\,(1-q)\,W_{22}.$$

In this equation the first index refers to the first or second niche, i.e., to early- or late-flushing oaks; the second index refers to the two phenotypes, i.e., to early- or late-emerging caterpillars. The fitness values of the phenotypes are 1 and 0, or 0 and 1, respectively, in the two niches: early caterpillars cannot survive on late oaks, and late caterpillars will die on early oaks. For this reason

$$\overline{W} = pq \times 1 + (1-p)\,q \times 0 + p\,(1-q) \times 0 + (1-p)\,(1-q) \times 1$$

$$= pq + (1-p)\,(1-q)\,.$$

This equation indicates particularly well that $\overline{W}$ depends on phenotype frequency q, niche frequency p, and also the fitness values of the phenotypes in both niches. Table 5 summarizes the dependence of mean fitness on niche and phenotype frequencies.

Apparently the best "strategy" of the *Quercus* population is to develop a polymorphism for flushing. The two flushing types should be represented with equal frequencies and distributed in random mixture. However, special "adaptation" may also be available to reduce the annual damage caused by *Tortrix viridana*, as the studies of SATCHELL (1962) indicate. In the Roudsea Wood National Nature Reserve, Great Britain, late-flushing types were resistant to *T. viridana*, but most of the early-flushing trees were also resistent. This phenomenon is probably caused by rapid shoot growth that apparently outpaces caterpillar feeding and the damage associated with it in spite of equal caterpillar frequency.

ALTENKIRCH (1966) reported on different "races" of *T. viridana* in Portugal. There *Quercus suber* and *Q. ilex* occur sympathically over large areas. *Q. suber* flushes early and pupae are already found at a time when the caterpillars on *Q. ilex* are just hatching. Obviously *T. viridana* has developed an adaptation that enables the insect to distinguish between the two species. No distinct morphological differences exist between the population feeding on *Q. suber* and that feeding on *Q. ilex*; ALTENKIRCH described them as races. Investigations of the nature and degree of isolation have not been carried out, yet the gap between the "races" is apparently large enough to classify them as subspecies.

BOVEY and MAKSYMOV (1959) described the similar case of *Zeiraphera griseana*, which occurs in the high Alps on *Larix decidua* and *Pinus cembra*. Here the two "niches" consist of two species in separate genera. Two populations of *Zeiraphera* exist whose members usually divide into *Larix* and *Pinus*. Crosses can still be made and the hybrids then feed on both host species. In the example dealing with the resistence of early- and late-flushing oaks against *Tortrix viridana,* for every niche frequency p there exists a gene frequency q that provides the population of *T. viridana* with maximum fitness. If $p = 1$ or 0.75, then maximum fitness is at $q = 1$, and if $p = 0$ or 0.25 it is at $q = 0$. Conditions are different only at $p = 0.5$; here the fitness value of the *T. viridana* population is entirely independent of q (the same argument holds for the fitness value of the *Quercus* population in relation to p). The optimal population of *Quercus* should be found here in consequence of the smallest damage from *T. viridana*. An exception would be the development of a situation similar to that described by ALTENKIRCH in Portugal — a not impossible occurrence in view of the short generation interval of the insect.

We shall now consider a case where the fitness set cannot be represented along a straight line, but by a curve. It is assumed that the fitness value of the population in two niches is determined exclusively by the frequencies of the three genotypes of a locus with two alleles. There is random mating in the population. The mean fitness of the population will again depend upon the niche frequencies p and $1 - p$ and the gene frequencies q and $1 - q$.

In the *Tortrix-Quercus* example, the fitness values of the two phenotypes in the first and second niche were 1 and 0 and 0 and 1, respectively. This need not be so, and generally the mean fitness of a population over two niches is given by its "adaptive function" (LEVINS, 1962):

$$A(W_1; W_2) = pW_1 + (1 - p) W_2 = K.$$

Now assume that in every generation the population will be distributed over both niches in the following way: the probability of any one individual to grow up in the first niche is p, and in the second niche $1 - p$, and this is independent of the parental niche. The highest mean fitness is achieved by the population with the largest value for K. When writing the equation

$$W_2 = \frac{K}{1 - p} - \frac{p}{1 - p} W_1$$

and using a coordinate system where W_1 is the abscissa and W_2 the ordinate, a straight line is obtained. Its slope depends upon niche frequency p, and its position in the coordinate system upon K, i.e., the value of the adaptive function.

We now again consider the population with three phenotypes representing the three genotypes of a locus with two alleles. Frequencies of the three phenotypes are equal to those expected from random mating. The fitness values of the three phenotypes in both niches will be

	W_1	W_2
AA	0.8	0.6
Aa	1.0	1.0
aa	0.5	0.9

The mean fitness of the population in each of the two niches depends upon the frequency of A, i.e., q, and is given by

$$\overline{W}_1 = q^2 W_{12} + 2q(1-q) W_{11} + (1-q)^2 W_{10},$$

$$\overline{W}_2 = q^2 W_{22} + 2q(1-q) W_{21} + (1-q)^2 W_{20},$$

where the second index of W_{ij} stands for the number of A alleles of the genotype (phenotype). The extended fitness set F' is obtained from these equations. (See Fig. 16.) The "adaptive strategy" of the population should lead it to strive for a composition providing it with a maximum value of the "adaptive function" $A(W_1, W_2)$. However, this is valid only on condition of our premise that the mean fitness of a population is a good measure of its adaptive value.

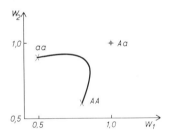

Fig. 16. "Extended fitness set." F' of pure and mixed populations if three phenotypes with different fitness are distributed over two niches. The frequencies of the three phenotypes (genotypes) correspond to those expected from random mating in a population having one locus with two alleles. The genotype Aa exhibits superior fitness in both niches (see text). Mean fitnesses of the populations lie along the curved line connecting the pure populations, namely, the points $W(1, AA)$, $W(2, AA)$, and $W(1, aa)$, $W(a, aa)$

2. Fitness Set and Optimum Population

Let us consider again the population giving rise to Fig. 16, this time by taking note also of niche frequencies. The highest value K of the adaptive function at a given niche frequency p should be associated with that population at the point of contact with the straight line, which is the only tangential line of the group of

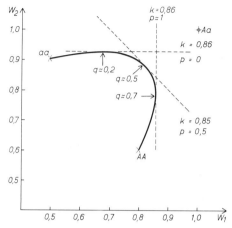

Fig. 17. Determination of the optimum populations (populations with highest K) for the same population given in Fig. 16 when $p = 0$, $p = 0.5$, and $p = 1$. Optimum gene frequencies are at 0.2, 0.5, and 0.7, respectively

possible (parallel) straight lines with the same p value. This particular straight line is the one with the highest value of K, having one point in common with the extended fitness set F' (See Fig. 17).

Yet one should note that the proceeding statement is valid only in a fine-grained environment (LEVINS and MACARTHUR, 1966), i.e., an environment the components of which are utilized by the population in the same proportion as they occur. In a coarse-grained environment the fitness function assumes a different form. This is most easily seen when examining the example of a population having successive generations in random order in niches 1 and 2. Let the number of generations be n, let the proportion of generations spent in niche 1 be p, and let the proportion spent in niche 2 be $1 - p$. Population growth after n generations is then given by the factor

$$W_1^{np} \ W_2^{n(1-p)}.$$

The mean growth rate for each generation was

$$W_1^{p} \ W_2^{(1-p)} = A(W_1 \ W_2) = K .$$

In fitness space, instead of having groups of straight lines we now obtain for K, in dependence of p, groups of parabolic curves of the form

$$W_2^{(1-p)} = \frac{K}{W_1^{p}}$$

or in logarithms

$$\log K = p \log W_1 + (1 - p) \log W_2 ,$$

$$\log W_2 = \frac{\log K}{1 - p} - \frac{p}{1 - p} \log W_1 .$$

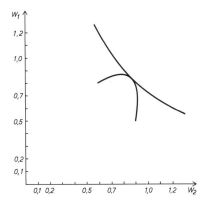

Fig. 18. The form of the adaptive function for $p = 0.5$ when the environment is coarse-grained. The population is that given in Fig. 16

Figure 18 is a graphic representation of this relationship when $p = 0.5$, and of the example given in Fig. 16.

Adaptive functions thus differ in fine-grained and coarse-grained environments, but the procedure of deriving them is the same in both cases. Of course in nature there will be transition between these two extremes. This means that the adaptive function is described by any intermediate curve, ranging from the straight line for a fine-grained environment to a parabole-like curve for a coarse-grained environment.

The example in Table 5 described a situation where no polymorphic optimum populations could be expected in a fine-grained environment, even with pronounced heterogeneity of the niches. The fitness set F' was represented by the straight line that connected the two monomorphic populations. Polymorphic populations would be possible, although not likely, only if $p = 0.5$. In this case the straight line with highest K corresponded to the straight line of the fitness set. Now let us consider a similar population in a coarse-grained environment. Here we will expect from the beginning that, as a rule, polymorphic optimum populations may be possible.

LANE and ROTHSCHILDT (1961) have found larvae of *Zygaena lonicera* requiring one and two years for development. They interpret this polymorphism as a defense strategy in a coarse-grained environment. The strategy ensures the existence of the population in those years when the survival of adults is in jeopardy for some reason or another, e.g., due to lack of food. To cite a forestry example, in a population of the seed wasp *Megastigmus*, LESSMANN (1971) found forms requiring one, two, or three years for development. It is interesting to note that the observation concerned a species associated with Douglas fir, which in Europe varies much in its annual seed production. This polymorphism was lacking in another species of *Megastigmus* having greater security of food supply. On the other hand, in years completely without food, the population having a one-year cycle would disappear completely provided it cannot again evolve from the second phenotype. This would require dominance of the gene controlling development in two years.

When considering the complexity of this situation, it is difficult to assume realistic fitness values for the phenotypes in the two niches. Yet the principle is clear. The fitness values cannot alternately be 0 and 1, for this would lead to the loss of the entire population in two successive years without food. (With separation of niches in space this catastrophic event would be avoided since the reestablishment of the population by immigration is then still possible.) We therefore arbitrarily substitute the following values:

Niches	Phenotypes	
	1	2
1	$W_{11} = 1$	$W_{12} = 0.1$
2	$W_{21} = 0.1$	$W_{22} = 1$

These values should be put into the equation for derivation of mean fitness that was given in the previous section.

Table 6 summarizes the values for K corresponding to those in Table 5. Monomorphic optimum populations can be expected only in the trivial cases where $p = 1$ and $p = 0$. With intermediate values of the niche frequency p, polymorphic populations will have the largest K values and therefore the highest mean fitness.

Up to this point we have dealt with phenotypes that could be represented by a single point in fitness space. This was based on the assumption of distinct phenotypic classes that result from inheritance through major genes (or gene blocks). In this case each distinct phenotype possesses a certain fitness in both niches (assuming that niche differences were equally distinct). In contrast, in nature we will find much more often continuous variation in the coördinates of niche space as well as in the scale of the phenotypes. To begin with, let us first consider the fitness of two phenotypes in relation to the continuous scale of a niche factor, perhaps temperature. Both phenotypes will probably reach optimum fitness at different values of the niche dimension, and fitness will decrease above and below these values. We assume that this curve of optimum values is symmetric and described by a normal distribution (the last assumption facilitates numerical description). Let y stand for optimum value and s for the distance of a certain niche from this optimum. Phenotype fitness will then decrease with the increasing absolute distance $|s - y|$. A factor may be introduced that measures the degree of

Table 6. Values of $\overline{W}$ (mean fitness) as determined by niche frequency p and phenotype frequency q. Example constructed for a coarse-grained environment

		Niche frequency, p				
		0	0.25	0.50	0.75	1
Phenotype frequency, q	1	0.1	0.178	0.315	0.562	1
	0.75	0.325	0.404	0.502	0.624	0.775
	0.50	0.55	0.55	0.55	0.55	0.55
	0.25	0.775	0.624	0.502	0.404	0.325
	0	1	0.562	0.316	0.178	0.1

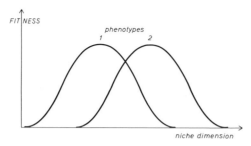

Fig. 19. Distribution of the fitness values of two distinct phenotypes over one niche dimension measurable by means of a continuous scale

reduction, and one could therefore write $C|s - y|$. This will account for the "tolerance" or other characters of the phenotype (LEVINS, 1962), at least in a formal manner. Figure 19 illustrates this for two phenotypes in relation to the continuous scale of one niche dimension.

Because the terms "phenotype" and "niche" are interchangeable, the relationship can be presented by drawing the fitness values in distinct niches over the continuous scale of phenotypes. The fitness curves obtained in this way can be transferred to a two-dimensional fitness space applicable to two niches, as shown in Fig. 20. It is seen that the fitness set F depends upon the degree of overlap of the distribution curves of the fitness points for both niches: if the distribution curves intersect above the turning points of the curve, the fitness set assumes a convex configuration (Fig. 20a); if intersection is near or just above the turning points, it is flat-convex (Fig. 20b); with intersection below the turning points the set becomes concave (Fig. 20c).

We are interested of course only in the form of the fitness set in the range where we suspect populations with maximum K values, i.e., the right upper boundary range. As we have seen, the relative distances between niches, i.e., the niche distance measured perhaps in standard deviations as in the preceding graphs, is decisive for the form of the fitness set. Therefore, the absolute distance between niches as well as the tolerance of phenotypes measured by the value of C determine the form of the fitness set within the range of interest here.

Thus the optimal adaptive strategy of a population is determined primarily by the form of the fitness set and the kind of environmental heterogeneity. Environmental heterogeneity will be discussed in the next section. In addition to the factors already mentioned, the form of the fitness set is also determined by the mating type. For example, with inbreeding the extended fitness set F' in Fig. 16 will become flatter, i.e., will be less convex than in the above situation where random mating was assumed. Of course, the reason is the increase in the number of homozygotes associated with inbreeding. Therefore, the form of the fitness set has been modified in spite of the fact that the fitness values of the phenotypes in both niches and also the environment had remained the same. The only difference had been introduced by the breeding system. As a result, the form of the fitness set is not fixed and can be changed by selection, and also by factors other than those associated with the mating system.

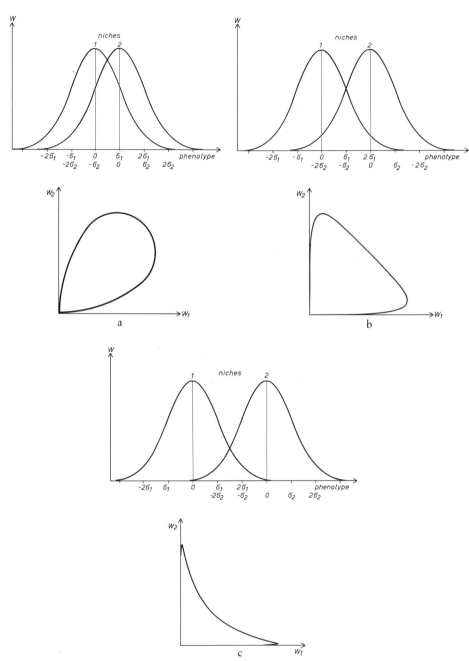

Fig. 20 a–c. Dependency of the form of the fitness set F upon niche differences. Two niches are assumed with one optimum phenotype in each. a Niche differences relatively small; the fitness set for both phenotypes is convex. b Niche differences are so large that the fitness curves cross near the turning points; the right, upper boundary of the fitness set is a straight line. c Niche differences are so large that the fitness curves cross below the turning points; the fitness set is concave

In this case, selection for self-fertility could be an indirect reason for such a change. This could apply in regions where population density is inadequate to supply enough foreign pollen to a normally cross-fertilizing population. The switch to self-fertilization in the strict sense, such as could be expected from selection for a high probability of self-fertilization, would eventually lead to the evolution of only two phenotypes. The hybrids arising from a low rate of cross-fertilization would mainly serve to create new variability. The consequence would be an elongation of the fitness set F' to a straight line connecting the fitness points of the two homozygotes. Simultaneously, there would be a decline in the mean fitness of the optimum population, since K becomes smaller when the contact point of the curve of the adaptive function with largest K in the fitness set moves closer to the origin. The optimal rate of cross-fertilization, a property of the genetic system, is then a compromise solution. This would resolve the conflict between loss of fitness resulting from inbreeding, not only partial or total loss of heterozygote advantage but also other detrimental effects of inbreeding upon fitness, and the loss of fitness with obligate cross-fertilization under conditions of an inadequate pollen supply.

Of those factors influencing the fitness set and thereby the adaptive strategy of a population in a heterogeneous environment, we have become familiar with the following: degree of dominance for the character "fitness," environmental tolerance of phenotypes, and mating type.

3. Optimum Populations in Heterogeneous Environments

Populations, made up of individuals developing in the same and absolutely stable environment, and in environments remaining equal from generation to generation, are required to produce only one — the optimal — phenotype. Under the premises stated earlier, this was the phenotype with maximum fitness as measured by its reproductive rate. In a heterogeneous environment similar conditions may apply provided the environment is fine-grained, or, in a coarse-grained environment, the form of the fitness set F admits a single optimum phenotype. We would now be in a position to imagine the qualities of optimal populations in all possible variations of the environment that occur with forest trees. Of course in doing so we would have to regard the premises fixed by the populations, such as their breeding system, form of the fitness set, etc.

However, this appears to be a rather difficult approach. We would be required to look at too many variables simultaneously. It is conceivable, for example, that the method of seed dispersal contributes as much to the maintenance of genetic variability and the outcrossing structure of the population as a genetic system marked by a high recombination rate. Adaptations are at least partially exchangeable, and in a new situation it is likely that the evolutionary history of a population will determine which of the possible adaptations will be acquired. VAN DER PIJL (1969), for instance, compares trees of the second-growth rain forest with those of the climax stage. To colonize new areas rapidly, the former species must achieve rapid seed distribution over considerable distances. The seeds are therefore light and dry for dispersal by wind, still the best way of moving large numbers of seed during a short time period. For the species of the climax forest, animals undertake the distribution. The seeds are large, to produce seedlings capable of "waiting" in shade, and possess qualities attractive to animals. Seeds of the pioneer species often remain dormant so that after destruction of the climax forest regeneration may be initiated from dormant seed alone. Pollen vectors of the pioneer species are also better equipped for distant transport.

This means that species of the second-growth forest, at the beginning of succession, tend to population structures where genotypes mingle continually beginning with the manner of seed dispersal; one might say by continuous recombination in space.

For these reasons we will not become involved in a discussion of specific optimum populations but instead ask how the so-called variation patterns have arisen that are known from forest genetics literature. In doing so we shall not limit ourselves to the conventional question as to whether a variation pattern is continuous, clinal, or discontinuous, sometimes erroneously termed "ecotypic," but rather consider the aspects under which clinal or discontinuous patterns should be seen; examine the role of selection and migration; discuss estimates of selection intensities and migration rates; investigate whether genetic drift plays a role and how its effects are estimated; and study the function of linkage and recombination. In this way we shall try to go beyond the merely descriptive stage and arrive at explanations such as have been intimated in an earlier work (STERN, 1964). Methodical aspects will be emphasized, especially the methods useful when working with forest trees. Unfortunately, not many of the methods developed for ecological genetics are applicable here. FORD (1964) sees the methods of ecological genetics as a combination of field observation, outdoor trials under realistic conditions, and laboratory experiments. Every experimental biologist will realize at once that the second category of methods especially is applicable to forest trees only with extreme difficulty. Field observations, too, become very problematic in view of the long time period included in the study of only a single generation of a forest tree species. It is evident therefore that on occasion we must draw conclusions by analogy to results from similar evolutionary situations in "simpler" organisms. This method, which is disposed of by many colleagues as *Drosophila* mythology, cannot be disregarded here. It is well to remember that most of the well-founded knowledge in forest-ecological genetics is based on the methods of the *Drosophila* mythologians.

a) Continuous Clines

The clinal variation of population means extending over large areas, first discovered by LANGLET (1936 and earlier), can be explained in the simplest way as the product of selection for optimal characters that vary continuously along geographic gradients. This selection is conceived as selection for phenotypes that are adapted to the annual rhythm of weather at a certain locality in an optimum way: they initiate annual growth at the optimal date, grow daily at a maximum rate during the optimal time period, switch over to latewood production, and cease growing in the fall again at the optimal date. A similar optimal timing should be achieved for the successive phases of the generative cycles. Within the local populations, selection for the characters of importance for adaptation to the geographic gradients should be mainly stabilizing selection, aiming to maintain local populations around their optimal mean values. Migration would then counteract by continually introducing new genetic variants from neighboring populations. Directed selection within local populations should operate only at the edge of a species' range; the goal here is an ever-better adaptation to extreme conditions, and the optimal phenotypes are those of the one extreme. In a schematic way the situation can be presented as shown in Fig. 21.

Optimal conditions for adaptation to such situations would be provided by the population having many loci with many alleles and with small effects of each allele. Allele frequencies would increase or decrease along the cline in a systematic manner; frequencies deviating from 0 or 1 would initially be accounted for only with reference to migration. We therefore shall investigate first the interplay of selection and migration in a section of such a cline under the simplest conditions.

HOLGATE (1964) examines a population with continuous distribution in one dimension, e.g., along a river course. At a certain place, a new beneficial allele A_2 appears and begins to diffuse into the population. The three genotypes A_2A_2, A_1A_2, and A_1A_1 have the fitness values $1+2m$, $1+m$, and 1, respectively (i.e., there is intermediate inheritance). Let x be the distance of any point from the place of origin of the allele and $p(x, t)$ the frequency of the allele A_2 at this point at time t. We will also assume discrete generations and random mating in the locality as well as normal distribution of the migration of progeny with a standard deviation k. Then the diffusion of A_2, with generation time approaching 0, will be

$$\frac{dp(x, t)}{dt} = k \frac{d^2p(x, t)}{dx^2} + mf(x, t)\,[1 - p(x, t)]\,.$$

If the selection coefficients of the alleles apply only to a final segment of the linear habitat, with other conditions prevailing outside of this segment, then frequencies may remain equal over several generations and will depend only upon x. In this case a second derivation using x would describe this "cline,"

$$kp''(x) = -mp(x)\,[1 - p(x)]$$

where t can be neglected.

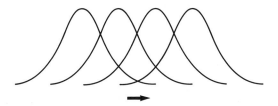

Fig. 21a. Schematic diagram showing subpopulations along a cline with continuous change of the population optimum

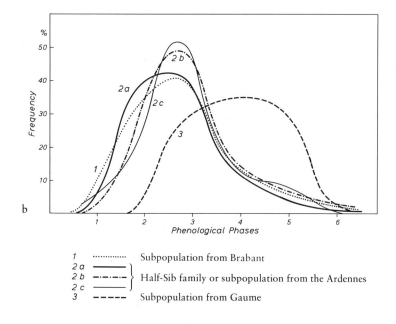

b

1 ⋯⋯⋯⋯⋯⋯ Subpopulation from Brabant
2 a ━━━━━
2 b ━·━·━ } Half-Sib family or subpopulation from the Ardennes
2 c ─────
3 ─ ─ ─ ─ Subpopulation from Gaume

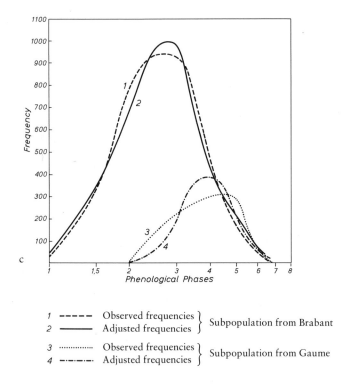

c

1 ─ ─ ─ ─ Observed frequencies }
2 ━━━━━ Adjusted frequencies } Subpopulation from Brabant

3 ⋯⋯⋯⋯⋯ Observed frequencies }
4 ─·─·─·─ Adjusted frequencies } Subpopulation from Gaume

Fig. 21b and c. Flushing time in two different subpopulations of beech of different origin in
Belgium. (From GALOUX, 1966)

If it is assumed that the preceeding equation is valid for the interval $[0, b]$, then the frequencies of the three genotypes within the interval is obtained from

$$\bar{p} - \frac{k}{bm}[p'(0) - p'(b)] \quad \text{for} \quad A_2 A_2$$

$$\frac{2k}{m}[p'(0) - p'(b)] \quad \text{for} \quad A_1 A_2$$

$$1 - \bar{p} - \frac{k}{bm}[p'(0) - p'(b)] \quad \text{for} \quad A_1 A_1$$

where $\bar{p}$ represents the mean frequency of A_2 over all partial populations of the cline (here interpreted as a continuum).

FISHER (1937) has also discussed the distribution of a beneficial gene using a diffusion equation. His solution is based on the assumption of a stationary wave that advances at a constant rate. In a subsequent paper he treated the case of linear variation of the selective advantage of A_2, assuming intermediate inheritance as well (FISHER is of the opinion that the presence of dominance would indicate a genetic polymorphism of different origin in most cases). His results are outlined below.

To describe the distribution of gene frequency, first a "neutral" line is found where selection and diffusion produce a situation where $p = 1/2$. The distribution is then measured above the distances from this line. The distribution is first of all determined by the selective gradient — that is, changes of the selective value along the cline — and by the migration rate. According to FISHER (1930), the selective advantage i of the gene with frequency p is measured by the rate of change of the logarithm of the gene ratio. If i is proportional to x, and x varies in both directions, then the rate of increase of gene frequency resulting from selection is

$$\frac{dp}{dt} = pqi = pq\,g\,x$$

where q is $1 - p$ and g the gradient of selective advantage. The increase resulting from diffusion is then

$$k\frac{d^2p}{dx^2} = -k\frac{d^2q}{dx^2}$$

and an equilibrium exists at

$$k\frac{d^2q}{dx^2} = pq\,g\,x.$$

Putting $g = 4k$, the relation between q and x is given by

$$\frac{d^2q}{dx^2} = 4k\,x\,pq$$

with boundaries $x = 0$; $q = 1/2$ and $x = \infty$; $q = 0$.

The author provides a table of standardized deviations for the range of gene frequencies 0.5 to 0.1, which he terms "legits," as well as a solution of the preceding differential equations for the range $q = 1/2$ to 0 given to eight decimals.

From the normal distribution, a standardized measure of distance is derived,

$$a^3 = \frac{4k}{g}$$

where a measures the distance between points (lines) with $q = 0.5$ and $q = 0.1004$ and $q = 0.8996$ respectively.

When the equations are applied, the weighted regression

$$X = b_0 + b_1 x_1 + b_2 x_2$$

is first calculated for the geographic coordinates x_1 and x_2. A first approximation of the "neutral" line is obtained when this equation is set equal to 0; the solution for $+1$ and -1, respectively, provides a first estimate of a. The tables make it possible to provide further information. A test for goodness of fit can also be made.

The usual condition encountered in nature is probably the occurrence of pairs of alleles, one of which may be more or less dominant. Exceptions are biochemical polymorphisms, some blood and serum groups, isoenzyme polymorphisms, etc., but systematic investigations of these are of recent origin. The more common situation, the basis for a frequency cline with dominance of one of the alleles, was investigated by HALDANE (1948). His specific assumptions include the presence of a continuous population across a plain divided by a sharp boundary into two halves. The distance of every point from this boundary is measured by x. Within the range of positive x values, the homozygous recessive genotypes, aa, possess greater fitness in the amount of $1 + K$ in comparison with the two other genotypes; these in turn have superior fitness in the range of negative x values by the amount $1 + k$. The plain must be large enough to allow monomorphic populations in its marginal areas. The species concerned is an annual. Migration is a random process; the migration rate t is measured by its standard deviation. Selection becomes effective at the place where the individual lives. There is no assortative mating and the population is in equilibrium. The frequency of the gene a in adult plants at the distance x from the boundary is q, the frequency of the homozygotes of this allele q^2.

Migration changes the gene frequency q_0 to approximately

$$q_1 = q_0 + \frac{t^2}{2!} \frac{d^2 q_0}{dx^2}$$

which can be simplified if t is normally distributed; t^2 then becomes 1. The influence of selection in one generation within the range $x > 0$ results in a change of

$$(1 - q)^2 AA : 2q(1 - q) Aa : q^2 aa$$

to

$$(1 - q)^2 AA : 2q(1 - q) Aa : (1 + K) q^2 aa$$

and subsequently approaches

$$q_b = q_0 + K q_0^2 (1 - q_0) .$$

The value of q at the boundary q_b under equilibrium conditions is obtained from

$$q_b = \left[(1 + k/K) \left(\frac{4}{q_b} - 3 \right) \right]^{1/4} .$$

The author gives a table showing the dependence of q_b on the ratio K/k. For equilibrium conditions, expected values are found for every distance x from the boundary line. The derivation will not be given.

HALDANE has applied his model to some data from the literature. KETTLEWELL and BERRY (1961) introduced a particularly instructive example based upon a cline of the melanic form *edda* of the moth *Amathes glareosa* on the Shetland Islands (see also KETTLEWELL and BERRY, 1969 a, b). The north–south range of the species is approximately 54 miles. The frequency of *edda* increases from 1 to 2% in the south to nearly 100% in the north. For marked moths, the mean distance between the localities of release and recapture was recorded as about one-half mile. A valley in the southern part of the main island constituted an obstacle that was rarely overflown. In spite of this, gene frequencies north and south of the valley were approximately equal.

The selective values of both morphs were very small, both those estimated from the equation for the cline and those obtained from the recapture experiments of marked moths. There appeared to be no differences, especially in the southern part of the main island, but the melanic form was slightly superior in the north. Recapture experiments demonstrated higher survival rates of the marked animals near the locality of capture and thereby local adaptation of both morphs. The flying ability particularly of northern populations was reduced; in these populations the dominance of *edda* was expressed most clearly. This result indicates clearly the difficulty of carrying out field studies when dealing with complex problems.

Additional models for the description or explanation of morphoclines are given by HANSON (1966). In general, these constitute a generalization of HALDANE's (1948) results. HANSON's model population is realized as follows: it is a closed population; its range includes localities with particular requirements upon the fitness of individuals; pairing within each generation is synchronized; the environments of successive generations are autocorrelated, and this applies to all subpopulations; the size of the subpopulations remains constant; all subpopulations are of equal size and are distributed in a regular way across the range; the subpopulations are large enough to estimate gene frequencies fairly realistically; migration can be described by a symmetric distribution; the pair of alleles being considered is independent of other loci in every respect; the species is monoecious. The symbols to be used in the two subsequent equations dealing with steady-state frequencies are summarized in Table 7.

Table 7

Genotypes	Frequencies in infinite groups	Initial frequencies in subpopulation (i)	Frequencies in sample n_i	Relative fitness
aa	$[q_i(t)]^2$	u_{1i}	v_{1i}	$1+k_{1i}$
Aa	$2q_i(t)[1-q_i(t)]$	u_{2i}	v_{2i}	$1+k_{2i}$
AA	$[1-q_i(t)]^2$	u_{3i}	v_{3i}	1

Furthermore, it is

$$u_{4i} = u_{1i} + u_{2i}/2,$$

$$v_{40} = v_{1i} + v_{2i}/2.$$

The coordinates x_1 and x_2 of every individual in the area are to be measured in units of

$$\sigma = \sqrt{(1-P)}\,\sigma_w$$

where $(1-P)$ is the probability of emigration and σ_w the standard deviation of the distribution of migration.

Two problems are considered:

(a) The contour lines of equal $k(x_1, x_2)$ may be represented by lines vertical to the axis of x_1. Under equilibrium conditions x_2 is of no significance and need not be given. This is the case of a latitudinal cline, for example. Equilibrium conditions are obtained as usual, by setting the equation for a change of gene frequencies in time equal to zero:

$$0 = 2kq(1-q) - 2k(dq/dx_1)^2 + (d^2q/dx_1^2)\,[2 + k(1-2q)]\,,$$

which is valid for partial dominance, and

$$0 = 2kq^2(1-q) + 2k(dq/dx_1)^2\,(1-3q) + (d^2q/dx_1^2)\,[1 + k(2q-3q^2)]$$

for complete dominance.

(b) The contour lines of equal $k(x_1, x_2)$ may be represented as concentric circles around a midpoint (x_1', x_2'). A transformation into polar coordinates $(r, 0)$ is first performed. At equilibrium all partial derivations for the second coordinate (0) become zero; therefore,

$$0 = 2kq(1-q) - 2k(dq/dr)^2 + [(dq/r\,dr) + (d^2q/dr^2)]\,[2 + k(1-2q)]$$

for partial dominance and

$$0 = 2kq(1-q) + 2k(dq/dr)^2\,(1-3q) + [(dq/r\,dr) + (d^2q/dr^2)]\,[1 + k(2q-3q^2)]$$

for complete dominance.

An example will illustrate this relationship. LEDIG and FRYER (1971) estimated frequencies of trees with serotinous cones on the basis of samples. The result is shown in Fig. 22. *Pinus rigida* is closely related to *P. serotina*, which bears almost exclusively serotinous cones everywhere in its range. The transitory zones of the two species (introgression?), i.e., the southern part of the range of *P. rigida*, have been treated in a special way.

The observed clinal variation in the frequency of serotinous cones may be described as a fairly regular decrease in all directions from the area with maximum frequency (100%) in the pine plains of New Jersey. Here *Pinus rigida* regenerated exclusively after forest fires, and is a dwarf tree as a result of the frequent fires. Trees with cones that open normally apparently do not have a chance to establish progeny. Unfortunately there is little knowledge concerning forest-fire frequency in other areas; this may influence variation of the character.

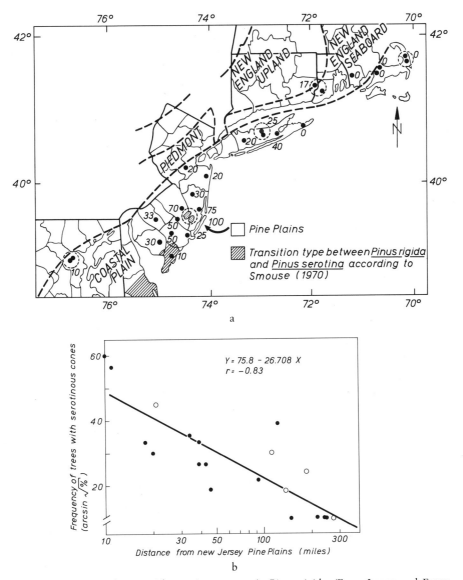

Fig. 22. Frequencies of trees with serotinous cones in *Pinus rigida*. (From LEDIG and FRYER, 1971). (a) Trees bearing 100% serotinous cones are found in the pine plains of New Jersey. From this center frequencies decrease in all directions (b)

LEDIG and FRYER interpret this variation pattern in the sense of HANSON's pocket model. With distance from the fire pocket the frequencies of serotinous cones decrease steadily. It is feasible that the clinal pattern developed solely through migration from this narrow zone of differential fitness, especially since *Pinus rigida* has been a native species for many generations. The authors consider that the necessary conditions for this theory have been met: the diameter of the two centers with serotinous cones is 15–20 miles, and the mean migration distance of seed according to BANNISTER (1965, applicable to *P. radiata* in New Zealand) is assumed to be 11 miles. Good possibilities are seen that distant pollen may achieve fertilization. Should these assumptions really apply, then a similar variation pattern is to be expected. Since alternative explanations are possible, one should first of all obtain experimental data on mean migration rates of pollen and seed and forest-fire frequency over the whole area before drawing final conclusions.

When estimates are available of the coefficient of selection for the alleles and of the parameters of migration, it is possible to arrive at helpful interpretations on the basis of the above models designed to study certain types of frequency clines (morphoclines). Often such estimates can be obtained only with difficulty or not at all. This then limits a study to the search for the most likely model among the several that explain the origin and maintenance of clinal variation. CLARKE (1966), for example, investigates the evolution of such morphoclines on the basis of a model with one locus and two alleles. He arrives at the interesting result that selection along such clines not only affects the alleles themselves but also the modifiers. When presenting the results in the form of a regression on geographic or ecological variables, the modifiers codetermine the slope and the position of the regression line to a remarkable degree. In such cases migration does not check the fitness reaction of the genotypes to modifiers but may influence their order of magnitude. This may explain the existence of pronounced clinal variation in spite of small differences in selective values of genes along a morphocline, as in the case of the moth *Amathes glareosa* discussed by KETTLEWELL and BERRY (1961, 1969 a, b).

However, the interpretation of such experiments must be a very cautious one since estimates of selective values from such simplified experiments are not necessarily realistic. This is so because the fitness value of a genotype is a complex quantity of which only a part can be assessed in an experiment. It may be noted that the models of CLARKE also provide explanations of the so-called area effects, which were found especially in the snails *Cepea* and *Partula* and could satisfactorily be related to niche diversity within the larger areas (see earlier discussion). This cannot be considered here in detail but should be mentioned.

In this connection it is of interest to discuss some typical examples of frequency clines from the literature. BARBER (1964), who has reviewed them critically, is of the opinion that they demonstrate large differences in fitness of the genotypes along such clines. Unfortunately the related investigations are not specific enough in most cases, but could be undertaken with young plant material as BARBER indicates. This particularly applies of course to long-lived organisms such as tree species where population genetic experiments with older material are possible only in special cases. The same author (BARBER, 1965) gives examples from the Australian flora. Large differences in the coefficients of selection of single genes along clines are indeed found. The juvenile stages and neoteny observed by him may also be explained in this way. Convergence is observed primarily in vegetative characters whereas generative characters remain unchanged.

The first of the examples to be discussed here is the morphocline of waxy stem covering of *Rhizinus communis* in Peru studied by HARLAND (1947). This example is chosen because "blue coloration" of needles and shoots in trees shows similar trends. Many species, both angiosperms and gymnosperms, reveal altitudinal clines such as *Rhizinus* in Peru. In this species the waxy covering, and thereby blue coloration, is determined by a dominant allele *B*. Its frequency in the high Andes Montains at about 2400 metres is close to 1; it decreases with lower elevation of the place of origin to 0.01 in the vicinity of Lima. The basis of the cline is a systematic change of the selective value of *B* with diminishing elevation, which in this case implies reduced sun radiation and cool summer months because of frequent fogs. In Lima, for example, the genotype *BB* is incapable of producing fruits, *bb* forms fruits throughout the year, and the reaction of *Bb* is intermediate. Thus the systematic change to the selective value appears to be the actual reason for the cline. The role of migration is unknown.

According to other authors similar explanations hold for frequency clines in *Justicia simplex* (JOSHI and JAIN, 1964) and a latitudinal cline of *Trillium* chromosome types at the northeast coast of Hokkaido (YOSHIMICHI and KURABAYASHI, 1960). The latter involves "supergenes" (whole chromosome sections) and not alleles of a locus. Many similar examples, but often not as well investigated, are found in the literature (see STEBBINS, 1950, BRIGGS and WALTERS, 1970 for summaries and interpretations).

An especially interesting example is found in the polymorphism for content of hydrocyanic acid in several species of different genera, which can also be explained by gene frequencies. Frequency clines have been determined in all cases investigated for this character. DADAY (1954) gave a detailed description for *Trifolium repens*. Two alleles of two loci are responsible for the production of HCN and both are dominant. The first controls the production of the glucosides linamarin and lotaustralin; the second controls production of the enzyme linamarase. HCN is found only in those genotypes containing at least one of each the two dominant alleles.

The frequency of both dominant alleles in populations of *Trifolium repens* decreases systematically from the Mediterranean Sea in a northeastern direction, from 1 to 0. A similar decrease is observed with increasing elevation. The cline is fairly well described as a dependence of gene frequency upon the mean temperature regimes of the source locations. A 1° F decrease in mean January temperature corresponds to an average frequency reduction of the first dominant allele by 3.2% and of the second by 4.2%. The author considers HCN in the leaves and stems as a possible retardative agent that slows down metabolism and increases the frost resistance of the cyanogenous plants. In his opinion there are also few active isoalleles in areas with low temperatures. This brings us again to the model of CLARKE (1966), for modifiers affect selective values in a similar way as isoalleles of the same function if dominance modifiers are neglected; but theoretically even the dominance relations can be changed through substitution of isoalleles.

BISHOP and KORN (1969) investigated the different attraction of cyanogenous and acyanogenous plants of *Trifolium repens* to snails and slugs that was assumed by other authors. They found that these leaf eaters do not distinguish between morphs. There is also no growth difference. Thus the polymorphism is maintained by causes other than the selective action of animals.

JONES (1966) undertook similar experiments with cyanogenous and acyanogenous forms of *Lotus corniculatus*. Three different cyanogenous forms were offered as sole food to animals kept in cages: the caterpillars of two species of *Lepidoptera*,

slugs, snails, and voles. All animals preferred the weakly cyanogenous forms but
the voles only when other food was given at the same time. The author therefore
supposes that the feeding habit of animal opponents is at least one of the mechanisms initiating and maintaining this polymorphism. Later JONES (1968) examined
the dependence of the polymorphism of cyanogenesis in *Lotus corniculatus* upon
that of *Trifolium repens*. In the study area both species occur in pure and mixed
populations. He concludes that there is in fact a correlation between the frequency
of cyanogenous forms in mixed populations and the same frequency in the admixture of *Trifolium*, and also between the frequency of cyanogenous forms and the
distance between pure and mixed populations. This result is of interest not only in
relation to the subject being discussed here but also as an example for coevolution
in ecosystems.

The variation pattern of cyanogenesis apparently has at least three causes: temperature dependence of the selective values, animal feeding, and the same polymorphism in
the companion population or neighboring population of other species.

Most of the "ecoclines" (clines developed and maintained by natural selection)
extending over large geographic areas supposedly result from a systematic change
in gene frequencies caused by variable selective values of the genes, along parallel
ecological gradients. Climatic factors are the main sources of large area clinal
variation. It is possible, therefore, to demonstrate clinal variation by regression on
certain climatic factors, most often on the long-term means of temperature, precipitation, length of the growing season, etc. Yet often annual values fluctuate considerably and large deviations are observed around these means; in the temperate
and boreal zones this is always so. The distribution of late frosts in northwestern
and southeastern Schleswig-Holstein may illustrate this (Fig. 23).

The long-term averages of late frosts correspond fairly well to one side of the
normal distribution. In northwestern Schleswig-Holstein, frosts are very rare after
the first day in June as a result of the influence of the North Sea. In the southeastern
part at Woltersdorf they occur more often, at least once in ten years. From a
Norway spruce experiment, the distribution of flushing dates in one provenance,
which is average with respect to this character, is drawn into the same chart. The
different probabilities that spruce will encounter late frost in these two areas are
obvious. As is known from many experiments, the mean date of late frost is the

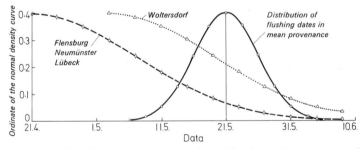

Fig. 23. Frequencies of late frosts in maritime (stations Flensburg, Neumünster and Lübeck)
and continental (station Woltersdorf) parts of Schleswig-Holstein, in comparison with the
distribution of flushing dates on the average of local provenances of *Picea abies*

niche dimension that is best correlated with the mean date of provenance flushing. It now becomes evident that actually the frequencies of certain late frost occurrences, i.e., values of the niche dimension, change. A similar explanation may be found for the example of clinal variation of serotinous cones of *Pinus rigida* given by LEDIG and FRYER (1971) by assuming changing forest-fire frequencies along the cline.

If this is actually the case, we may again apply LEVINS' method of presentation as described in Sections 1 and 2 of this chapter. We shall first consider a convex fitness set and changing niche frequencies (only two niches) along a geographic gradient. In Fig. 24a the fitness set is strongly convex. Here the optimum phenotype changes little even with considerable divergence of niche frequency along the geographic cline. This is different in the case of Fig. 24b, where the fitness set is flat-convex: small alterations in niche frequency along the gradient are sufficient to cause pronounced differences among the optimum phenotypes.

A concave fitness set introduces basic differences. As Fig. 24c indicates, if the frequency of niche 1 increases, initially the optimum phenotype is not modified or is only slightly modified. However, as this niche exceeds a certain value, here indicated by the straight line with average slope, the optimum phenotype is altered rapidly: populations of the first phenotype would be followed by monomorphic populations of the second phenotype after a relatively narrow transition zone, the

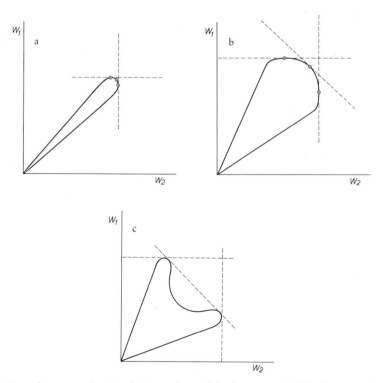

Fig. 24. Gene-frequency cline in relation to form of the fitness set: (a) The fitness set is strongly convex. (b) The fitness is weakly convex. (c) The fitness set is concave

width of which is determined by rate and distance of migration. Both figures are valid for fine-grained environments, where straight lines may be used to represent the adaptive function with different niche frequencies (see Section 2 of this chapter).

In the case of a coarse-grained environment the adaptive function describes parabolic curves. With a convex fitness set the optimum phenotype changes systematically, depending upon the degree of convexity along the cline (Fig. 25a). Yet if the fitness set is concave, extreme niche frequencies create monomorphic optimum populations and within a broad band of variable niche frequencies, polymorphic optimum populations (Fig. 25 b).

The two most prominent examples for abrupt character change after a narrow transition zone of polymorphic populations are probably provided in Scandinavia by the shift from *Pinus sylvestris* to *Pinus sylvestris* var. *lapponica* with extremely narrow and long crown, and from *Picea abies* to *Picea abies* types with their *obovata*-like cone shapes (Fig. 26). Both were formerly classified as separate species, *Pinus lapponica* and *Picea obovata*. East of the transition zone all spruces bear *obovata* cones, to the west of it exist a polymorphism of cone shape and *"typical"* variants were given very imaginative names. Yet the presence of *obovata* types in mountains far west of the transition zone is proof that *obovata* cones are nothing but morphs and components of the general Norway spruce cone polymorphism. However, these mountain populations possess selective advantages in geographically separate areas and extreme ecological conditions (see part c of this section). SCHOENICKE et al. (1959) present a good example of clinal variation of cone serotiny in *Pinus banksiana* (Fig. 27). In the northern part of its range trees with serotinous cones predominate. In the south-central and western parts there is a narrow transition zone with 40–70% of serotinous cones, and finally in the extreme south trees prevail with even lower percentages.

As the same authors point out, *Pinus banksiana* has developed a new adaptation in the northwest, i.e., in the areas with largest percentage of serotinous cones. Populations from this area possess high proportions of trees that bear both serotinous and normally opening cones. In many situations this is certainly the best reaction to the environmental exigency of many forest fires. *P. clausa* also is reported to produce both cone types in areas with many forest fires. NEUMANN et al. (1964) have investigated the effects of cone serotiny upon cone opening in *P. banksiana* in some detail. The cone opening mechanism appears to be more complex in nature in this species.

A dimorphism of the generative apparatus borne by the same individual, which is comparable in principle, is also found in other plant species, for instance, the seeds of *Nicandra* (cited by FORD, 1964).

We shall now consider the advantages or disadvantages of the fitness of the heterozygote as contingent upon gradients in the frequencies of two niches (LEVINS and MACARTHUR, 1966, giving a generalization of the model of LEVENE 1953). Figure 28a depicts a situation where the heterozygotes $A_1 A_2$ are almost equal in fitness value or superior to one of the homozygotes in each of the two niches. The fitness set, shown as a curve arching toward the fitness point of the heterozygotes, is convex. The adaptive function with maximum K for a coarse-grained environment is moved to the right and down with increasing frequency of the second niche. There is a frequency cline of the two alleles. This cline is continuous.

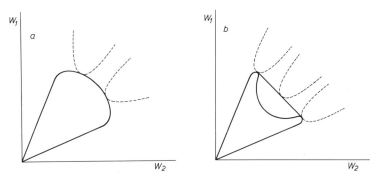

Fig. 25. Gene-frequency clines in a coarse-grained environment is contingent upon the fitness of set: (a) With a convex fitness set the optimum phenotype changes systematically with niche frequency. (b) A concave fitness set produces monomorphic and polymorphic optimum populations, again in dependence upon niche frequency

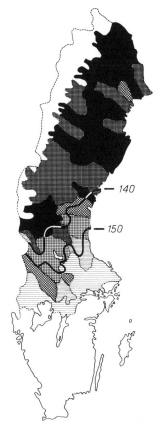

Fig. 26. Distribution of "Norrland pine" and "Sörland pine" in Sweden, based upon an index developed from needle length, cone color, thickness of the cone scale, and crown width. These characters are to a large extent under genetic control. The two lines in the center indicate the number of days in the growing season with $+6°$ C or above. (From LANGLET, 1959)

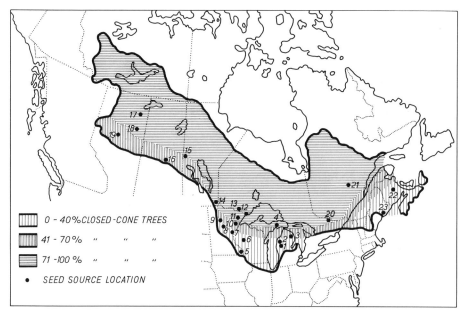

Fig. 27. Frequencies of trees with closed and open cones of *Pinus banksiana*. Numbers on the map refer to the sequence of seed-source numbers. (From SCHOENIKE et al., 1959)

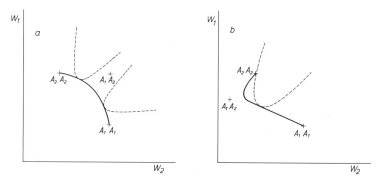

Fig. 28 a and b. Optimum populations for one gene locus with segregating alleles with heterozygote advantage or disadvantage, respectively, in a coarse-grained environment. a Along the convex fitness set the composition of the optimum population changes systematically depending upon niche frequency. b In consequence of the form of the fitness set there are monomorphic and polymorphic optimum populations. This drawing shows two adaptive peaks, one with a polymorphic, the other with a monomorphic optimum population

Figure 28 b represents a different situation. The fitness set is concave. In the special case shown here two optimum populations exist, one monomorphic and the other polymorphic. When the fitness point of $A_1 A_2$ is moved closer to the origin of the ordinate, the adaptive function is no longer arched sufficiently to touch the concave fitness set at the inside. The population will tend to become monomorphic for one of the two homozygotes. A continuous cline of heterozygote fitness or niche frequencies leads to an abrupt alteration in the composition of the population (regarding the role of migration in such situations see DEAKIN, 1968).

The last situation is at the same time an example for multiple adaptive peaks in the sense of WRIGHT for one locus with two alleles: populations of different composition may possess equal adaptive values. Introduction of additional loci (alleles) and niches may lead to situations of increasing complexity, i.e., more adaptive peaks. To evolutionary history of the population or genetic drift in the wider sense, respectively, determine which peak will eventually be reached (see WRIGHT, 1960 for comprehensive description). The obscure results of the investigations concerning frequency of morphs of the cone polymorphism of Norway spruce, for example, may have their origin here.

Figure 28 a and b is also an example of the importance of the mating system for at least some cases of clinal variation. Increasing self-fertilization would augment the frequency of both homozygotes, and reduced weighting of the heterozygotes would make the fitness set flatter. This in turn would have the consequence of increasing the frequency of polymorphic optimum populations. A simple deduction of this kind indicates the necessity of accounting for the mating type when investigating and interpreting clinal variation patterns. Its influence upon the results may be significant. Similar deductions lead to the conclusion that the kind of environmental heterogeneity (in addition to fine- and coarse-grained categories, which have already been discussed) should also be included in the investigation. The next section will discuss this aspect.

Thus the component "clinal variation" of the genetic variation pattern turns out to be a complex phenomenon. The origin and shaping of a cline may be due to the joint action of such diverse factors as geographic trends of the niche frequencies, type of environmental heterogeneity (fine-grained or coarse-grained, etc.), geographic trends of the selection coefficients of single alleles, form of the fitness set, mating type, and migration. The existence of opposite variation patterns in the presence of equal types of environmental heterogeneity is a paradox only if one regards the environment alone and disregards the factors of the genetic system. One case with a continuous cline, the other with abrupt shift from one to the other monomorphic population, are instances of such apparent inconsistencies. PEACOCK and MCMILLAN (1968) discovered very different strategies of adaptation to the same climatic trends in the two sympatric legumes Prosopis juliflora and Acacia farnesiana of North and South America. In the phytotron, P. juliflora subpopulations exhibit pronounced genetic differentiation along the gradients but in A. farnesiana there is only a weak indication of this.

We shall now consider what is known about clinal variation in forest tree species, and first, about large-scale clines in the sense of HUXLEY (1942). LANGLET's fundamental investigations in Norway spruce and Scots pine have already been mentioned. These provided the impulse for the many investigations found in forest genetics literature. Of course, only a selection can be discussed, taking note of the manifold objects and environments. The question as to whether clinal variation can be equated to ecotypic variation (this was the subject of an extensive controversy between LANGLET and TURESSON that lasted until recently) will not be considered. The point made by CLAUSEN et al. (1940) that the ecotype can be regarded as one of a series of distinct genetically different units the boundaries of which coincide with the distribution of an environment, may be correct if the environmental factor actually shows such distinct limits of distribution (see next section). CRITCHFIELD (1957) was probably confronted with a similar situation in *Pinus contorta*. Its range is made up of parts that are isolated from each other. He may describe them as ecotypes and probably will find clinal variation within each ecotype, and perhaps also among group clines in the sense of HUXLEY (1940) because selection for optimum populations may exist within the local populations. Overlooked by many authors, BONNIER (1890) and KERNER VON MARILAUN (1891) had already performed the critical experiments related to this problem in the last century, even using the modern method of testing the characters of individual genotypes by means of clones.

The frequent thesis that clines result from migration between two ecotypes (in extreme niches) had already been repudiated by SINSKAJA (1942), who found clinal variation in typical self-fertilizing species with very low migration rates. She interpreted this as proof against the dominance of the "oecotype" and then, in contrast to earlier studies of single characters, emphasized the importance of character combinations that are borne by what she called "oeco-elements." She also indicated the importance of the variance within demes. This will be discussed later. Still, the whole subject matter, explored and interpreted with intuition and sagacity over the last decades, has been expressed in the form of mathematical models only recently — and these are anything but satisfactory even now. Even the genetic background of some large-scale clines was investigated reliably at an early date, such as the clinal variation of some characters of *Lymantria* (GOLDSCHMIDT, 1934). Depending upon the character, these were controlled by multiple alleles or by loci with alleles that segregate independently whereby cytoplasmic inheritance (MICHAELIS, 1954: HAYWARD and BREESE, 1968), or epistasis, also may have played a role. Again, this part of the literature cannot be considered.

Typical examples of clinal variation over great distances were found by BURLEY (1966) in many characters of *Pinus taeda* in a review of provenance experiments with this species. The cline is considered the probable result of geographic trends in precipitation, temperature, and perhaps also photoperiod. LACAZE (1967) describes a relatively simple case, the clinal variation in *Abies grandis*, a tree species with nearly linear north–south distribution along the west coast of North America. His investigation, like most "seed source studies" in forestry, was aimed at finding the economically best provenance. In contrast, the clinal variation of *Acer saccharum* described by KRIEBEL (1957) is two-dimensional, and three-dimensional if elevation is included as a third ecological variable. PRAVDIN (1964) summarizes the literature on *P. sylvestris*. He distinguishes, as CRITCHFIELD does for *P. contorta*, ecotypes or subspecies. In addition to the typical subspecies he recognizes: subsp. *hamata* on the Crimean Peninsula and in the Caucasus, *lapponica* in northern Europe and Asia, *sibirica* in Asia, and *kulundansis* in the zone of Siberian steppes. With effort, one could probably find additional populations that are distinguishable and geographically isolated, e.g., in the Pyrenees. However, as LANGLET has shown, the clinal variations of at least some physiological characters does not reflect unevenness at the subspecies boundaries (stepped clines). SCHRÖCK (1967a, b) investigated the geographic variation of the phototropic reaction of germinant seedlings of *P. sylves-*

tris, a character that had been recommended for early testing and provenance identification by WERNER SCHMIDT 30 years earlier. He found a clinal pattern depending upon latitude, longitude, and elevation. A southwest–northeast trend of the character is predominent. SQUILLACE and SILEN (1962, see also WELLS, 1964), working with *P. ponderosa*, found an east–west and a north–south cline only in the eastern part of the range. In the western part the latitudinal cline is entirely absent. SQUILLACE (1966) also gives a literature review of the variation pattern of *P. elliottii*. Clinal variation, controlled by climate, is its main feature. WAKELEY (1961) found primarily clinal variation in a paper summarizing results of the southwide, seed-source study involving several pine species native to the southeastern United States. At different test sites the clines may have opposite trends as a result of faster growth of southern sources in the south and northern sources in the north (see also SNYDER et al., 1967). Susceptibility to disease, too (here to the fusiform rust, *Cronartium fusiforme*), varies clinally and the native source or source nearest to the test site is not always the best. These examples may be sufficient to indicate the multiformity of clinal variation and its practical importance in forestry.

It would be interesting, of course, to know how many of the polymorphic gene loci have an associated clinal variation of allele frequency. Among the experiments performed to answer that question are those of F. BERGMANN, Göttingen, dealing with Norway spruce. These have not yet been finalized or described. Such investigations are rare. One example (PRAKASH et al., 1969) from natural populations of *Drosophila pseudoobscura* may therefore be sufficient. Twenty-four loci were investigated. Concurring with expectations, the degree of polymorphism was highest in the central and lowest in the small marginal populations (see part f of this section). When disregarding two loci on the third chromosome, these differences vanished. In the main population, 40% of the loci were polymorphic (as shown by isoenzyme techniques), per individual an average of 12%. For five loci, equal gene frequencies were found in the North American population; one locus varied clinally; the polymorphisms of three loci were limited to certain populations; extremely fluctuating allele frequencies were discovered among local populations at two loci, probably as a result of association with adaptive gene blocks. Balancing selection appeared to be the prevailing kind, and agreement with WRIGHT's theory of shifting balance (see WRIGHT, 1970 for summary) is evident. If it is desired to evaluate the significance of clinal variation for frequencies of polygenes, it should be realized that although clinal variation of the population means may be present, parallel frequency clines must not necessarily exist in the underlying distribution pattern of the specific polygene. Optimal population phenotypes along the cline may be obtained from very diverse gene complexes when different forms of selection are superimposed over the specific cline that is being investigated, accidents of sampling are involved, and other factors.

Estimates of the intensity of selection along a cline of the population means for a given quantitative character encounter difficulties similar to those in estimates along frequency clines. One should know the migration rates along the cline, and heritability of the character or its heritabilites (since these are not necessarily the same everywhere), and not merely the selective values of genes. Glimpses of selection intensity are also obtained from the "realized heritability" when comparing neighboring populations in the cline.

STERN (1964), for example, estimated the clinal variation of several characters in *Betula japonica* and *B. maximowicziana* in an experiment in which every local population was represented by several half-sib families (progeny of individual trees following open pollination). The ecological variables were latitude, longitude, elevation above sea level, mean annual temperature, and mean annual precipitation. The effect of each of these variables upon clinal variation was measured by its partial regression coefficient. In *B. japonica* for the character initiation of growth, this coefficient was -0.2871 for latitude measured in whole degrees. The character was measured in three-day intervals. Therefore, the partial regression on latitude implies that, in the experiment, of any two populations chosen for comparison, the population that originated one degree of latitude further north flushed $(-0.2871) \times$ three days, i.e., approximately one day earlier.

The additive genetic variance of the same character, estimated from the covariance of half sibs, was 0.9602 and its standard deviation therefore 0.9799. If this partial regression on latitude is measured by the genetic standard deviation, it is found that the population that originated one degree of latitude further north flushed earlier by $0.2871/0.9799 = 0.2930$ standard deviations.

We may now ask: what selection intensity is required to change the mean flushing date of one population in one single generation of selection into that of a population located one degree of latitude further north? Here we remember that the response to selection (the genetic gain) is given by the relationship

$$R = h^2 \, \sigma \, \bar{\imath}$$

where R is the response to selection (here understood as realized heritability), σ the phenotypic standard deviation, and $\bar{\imath} = z/p$, the ratio of the height of the normal distribution ordinate at the point of truncation by selection (z) to the proportion of the individuals found above this point that will be selected. For the purposes here, one may suppose that there is no environmental variation, i.e., the phenotypic standard deviation equals the additive genetic variance, hence that heritability assumes the value of 1.

From the data and the value of *i* obtained from statistical tables, it is seen that the 83% earliest flushing individuals must be selected to achieve the stated response. The selection intensities for other characters, which also varied clinally, were of the same order, averaging 85%. Yet it must be realized that the results were obtained under two conditions: (1) there was no environmental variance, which is probably not the case in nature; and (2) selection was considered only in relation to one character that varied parallel to the chosen geographic gradient. With simultaneous selection for several independent characters, the picture changes quickly (see Table 8). Thus, with only 15 characters, selection intensities equal those used by plant and animal breeders who aim at a rapid change of quantitative characters.

This simple method gives only an indication of the intensity of selection responsible for the inception and maintenance of clinal variation. Its objectives and design resemble the method of LUSH (1953) applied to compare a population subjected to longterm selection with the initial population. This, too, has certain weaknesses that will not be discussed here.

Table 8

Number of characters	% of individuals to be selected
1	85
5	44
10	20
15	9
20	4
25	2
30	1

The study of HENDERSON et al. (1959) should be mentioned in this respect. HALDANE (1949) has suggested measuring the evolutionary rate in a similar way (see also the critique of SIMPSON 1949). STEBBINS (1949) has reviewed this problem in the higher plants. As a rule, the evolutionary rate must be related to quantitative characters, making assumptions much like those in the preceeding example of clinal variation. The only difference is that the change of population means is a matter of time and not of space. HALDANE concurs with us and concludes that evolutionary rate may be measured in standard deviations. In his case it is the phenotypic standard deviation, for its genetic component, of course, cannot be measured back in time.

Unfortunately, in forest genetics, conditions are practically never suitable for the measurement of selection intensities in successive generations. Generation intervals are simply too long for reasonable experimentation. This is a serious handicap, for MORLEY's (1959) remarks concerning genetic-ecological investigations are equally valid for forest genetics. His opinion is that the relevant experiments should measure selection intensities particularly to reveal the pattern of adaptation. The effect of selection should be related not only to the mean values of populations but also to the distribution of populations in space, and their structure and size (number of individuals). He considers means and variances and their development in space and time to be key parameters as a comparison of means in the sense of the old genecology is no longer sufficient. Diversity within the populations is as informative as uniformity. Finally, in modern times the influence of man becomes an ecological variable that must also be considered. Man exerts strong influences upon the environment of plant and animal species through animal husbandry, agriculture, destruction of forests and prairies, elimination of competitors, and application of fertilizers and protective chemicals. To this we must add that man influences his own environment also. The idea of protecting the environment, of course, was not as topical in 1959 as it is today.

The review of MORLEY cited above, which is still readable, gives priority to the measurement of selection intensity, among the criteria needed to characterize ecological-genetic relationships. This is probably justified according to what we know about evolution. Selection is the driving force in micro- as well as in macro-evolution. For this reason it is appropriate to consider in a very general way the methods of measuring the kind and intensity of selection in natural populations.

We shall begin with methods of measuring stabilizing selection (here identical with selection for an optimum, as long as the population is not far from this optimum).

HALDANE (1954 a, b) shows that in the presence of an optimal phenotype and selection, a proportion $w_0 - \bar{w}$ of the population survives, where w_0 is the proportion of optimal phenotype survivors and $\bar{w}$ the proportion of the whole population. Given these premises, he defines the intensity of selection by

$$I = \ln w_0 - \ln \bar{w} .$$

The application of this relation in practice requires a knowledge of the proportions w_0 and $\bar{w}$. Even when $\bar{w}$ is unknown, but the distribution of the quantitative character x is known from two successive generations, I can be estimated from

$$I = \ln f_2(x_0) - \ln f_1(x_0)$$

where f_1 and f_2 are the functions of the distributions. For details, readers must consult the original article. For the same purpose VAN VALEN (1965, 1967) uses the relation

$$I = 1 - \bar{w} = \frac{w_0 - \bar{w}}{w_0} .$$

He gives a number of functions to estimate selection intensity under different conditions, as well as graphic presentations of these functions, which facilitate their application to concrete situations. O'DONALD (1968) criticizes the manner in which these functions have been used in specific cases and introduces a variant that is valid. The maximum-likelihood estimators of DUMOUCHEL and ANDERSSON (1968) appear to be useful, at first, primarily for experimental populations.

These estimators can be applied in comparisons of gene frequency of only one locus. MORAN (1963) introduces a model that facilitates the estimation of selection intensity in certain cases, even when more than one locus is involved.

Among the methodological studies devoted to genetic polymorphisms are those of KIMURA (1956) and TURNER (1968). However, we must realize the difficulties that confront experiments designed to measure selection intensities and/or fitness values of certain genotypes in natural populations. One difficulty is that inbreeding changes the frequencies of genotypes, often very drastically; therefore, estimates of the degree of inbreeding should be available. A number of methods to accomplish this are given by LI and HORVITZ (1953), and JAIN and WORKMAN (1967) present a general description of the relationship between selection and inbreeding using WRIGHT's coefficient of inbreeding F. The second difficulty, which is particularly serious in natural populations, is the dependence of relative fitness values of the morphs in a polymorphism upon their frequency in the population (PROUT, 1965 and others).

An exhaustive or even satisfactory discussion of these problems is not possible here; however, because they are of particular importance for genetic work with forest trees and forest ecosystems, at least reference will be made to some additional literature. CROW (1963) investigates possibilities of measuring selection intensities in man at the three levels of total intensity, phenotypic, and genotypic selection. The work of ALLARD et al. (1966) should be of particular interest because of its instructive examples demonstrating how to estimate selection values and apply them in the prediction of population change. This work has an experimental orientation and conveys impressions of the frequently drastic effects of selection in natural populations. Finally, COOK (1970) has devoted a monograph to this topic but has given the special problems of plant populations a rather brief treatment.

With regard to experimental studies of clinal variation in forest trees, HUXLEY (1942) distinguishes "short clines" and "large-scale clines." The former are thought of as being more or less temporary adaptations, whereas the latter are considered more permanent characteristics of a species' variation pattern. Yet it is difficult to find a distinction between the two. Not only are there all possible transitions, but the short clines, for example, on mountain slopes appear to be permanent unless the species considered is subjected to severe environmental fluctuations in time. BRINAR (1968), for example, finds in beech dependencies between crown types and topography (elevation and slope aspect). The most instructive and methodically advanced investigations have been undertaken in several tree species by GALOUX in Belgium (GALOUX and FALKENHAGEN, 1965; GALOUX, 1966, 1967). To explain the observed variation pattern, the author attempts to use simultaneously the vegetation type and the position of the species in succession. According to GALOUX (1967), the relationships between the general environment (soil and climate), the vegetation, the species, and subpopulations (demes) can be considered at the levels shown in Table 9. Within this framework the genetic variation pattern needs to be described and explained. The special properties of the

Table 9. Levels of relationship: environment-vegetation species-subpopulations
(for investigations in Belgium)

Ecological territory	Flora	Vegetation	Syngenetic status of vegetation types	Genecological differentiation
Dominion	Specific endemism	Typical for the dominion	Pioneer communities, stages of succession, climax	Genetical differentiation possible, clines, races
Sector	Preferred or avoided by the species	Regional groupings with centers on specific sites	Pioneer, position in succession, climax	As above
Subsector (warm or cold)	Specialized species	Specialized communities	Permanent communities, variants of the climax	
District	Presence or absence of the species	Ecological groups		
Station	Frequency and dominance		Variants of the climax	Differentiation not proven

genetic system of the species must also be considered as far as they influence the variation pattern.

A second consideration concerns geographic distribution as shown in Table 10.

Clinal variation was measured in the customary manner by means of simple and multiple regressions. Regressions were also found between morphological and/or physiological characters; the correlations resulted from parallel selection along the cline or the characters were functionally dependent. The regression coefficients in *Fagus sylvatica, Acer pseudoplatanus,* and *Alnus glutinosa* were marked by the same sign, with one exception. When the data were analyzed hierarchically by regions (termed ecological sectors) and half-sib families in regions (open-pollinated families), considerable differentiation was found among regions and among subpopulations within regions. In *Acer,* 70–90% of the genetic variation in flushing was contributed by regions and the remaining 10–30% by half-sib families; in *Fagus* the corresponding figures were 50 and 50%. For other characters the results were similar: in spite of large differences among regions, a large amount of additive genetic variation remained within subpopulations (the variance of the means of half-sib families accounts for only 1/4 of the additive genetic variance; for comparison with other quantities it should be multiplied by four).

For *Acer pseudoplatanus,* a specialized species with considerable rate of inbreeding, seven estimates of the inbreeding coefficient F are available that range from 0.09 to 0.72 and they are much higher than the mean inbreeding coefficient for *Fagus silvatica* (0.11). Estimates of the intensity of selection were also obtained but these are not directly comparable.

The work of GALOUX reflects his endeavors to advance toward an explanation of the deeper causes determining the variation pattern of his species. His approach is also characterized by a skillful selection of his study material in a restricted, ecologically well-known area (see also GALOUX, 1969).

The experiments of GALOUX sketched here represent special applications of the strategic analysis developed mainly by LEVINS. A general model is used in an attempt to describe the adaptive strategy of a species in a complex environment, and to explain the differences between the strategies of different species. LEVINS (1962, 1963) also has given the criteria that were used by GALOUX, and STERN (1964). Concerning the analysis of biota, LEVINS (1963) states that

"The plants and animals which live together in the same region do not necessarily experience the same kind of environmental heterogeneity. For example, the meadow environment is more variable than the forest and less variable than disturbed ground. The average conditions over a longer life span have a smaller variance than over a short life span. The seed of climax forest trees usually fall near the parent and are more likely to grow under similar conditions than would the seed of weeds. Thus, from the general ecology of a species it is possible to derive a quasiquantitative measure of environmental heterogeneity."

The author then proposes to develop an index of heterogeneity in time, perhaps in the following manner:

1. Habitat, applying the numbers -1 of forest, 0 for meadow or savanna, 1 for a disturbed habitat.

2. Position in the succession: -1 for climax, 0 for subclimax, 1 for colonizing species and weeds.

3. Seed dispersal: -1 for short-distance distributors, 1 for long distance.

Table 10. Types of geographic distribution, syngenetic status, and genetic parameters

Distribution	Plains	Hills, mountains	Along river courses	
Topography		Curves and gradients	Curves	
Number of dimensions of variability	Longitude, latitude	Longitude, latitude, elevation a.s.l.	Longitude or latitude	Two or three dimensions
Primary ecological factors				
(a) Radiation	Subconstant	Variable	Subconstant	Variable
(b) Water	Subconstant	Variable	Constant	Constant
(c) Minerals	Variable	Variable	Subconstant	Subconstant
Ecological territories	One sector	Several sectors and subsectors	One sector	One or more sectors
Climax	Optimal distribution, ecological optimum, weak clinal variation, small coefficient or relationship, weak regression (*Fagus sylvatica, Quercus robur*)	Optimal distribution, ecological optimum, variable clinal variation, sometimes abrupt, subpopulations differentiated, weak relationship, variable regressions (*F. sylvatica*)		
Specialized species (in subsectors)		No ecological optimum, no optimum distribution, subpopulations at the edge of the range, sometimes relicts, discontinuous clinal variation with clinal component when special habitat includes ecological gradients, sections of special clines within a general cline, regression coefficient variable. (*Acer pseudoplatanus, Quercus pubescens*)	Clinal variation weak, regression coefficient weak (*Alnus glutinosa*)	More clinal variation with respect to elevation, larger regression coefficients (*Alnus glutinosa*)
Pioneer species	Variation result of accidents, not oriented, occasionally closer relationships (*Betula verrucosa*)	Variation result of accidents, not oriented, occasionally very much closer relationships (*B. verrucosa*)		

4. Life expectancy: −1 for perennial species, 0 for annuals, and 1 for ephemeral species.

Similarly, the effective heterogeneity in time depends upon the area occupied by a deme (better, panmictic unit), the occurrence of microniches large enough for full development and reproduction of an individual of the species concerned, and the number of occupied niches (or the width of the niche). Here again an index may be constructed where positive values indicate greater heterogeneity:

1. Frequency of occurrence: −1 rare, 1 frequent.

2. Size: −1 for large trees, 0 for shrubs, 1 for small herbs, lianas, epiphytes.

3. Pollination: −1 for self-fertilizers, 0 for bees, and 1 for fertilization through less specialized animals and wind.

Furthermore, optimum structure depends upon individual homeostasis (see LEVINS, 1967). This could be measured in experimental plantations or derived indirectly from properties of the habitat (see also LEVINS, 1964, and for summary 1968). Comparisons of this kind should produce good results when they involve pairs of closely related but ecologically distinct taxonomic units (such as the two *Betula* species studied by STERN, 1964, one of which belongs to a section composed of species occupying near-climax positions in succession whereas the other species belongs to a section with pioneer species). Across geographic gradients, LEVINS expects:

1. The adaptive type of annual weeds characterized by high individual homeostasis; random genetic differences among subpopulations in the same region; less genetic differentiation along transects; every species occupies a large ecological range; little reaction to selection for microniches (which is often associated with polyploidy). The genera concerned may be large but are not subdivided into coenospecies in the sense of TURESSON.

2. An intermediate adaptive type, made up of the herb layer in forests, shrubby weeds, secondary growth, and subclimax species.

3. The adaptive type of climax vegetation, most distinctly expressed by forest trees; less individual homeostasis; adaptive genetic variation along transects; very specialized for a narrow ecological niche. There is little adaptive polymorphism within the population (although this is limited to certain types of genetic polymorphism, as will be discussed later); allopatric races or coenospecies exist; and there is pronounced reaction to selection.

These relationships are expressed even more clearly in Table 11 given by the same author (LEVINS, 1963). The environment consists of two niches.

Again, we recapitulate the statement that the variation pattern is determined by properties of the genetic system and the kind of environmental variation, and this is valid also for geographic gradients. It is therefore not surprising that we may find different variation patterns along the same geographic gradients within the same species but for different characters. These encounter specific components of the environment, and react to them with respect to fitness.

It remains to be added here that LEVINS' prediction was often confirmed. He stated that climax species are much more specialized than species at the beginning of succession: in field experiments on several sites with the climax species *Betula maximowicziana* and the pioneer species *B. japonica,* the former species failed on all soils that did not correspond approximately to the soil types of climax forests of

Table 11. Optimal phenotype, optimal population structure, and type of variation along geographic gradients

| | Difference between niche optima small in comparison with individual homeostasis | Difference between niche optima small in comparison with individual homeostasis | |
| | | Environment | |
		Heterogeneous in space	Heterogeneous in time
Optimum phenotype	Intermediate between the optima in the two niches	Optimum in the more common niche (specialization for one niche)	Specialized for one of the two niches
Optimal population structure	Monomorphic, moderate fitness in both environments	Monomorphic, specialized	Polymorphic mixture of specialized types
Variation along geographic gradients	Continuous cline in the phenotype	Discrete races, separated at any critical value of niche frequency	Frequency cline of the same morphs

the temperate zone. The related subject of the width of a niche — specialization — will be discussed in the next chapter.

If similar predictions are desired for animals, characteristics of behavior particularly would have to be taken into account, for instance, the capacity for niche selection (see LEVINS, 1963). In general, however, the same principles are valid as is known from many experiments with animal species, especially with *Drosophila*. The degree of specialization resulting from the capacity of niche selection is indicated not only by the investigations of host–parasite systems (Chapter V), but also those of other forms of coevolution, e.g., of wasps *(Agaonidae)* and *Ficus* species. RAMIREZ (1970) studied 40 species of *Ficus* in Central America. Every one of these species has its own pollen vector; only one species has two. Of the two groups of *Ficus*, each possesses its own group related pollinator species from *Agaonideae*. Introduced species of *Ficus* do not produce seed as long as the pollinator is missing. The ability of the pollen vector for niche selection supports specialization and leads to monotropism of both participants.

The author concludes that the improbably large number of *Ficus* species, a total of 900, is a result of this specialization. In the Old World conditions appear to be similar. Another prerequisite, of course, is the capability of certain genotypes of *Ficus* of specializing for certain environments (see next section). A comparison of the ability of animal and plant species to develop adaptive strategies always favors the animals because of their possibility to bring behavior into play.

Selection along geographic gradients nearly always involves: 1. the effect of the same environmental factor upon several characters and thus groups of genes, and consequently 2. the variation of a whole character complex along such gradients.

Hence the correlation of characters with individual environmental factors (which in turn may be correlated among themselves along such gradients) is not the only one for measurement of clinal variation: the correlations among the characters along the gradient (DAVIDSON and DUNN, 1967) could also be utilized for this purpose, at least if it is intended to demonstrate clinal variation. The optimal procedure in such cases would include both source of information, i.e., the intercorrelation of the environmental variables and of the characters, as well as the correlation of environmental variables with the character complex (JEFFERS and BLACK, 1963; ANDRESEN, 1966). BERG (1960), for example, uses the so-called "correlation pleiads," i.e., sets of correlation coefficients, to analyze the interplay of the environmental components, the components of the genetic system (see GALOUX, 1967, 1969), and finally the interplay of both groups.

The "genetic coherence" (CLAUSEN and HIESEY, 1959) measured by sets of correlation coefficients also belongs in this group. In fact a number of procedures have been suggested and tested to explore these relationships, in forest genetics as in other fields. Still missing, however, are critical, comparative investigations of methods to find an optimal procedure of multivariate analysis, such as has been undertaken by NAMKOONG (1966) for the analysis of introgression, a closely related problem. SQUILLACE (1966), for example, suggested using MAHALANOBIS' generalized distance D, a measure derived from FISHER's discriminant analysis. Using this procedure he succeeds in a satisfactory description of the geographic and predominantly clinal variation pattern of *Pinus elliottii* on the basis of provenance experiments. It is still an open question, however, in regard to the description of those characters or their underlying gene complexes that do not follow the prevailing trends of selection. MISRA (1966), who investigated the geographic variation of a *Drosophila* species, suggested applying a variant of vector analysis for this purpose. Dealing with the same problem, STERN (1964) employed factor analysis, which requires at least a general knowledge of the causal relationships. Parallel variation of the environmental factors leads to parallel variation of character complexes, either as a result of pleiotropism or of linkage (WRIGHT, 1931, 1965). A supplement of his analysis is therefore related to the question regarding the presence of "genuine" and "false" genetic correlations, i.e., the inquiry as to whether pleiotropy and/or linkage are involved. Comparisons of related species with the same or similar range again provide the answer. STERN found (see Section IV. 4g) his assumption confirmed, namely, that with parallel selection along a geographic gradient, linkage is more pronounced in climax species than in species occupying positions at the beginning of succession.

Of course, experiments with forest trees dealing with such questions cannot be carried out within reasonable time periods and expenditures unless they are confined to young plants. NIENSTAEDT (1961) therefore justifiably recognizes the need to combine laboratory, nursery, and field experiments, as O. SCHRÖCK suggested 15 years earlier. To arrive at optimal and conclusive results, certain principles of sampling theory must be followed (WILKINS, 1959, and others). To estimate linkage relationships, for example, it is necessary to study the population means as well as the genetic variances and covariances within the demes (MORLEY, 1958; STERN, 1964, and others.) An instructive investigation using some of the preceding methods is that of MORGENSTERN (1969). His study will indicate the kind of conclusions to be drawn from a multivariate analysis of clinal variation, if the experiments are planned appropriately beginning with an adequate system of sampling and concluding with an effectual analysis.

The experiments were based upon 24 geographic seed sources of *Picea mariana* covering a considerable part of the species' range. The sources could be assigned to nine regions (ecological territories in the terminology of GALOUX) and represented 24 demes. Seed from several trees was collected within each source and was kept separate in most cases. The experiments were conducted partly in the greenhouse and partly in the nursery. Table 12 gives results of the first experiment.

The matrix of correlation coefficients (Table 12) was subjected to a principal component analysis, the results of which are shown in Table 13.

The analysis makes it possible to recognize the influence of at least four "factors", which are interpreted at the bottom of the table. The first factor, termed "seed size," "loads" the characters cotyledon number and hypocotyl length. The second factor was designated as "vigor"; the lower the soil moisture regime at the place of seed origin, the higher the germination and survival percentages. The third factor shows that hypocotyl length depends not only upon seed weight but also upon geographic latitude and longitude. This factor therefore measures the influence of geographic gradients and so was termed "geographic origin." It could have been termed "cline" with the same justification. Finally, the fourth factor measures the result of competition among the young seedlings: the lower the germination rate and germination percent, i.e., the lower the density of the young seedlings, the larger was their size within the ten-month experimental period.

Subsequent tables (Tables 14 – 19) give results of the other experiments in this series. A discussion is not necessary and readers are referred to the original publication.

This investigation, and those of other authors designed to answer similar questions, indicate that in many cases a clearer exposition of the causes of clinal variation may be possible than with regression techniques.

Multivariate methods may also be successfully applied to define character complexes resulting from specific adaptation, i.e., complexes that are the consequence of functional dependencies or of parallel selection. MURTY et al. (1970), for example, investigated such dependencies in *Sorghum, Pennisetum,* and *Brassica* by means of factor analysis, utilizing the genetic correlations of 14 presumably adaptive characters. In *Sorghum,* two factors became evident that explained nearly the whole "communality." The first factor contributed "loadings" for characters of vegetative growth, the second for reproductive characters. In *Pennisetum,* characters of the spike were associated with the first factor, early growth and flowering with the second. Of particular interest was the authors' comparison of the correlation matrix of one population before and after several generations of disruptive selection. Disruptive selection had changed the correlation matrix completely. It would be interesting to examine relationships along a cline by means of similar investigations, to obtain a better idea of the kind and intensity of selection as well as of the changes taking place under selection pressure.

b) Discontinuous Races

Discontinuous geographic variation, also termed racial variation or ecotypic variation, is often set aside as a special case and considered the opposite of clinal-continuous variation. It is thought of as originating from the discontinuous distri-

Table 12. Correlation coefficients of Experiment No. 1 based upon pooled data from 115 single-tree progenies and three stands in population 1 to 9[a]

Variables		X_1	X_2	X_3	X_4	X_5	X_6	X_7	X_8	X_9	X_{10}
X_1	Germination rate	+1.00	+0.59***	+0.53***	+0.01	+0.31**	-0.15	+0.33***	-0.27**	+0.25*	+0.07
X_2	Germination percent		+1.00	+0.13	+0.01	+0.27**	-0.40***	+0.26**	-0.19*	+0.19*	+0.00
X_3	Survival percent			+1.00	-0.24**	-0.11	+0.12	-0.05	-0.27**	+0.21*	+0.07
X_4	Cotyledon number				+1.00	+0.62***	-0.08	+0.62***	-0.01	-0.44***	-0.25**
X_5	Hypocotyl length					+1.00	-0.01	+0.63***	-0.01	-0.52***	-0.27**
X_6	Ten-month height						+1.00	-0.15	+0.04	-0.08	+0.08
X_7	Seed weight							+1.00	-0.09	-0.25**	-0.12
X_8	Soil moisture regime								+1.00	-0.14	-0.09
X_9	Latitude N									+1.00	+0.62***
X_{10}	Longitude W										+1.00

[a] Significance levels in all correlation tables: * 5%; ** 1%; *** 0.1%.

Table 13. Results of the principal component analysis of Experiment No. 1. The three highest loadings for each axis have been marked to identify the underlying factor. Axis 5 is from the analysis before rotation

Axis	1	2	3	4	5
Eigenvalue	2.45	1.93	1.60	1.55	0.84
Percent of total variance	25	19	16	15	8
Variables			Loadings		
Germination rate	+0.23	+0.79	+0.11	-0.36	+0.25
Germination percent	+0.17	+0.40	+0.07	-0.74	+0.12
Survival percent	-0.23	+0.85	-0.05	+0.13	+0.13
Cotyledon number	+0.85	-0.16	-0.15	+0.02	-0.14
Hypocotyl length	+0.84	+0.13	-0.27	-0.07	+0.15
Ten-month height	-0.00	+0.14	+0.04	+0.88	+0.20
Seed weight	+0.85	+0.14	+0.01	-0.16	+0.03
Soil moisture regime	-0.06	-0.54	-0.13	+0.01	+0.81
Latitude N	-0.40	+0.22	+0.77	-0.19	+0.05
Longitude W	-0.09	+0.04	+0.93	+0.10	+0.05
Factor interpretation	Seed size	Vigor	Geographic origin	Competition	Soil moisture

Table 14. Correlation coefficients of Experiment No. 2 based upon 37 single-tree progenies from each of five moisture regimes in populations 4 and 5

Variables		X_1	X_2	X_3	X_4	X_5	X_6	X_7	X_8	X_9	X_{10}
X_1	Survival: Bartlett index	+1.00	+0.94***	+0.71***	+0.68***	+0.57***	+0.61***	-0.13	+0.58***	-0.14	+0.91***
X_2	Survival: days to 50%		+1.00	+0.70***	+0.65***	+0.57***	+0.53***	-0.15	+0.49**	-0.11	+0.98***
X_3	Root dry weight			+1.00	+0.90***	+0.80***	+0.79***	-0.28	+0.69***	-0.10	+0.70***
X_4	Shoot dry weight				+1.00	+0.62***	+0.83***	+0.15	+0.72***	-0.11	+0.64***
X_5	Root length					+1.00	+0.56***	-0.46**	+0.42**	-0.21	+0.57***
X_6	Shoot length						+1.00	+0.00	+0.56***	-0.12	+0.54***
X_7	Dry-weight shoot-root ratio							+1.00	-0.01	-0.04	-0.15
X_8	Seed weight								+1.00	-0.03	+0.47**
X_9	Soil moisture regime									+1.00	-0.06
X_{10}	Percent water loss of trays										+1.00

Table 15. Results of the principal component analysis of Experiment No. 2

Axis	1	2	3	4	5
Eigenvalue	3.14	2.87	1.37	1.09	1.05
Percent of total variance	31	29	14	11	10
Variables					
Survival: Bartlett index	-0.87	+0.33	-0.07	+0.24	-0.09
Survival: days to 50%	-0.94	+0.29	-0.09	+0.12	-0.05
Root dry weight	-0.43	+0.77	-0.27	+0.32	-0.04
Shoot dry weight	-0.40	+0.81	+0.15	+0.36	-0.06
Root length	-0.34	+0.64	-0.54	+0.08	-0.19
Shoot length	-0.29	+0.88	+0.06	+0.13	-0.05
Dry-weight shoot-root ratio	+0.07	+0.03	+0.98	+0.00	-0.04
Seed weight	-0.29	+0.39	-0.01	+0.87	+0.04
Soil moisture regime	+0.05	-0.06	-0.01	+0.03	+0.99
Percent water loss of trays	-0.93	+0.31	-0.09	+0.09	-0.00
Factor interpretation	Water stress	Plant size (vigor)	Relative root size	Seed size	Soil moisture

Table 16. Correlation coefficients of Experiment No.3 based upon 37 single-tree progenies from each of five moisture regimes in populations 4 and 5. Numbers in parentheses refer to measurements in the following stages of the experiment: (0) before first drought period; (1) after first drought period; (2) after second drought period; and (3) after third drought period

Variables	X_1	X_2	X_3	X_4	X_5	X_6	X_7	X_8	X_9	X_{10}	X_{11}	X_{12}	X_{13}
X_1 Total dry weight (3)	+1.00	+0.99***	+0.93***	+0.05	−0.34*	+0.15	+0.87***	+0.88***	+0.85***	+0.77***	+0.66***	+0.71***	−0.17
X_2 Shoot dry weight (3)		+1.00	+0.86***	+0.21	−0.43**	+0.10	+0.89***	+0.85***	+0.83***	+0.73***	+0.65***	+0.67***	−0.20
X_3 Root dry weight (3)			+1.00	−0.30	−0.13	+0.26	+0.74***	+0.84***	+0.82***	+0.78***	+0.62***	+0.72***	−0.10
X_4 Shoot-root ratio (3)				+1.00	−0.56***	−0.35*	+0.24	−0.00	−0.01	−0.14	−0.02	−0.13	−0.22
X_5 Terminal bud percent (3)					+1.00	+0.26	−0.61***	−0.38*	−0.35*	−0.01	−0.16	+0.01	+0.09
X_6 Terminal bud percent (2)						+1.00	+0.05	+0.12	+0.31	+0.09	+0.04	+0.23	+0.11
X_7 Shoot length (3)							+1.00	+0.91***	+0.86***	+0.56***	+0.57***	+0.44***	−0.16
X_8 Shoot length (2)								+1.00	+0.92***	+0.61***	+0.53***	+0.51**	−0.15
X_9 Shoot length (1)									+1.00	+0.57***	+0.51**	+0.46**	−0.28
X_{10} Total dry weight (0)										+1.00	+0.83***	+0.74***	−0.11
X_{11} Shoot length (0)											+1.00	+0.56***	−0.12
X_{12} Seed weight												+1.00	+0.03
X_{13} Soil moisture regime													+1.00

Table 17. Results of the principal component analysis of Experiment No. 3

Axis Eigenvalue Percent of total variance				
Variables	Loadings			
Total dry weight (3)	+0.76	−0.60	+0.08	−0.07
Shoot dry weight (3)	+0.76	−0.57	+0.21	−0.09
Root dry weight (3)	+0.71	−0.62	−0.22	−0.03
Shoot-root ratio (3)	+0.05	+0.13	+0.83	−0.14
Terminal bud percent (3)	−0.49	−0.10	−0.76	−0.10
Terminal bud percent (2)	+0.43	+0.17	−0.70	+0.09
Shoot length (3)	+0.85	−0.35	+0.33	−0.02
Shoot length (2)	+0.85	−0.38	+0.08	−0.06
Shoot length (1)	+0.91	−0.28	−0.03	−0.20
Total dry weight (0)	+0.33	−0.89	−0.09	−0.08
Shoot length (0)	+0.28	−0.81	+0.08	−0.09
Seed weight	+0.34	−0.75	−0.17	+0.11
Soil moisture regime	−0.12	+0.01	−0.11	+0.98
Factor interpretation	Needle surface	Initial plant size	Relative root size	Soil moisture

bution of the environments of a species such as the distribution of certain soil types, and if the genetic system of the species concerned is capable of reacting to this distribution by forming such discrete races. This is not necessarily always so, as was described in the preceding section. In the presence of high migration rates and small differences of the selection coefficients of the alleles concerned, continuous variation patterns may develop where a discontinuous pattern would be expected as the optimal one from the distribution of environments and without a knowledge of the relevant components of the genetic system. Therefore, continuous-clinal and discontinuous (racial) variation are not qualitatively different, but reflect adaptive strategies of the population, which depend primarily upon the distribution of the niches (in time and space), relative differences in the selection coefficients of the alleles concerned, and the migration rate.

The discontinuous component of the genetic variation pattern of a species also contains the differences among subpopulations resulting from accidents of sampling (genetic drift and others). Many authors regard this part as unimportant but recent investigations indicate that it is significant (e.g., SAKAI et al., 1971). It has been discussed in Section IV.3e of this chapter.

Across the range of many species, there are niche dimensions having continuous geographic gradients such as day length, which depends upon latitude and follows the same rhythm year after year, without any discontinuity along the gradient. The variation of other niche dimensions is strictly discontinuous, such as that of soil properties, or the frequency of late frosts, which depends upon topography and is greater in depressions than on slopes. Therefore, the variation pattern of a species should contain continuous as well as discontinuous components, depending upon

Table 18. Correlation coefficients of Experiment No. 4 based

Variables	X_1	X_2	X_3	X_4	X_5	X_6
X_1 Prolepsis, first year	1.00	+0.52***	+0.26**	+0.39***	+0.40***	+0.48***
X_2 Growth cessation, first year		+1.00	+0.79***	+0.92***	+0.88***	+0.88***
X_3 Growth initiation, second year			+1.00	+0.75***	+0.64***	+0.65***
X_4 Growth cessation, second year				+1.00	+0.99***	+0.88***
X_5 Duration of growth, second year					+1.00	+0.87***
X_6 Height growth until second year						+1.00
X_7 Seed weight						
X_8 Soil moisture regime						
X_9 Latitude N						
X_{10} Longitude W						
X_{11} Mean date of last spring frost						
X_{12} Length of growing season (days above $+6°$ C)						
X_{13} Length of frost-free season						
X_{14} Degree days above $+6°$ C						
X_{15} Day length June 21						
X_{16} Weekly reduction of day length, June 21 to September 21						

Table 19. Results of the principal component analysis of experiment no. 4. Axis 4 is from the analysis before rotation

Axis
Eigenvalue
Percent of total variance

Variables	Loadings			
Prolepsis, first year	+0.36	−0.26	+0.59	+0.31
Growth cessation, first year	+0.77	−0.57	+0.14	−0.04
Growth initiation, second year	+0.52	−0.67	−0.09	−0.15
Growth cessation, second year	+0.88	−0.44	−0.03	−0.04
Duration of growth, second year	+0.90	−0.36	−0.02	−0.02
Height growth until second year	+0.84	−0.35	+0.20	+0.09
Seed weight	−0.01	−0.20	+0.82	+0.24
Soil moisture regime	+0.16	−0.32	−0.46	+0.81
Latitude N	−0.86	+0.49	−0.02	+0.03
Longitude W	−0.93	−0.00	−0.00	−0.04
Mean date of last spring frost	−0.23	+0.95	−0.13	+0.07
Length of growing season (days above $+6°$ C)	+0.49	−0.85	+0.15	−0.04
Length of frost-free season	+0.27	−0.90	+0.07	−0.08
Degree days above $+6°$ C	+0.56	−0.74	+0.14	−0.07
Day length June 21	−0.86	+0.41	+0.06	+0.08
Weekly reduction of day length, June 21 to September 21	−0.86	+0.47	−0.07	−0.00
Factor interpretation	Day length	Temperature	Seed size	Soil moisture

upon 121 single-trees progenies in population 1 to 9

X_7	X_8	X_9	X_{10}	X_{11}	X_{12}	X_{13}	X_{14}	X_{15}	X_{16}
+0.33***	+0.07	−0.42***	−0.32***	−0.41***	−0.48***	+0.35***	+0.48***	−0.36***	−0.45***
+0.20*	+0.21*	−0.93***	−0.71***	−0.75***	+0.88***	+0.73***	+0.86***	−0.86***	−0.94***
+0.08	+0.23*	−0.74***	−0.53***	−0.72***	+0.78***	+0.71***	+0.75***	−0.73***	−0.72***
+0.05	+0.26**	−0.98***	−0.79***	−0.62***	+0.81***	+0.62***	+0.84***	−0.94***	−0.96***
+0.04	+0.25**	−0.96***	−0.80***	−0.56***	+0.76***	+0.56***	+0.80***	−0.92***	−0.95***
+0.26**	+0.22*	−0.88***	−0.76***	−0.55***	+0.73***	+0.54***	+0.75***	−0.83***	−0.88***
+1.00	−0.09	−0.12	−0.05	−0.26**	+0.26**	+0.22*	+0.21	−0.02	−0.16
	+1.00	−0.26**	−0.18*	−0.22*	+0.24**	+0.23*	+0.21*	−0.23*	−0.26**
		+1.00	+0.79***	+0.67***	−0.84***	−0.68***	−0.86***	+0.95***	+0.99***
			+1.00	+0.22*	−0.46***	−0.32***	−0.44***	+0.79***	+0.78***
				+1.00	−0.94***	−0.96***	−0.84***	+0.59***	+0.66***
					+1.00	+0.90***	+0.95***	−0.73***	−0.84***
						+1.00	+0.75***	−0.62***	−0.63***
							+1.00	−0.77***	−0.86***
								+1.00	+0.90***
									+1.00

the variation of those niche dimensions controlling selection, and the reaction of the genetic system.

Discontinuous variation in forest trees has usually been discussed in connection with questions about the occurrence of races on different soil types. On the basis of the literature, VON SCHÖNBORN (1967) has tried to answer this question but all of the experiments reviewed were based upon inadequate methods. However, recent experiments with forest trees confirm the existence of races on soil types for these organisms as well, as expected by most authors. LAND (1967) was able to show resistance to salt in subpopulations of *Pinus taeda* growing near the coast. SCHMIDT-VOGT (1971) found a relationship between rooting of *Alnus glutinosa* and the soil at the place of origin. In a provenance experiment the strongest root development was found in the provenance from an area with adverse soil conditions. PARROT (1971) discovered that in an experiment initiated with *Juglans nigra* in Canada in 1882, the genetic development of the populations replicated over several sites depended clearly upon soil characteristics. BURDON (1971) confirmed a high genetic variation within populations for the utilization of mineral nutrients and tolerance of nutrient deficiencies (here shortage of phosphorus) in a clone test with *P. radiata*, such as is known from other plant species. Therefore, for forest trees, too, the conditions are doubtless fulfilled that will allow them to become adapted to different soils, but in the absence of a sufficiently large number of experiments with forest trees it is justified to refer to results from other plant species.

We again begin (see Section II.1b) with the investigations concerning the formation of races on different soil types undertaken on extreme soils with high concentrations of heavy metals that are normally toxic, such as are found in the spoils of ore mines. Here the selection coefficients are extremely high, at least those of the tolerant genotypes. From this point of view, therefore, good conditions exist for

genetic differentiation. Such "preadapted" genotypes actually seem to exist in
many populations, and the mode of inheritance of the character is such that selection will accumulate these genotypes rapidly even in species that reproduce only
generatively. DESSUREAUX (1959), for example, found exclusively additive genetic
variation for the character manganese tolerance in alfalfa (see also McNEILLY and
BRADSHAW, 1968).

Some of the observations on tolerance concern the following species, for example: *Agrostis tenuis* against copper (ANTONOVICS, 1968; JOWETT, 1958; McNEILLY,
1968; McNEILLY and ANTONOVICS, 1968), nickel (JOWETT, 1958), and lead (JOWETT,
1964); *Anthoxantum odoratum* against copper (ANTONOVICS, 1968; McNEILLY and
ANTONOVICS, 1968); *Armeria maritima* against zinc (LEFEBVRE, 1970). Thus there is a
considerable number of examples with several species and metals. In all cases the
immigration into the spoil population was moderate to high. The results reported
by McNEILLY (1968) are typical for the joint effect of strong migration and selection. The population on the mine spoils is situated in an area where the wind
prevails strongly in one direction. Consequently, immigration is almost exclusively
from the exposed side, and emigration at the leeward side. The transition from
the tolerant to the nontolerant population took place on the exposed side in a
narrow strip that was only one meter wide. The frequency of tolerant types decreased slowly on the protected side where the mine spoil area changed into a site
free of copper. Therefore, on copper-free soil the selective values of tolerant genotypes should differ little from those of nontolerant genotypes; but on soil containing copper these values should differ considerably for tolerant and nontolerant
genotypes. In a thorough investigation, JAIN and BRADSHAW (1966) have applied
simulation methods to this problem, using various models with joint action of
migration and selection. Of course the same problem of estimating selection coefficients, etc., could be tackled using the more general methods of HANSON (1966) or
HALDANE (1948). McNEILLY and BRADSHAW (1968) found that the transition zones in
several old copper mines in Ireland and Wales are not wider than 15–16 m. On the
old lead mines investigated by JOWETT (1964) the situation was different. To begin
with, the degree of lead tolerance of *Agrostis tenuis* populations differed on individual sites. In addition, there was still considerable variation of the character lead
tolerance on the mine dumps, indicating lower selection coefficients than on the
copper mine dumps.

For the interpretation of the situation in *Agrostis tenuis* one should also note
that this species is a perennial. Once tolerant plants have become established, they
may produce seeds over many years; the process of adaptation is less risky, and,
other things being equal, more rapid.

Although the evolution of races on different soil types under extreme selection
also depends upon migration, one could expect a reaction of the genetic system to
selection and a reduction of the migration rate. Conceivably, this could occur
through development of incompatibility and seasonal barriers, or through the
breakdown of existing self-incompatibility mechanisms, i.e., a change to essentially
self-fertilization. McNEILLY and ANTONOVICS (1968) did not find incompatibility
barriers between populations on and beyond the mine dumps, but crosses of *Anthoxantum* seed parents from the dump with pollen parents outside yielded a low
seed set. Perhaps an incompatibility barrier is evolving here.

ANTONOVICS (1968a) found a higher degree of self-fertility of *Agrostis tenuis* and *Anthoxantum odoratum* on the contaminated than on the normal soil. This relative self-fertility was inherited; the degree of self-fertility was distinctly different for populations on different mines, and the differences were in part explained by surrounding populations. The author interprets the higher self-fertility of the mine populations as a means of reducing immigration. A simulation study indicated that self-fertilization and a reduced immigration rate was accompanied by substantial reduction of the genetic pressure on the population (under the conditions of the mine dump). The response to selection for higher self-fertility could also be shown by simulation. Linkage of the locus carrying the self-fertility allele to the locus carrying the gene for tolerance, strong selection pressure, great frequency of the favored gene, dominance of the self-fertility gene, and recessivity of the tolerance gene lead to a rapid increase of the allele for self-fertility. This model explains, in the author's opinion, many of the known cases of return to self-fertility. LEFEBVRE (1970), on the other hand, in his studies of zinc-tolerant populations of *Armeria maritima*, explains the same phenomenon, the increase of self-fertility in the mine population, as resulting from the necessity for repeated resettlement of the mine area. The existing resistant population fluctuates greatly in size and maintains itself only by a shift to more frequent self-fertilization. The rate of immigration into the mine dump population through pollen and seed was large in all cases, which was proven by the particularly impressive studies of McNEILLY and BRADSHAW (1968).

Differences between the flowering dates of the populations on mine dumps and surrounding populations were discovered by several authors. Interestingly, the mine populations always flowered earlier. For example, McNEILLY and ANTONOVICS (1968) found that mine populations of *Agrostis* and *Anthoxantum* flowered an average of one week earlier. The smaller the mine areas, the larger were the differences in flowering date. These differences in flowering were maintained in subsequent transplant experiments, and hence can be interpreted as the result of selection. On the larger mine dumps the largest differences of flowering date were observed at the edge of the area. According to the authors the beginning of sympatric speciation may be seen in such developments (the formation of any type of reproductive isolation may be interpreted in this way). In his investigations of lead-tolerant forms of *Agrostis*, JOWETT (1964) found not only earlier flowering, but also greater uniformity and reduced growth. Growth reduction is possibly the result of inbreeding depression following an increase in self-compatibility, such as has been observed by other authors.

The fact that perennial forms are less influenced by migration has been denoted by ANTONOVICS (1968b). His studies based upon simulation are also illuminating with respect to the conditions for development of modifying gene systems.

A final aspect of the origin of races on soil types contaminated by heavy metal ions concerns the effect of correlated selection. JOWETT (1964) found that forms tolerant of metals were also tolerant of *low* levels of calcium and phosphorus. Quite likely tolerance of heavy metal ions may be achieved physiologically in very different ways, for example, better possibilities for selection or higher resistance against ion concentrations in the plant itself (see Section II. 1b), and therefore these correlated changes of physiological properties could vary from population to population.

Race formation on extreme soil types involving not only one but several niche dimensions will be mentioned only. Races on serpentine soils (KRUCKEBERG, 1954) constitute an example that has already been pointed out in connection with adaptation to soil and with marginal populations. Such situations are more difficult to analyze, not only because of the multiplicity of niche dimensions to be considered simultaneously, but also because of the correlated reactions such as have been found in lead-tolerant forms of *Agrostis* as discussed above.

The reactions of higher plants to different concentrations of calcium ions have also been investigated relatively throughly (ELLENBERG, 1958). Of course, differences here are not as distinct as on sites contaminated by heavy metal ions, where selection coefficients of the different genotypes may assume the extreme values of 0 and 1. Yet here, too, suitable methods of investigation have produced fairly clear results. Thus BRADSHAW (1960) and SNAYDON and BRADSHAW (1961) found distinct differences among populations of *Agrostis tenuis* and of *Festuca ovina* on soils of high and low calcium levels. The two authors concluded that these intraspecific differences can be greater than those among the calciphytes and oxylophytes that are being used as plant indicators by phytosociologists. SNAYDON (1970) analyzed a population of *Anthoxantum odoratum* from an old fertilizer study. In this experiment, plots 20 × 35 m in dimension were given different quantities of lime for a period of 40–60 years. In a subsequent experiment, plants from the high-lime plots grew better on calcareous soil, and plants from low-lime plots on noncalcareous soil. The correlation between pH value of the original site and growth was 0.95. Although the author discovered only small rates of gene flow among plots, differences among the populations growing on them were smaller than differences among natural populations of this species growing on soils that are comparable with respect to calcium content. In the author's opinion, a narrow genetic base of the original population may explain this.

Some cases of genetic reaction to levels of phosphorus in the soil have also been well researched. For example, SNAYDON and BRADSHAW (1962b) found a correlation of 0.96 between phosphorus content of the place of origin and reaction to phosphorus fertilization in an experiment. Plants from soils with low phosphorus content could take up and store more of this nutrient. KUMLER (1968) observes similar conditions in a comparison of two races of *Senecio silvaticus*, one from dunes, the other from mountain soils. The latter reacts more strongly than the former at minimum levels of mineral elements.

Superiority on its own soil was demonstrated again and again in many species, e.g., by SNAYDON and BRADSHAW (1962a) for white clover when applying the traditional clonal method. Therefore, MELCHERS' (1939) prediction of a close adaptation of plant species to soil characteristics has been confirmed. There is no reason to assume different relationships for forest trees, especially since the experiments mentioned at the beginning point in the same direction.

Characteristics of the soil also constitute the background for the classical cases of selection for mutual exclusion not only at the species level (ELLENBERG, 1958, and others), but also the level of the population and the genotype (see STERN, 1969 for summary). To mention only two examples: SNAYDON (1961) found only small differences among populations of *Trifolium repens* from different soils when grown in pure cultures, but in mixed cultures each population was superior on its own soil. In another experiment with the same species, he confirmed this for the niche dimension calcium content (SNAYDON, 1962). Therefore, competition becomes the all important factor here, leading to selection coefficients that differ, if not in direction, then certainly in their order of magnitude. The importance of competition for adaptive values and their estimation, respectively, has already been emphasized by HALDANE (1932). The neglect of competition in provenance experiments with forest trees has led to the current situation where we know little about adaptation to soil.

The consequence of this is frequently seen in forestry literature, namely, in the attempts to designate the fastest growing provenance as the best adapted one of the new planting site. NAMKOONG (1969) points out that this corresponds to the assumption of a close correlation between growth and reproductive success. He cites a considerable number of examples from forest trees where this is obviously not the case, in such different species as *Pseudotsuga menziesii, Pinus silvestris, Picea abies, Juglans nigra,* and others. The interesting consequences he sees for tree breeding cannot be considered here, but another example may illustrate the point. In the Northwest German Plains, *Betula verrucosa* occupies open terrain almost exclusively since the destruction of forests. It grows in hedges, in pastures, along roadsides, etc., where there is usually no competition and where a large crown surface and early and abundant flowering determine reproductive success and with it the direction of selection. In the hill forests to the south, however, this birch must compete with other tree species. Those trees that keep up in growth with other species and maintain themselves long enough to place progeny in the first stages of succession possess the highest probability of reproducing themselves. In experiments of the writer (STERN, unpublished), this is reflected pronouncedly, and it is clear that it is not the very different soils or climatic effects but unequal selection for reproductive characters that determine differences among seed sources of the species from the Plains or neighboring hill country. The former grow relatively slowly, are generally bushy and of poor form, and flower early and abundantly; in contrast, the latter resemble their competitors more closely. They grow more persistently, possess a straight, continuous stem, and flower later and less prolifically. Other adaptations may exist, such as those to the soil or the necessity of tolerating shade and other conditions (see HOLMGREN, 1968 for an example with *Solidago virgaurea*). Therefore, even when studying adaptations to soil, all relevant niche dimensions should be included and not only those of the soil itself.

LEVINS (1964) chooses another approximation to solve the problem of the combined effects of migration and selection. Migration is not measured by migration rate but by autocorrelation of the environments of successive generations. Of course, this is $+1$ if every individual develops in the same niche as its two parents. It becomes 0 with random mating (50% migration rate); even negative correlations are imaginable, e.g., with cyclical changes of the environment, etc. The results of his reflections are principally the same as those obtained from the conventional model: niche differences in relation to tolerance of the optimal phenotypes, and the degree of autocorrelation of the environments, determine the process of differentiation and its result.

This model is mentioned only because it may possibly serve as a basis for the description and explanation of the genetic structure of large, continuous populations of higher plants. For example, a study of a tree's pollen and seed distribution may indicate the probability of rediscovering a certain gene in an individual at a certain distance from the parent in the next generation. This probability decreases with distance from the parent (see Subsections a, d, e). The results obtained by LEVINS by means of the autocorrelation coefficient are in general agreement with those obtained in similar situations with other procedures.

Another model that should be mentioned in this respect is that of disruptive selection, which was developed in its typical form by MATHER (1955). Here, again, there is simultaneous selection in more than one niche. In the ideal case an extreme phenotype possesses greatest fitness in each niche. Random mating takes place over all niches, and all niches occur with the same frequency. There should also be perhaps equal fitness superiority of each optimal phenotype in its niche. Experiments with disruptive selection have been conducted primarily against the background of sympatric speciation (MAYR, 1963). The starting point was MATHER's (1955) anticipation that, given suitable conditions, disruptive selection could develop crossing barriers and thereby lead to speciation. In LEVINS' terminology this means that the development of such barriers is the optimal adaptive strategy. This strategy is the one most likely expected if a population extends across two niches and is represented by two phenotypes, each of which possesses greater fitness in one of the niches.

Some observations in natural populations support this hypothesis but usually population history is not sufficiently known to discount all other explanations. CLARKE and SHEPPARD (1962) found a mimetic polymorphism in *Papilio dardanus* where all intermediate forms experience a selective disadvantage. They believed disruptive selection could be sufficiently effective to induce sympatric speciation. However, complete certainty is only possible with controlled experiments. THODAY and GIBSON (1962) and GIBSON and THODAY (1964, see also further literature cited there) have performed the first successful experiments. In populations of *Drosophila* kept under conditions of disruptive selection, they found not only an increase of genetic variance, more polymorphisms and divergence of the partial populations, but finally also genetic mechanisms that limited gene flow between both population parts. Different selection intensities and rates of gene flow were induced. In the experiments of MILLICENT and THODAY (1960), for example, a divergence of the population halves could be observed even in the presence of random mating (50% gene exchange). The characters under selection were relatively simply inherited in all cases, and major genes were always involved, e.g., WOLSTENHOLME and THODAY (1963) mentioned two loci on the third chromosome each with a dominant allele. Here disruptive selection led to an accumulation of linkage-phase gametes and thereby to rapid divergence of the population parts. Other authors used mutants in their experiments. For instance, HOENIGSBERG et al. (1966) created a population from the mutant lines ebony and dumpy of *Drosophila melanogaster*. Hybrids were removed in each generation. This resulted in an increase of mutant frequency with every generation, and therefore a limitation of gene flow between lines. The main reasons for progressive genetic isolation were the development of behavioral barriers in most cases and of ecological isolation (THODAY, 1969). Behavior patterns of animals may increase the autocorrelation of successive environments and thereby create better conditions for genetic divergence through disruptive selection, e.g., if

the two or more genotypes develop preferences each for an optimal habitat (Mac-
Arthur, and Levins, 1965; Maynard Smith, 1966). A review of the most important
experiments is given by Thoday (1964). Of course, barriers did not develop in all
experiments (e.g., Robertson, 1966).

As an adaptive strategy, the development of genetic isolation barriers, has often
been the concern of population genetics. Among the physiological mechanisms
involved are, for example, hybrid sterility (see Ehrman, 1962 for summary regarding *Drosophila*) in both animals and plants; barriers of behavior in animals (always
realized in experiments, see Kessler, 1966, although the methods applied and
interpretation of results are often controversial, see Merrel, 1950 and Stalker,
1942); seasonal isolation in animals and plants, and the incompatibility of plants.
Genetic isolation may also result from geographic isolation alone when this is
maintained long enough (see Wallace, 1954 and others for conditions in *Drosophila*); but the many known examples of its origin as an adaptive strategy through
selection in sympatric populations in nature indicate its general importance for
evolution (Ehrman, 1965; Magolowkin et al., 1965 and others for *Drosophila*,
Vasek, 1964 and Grant, 1966 for plants, to mention at least two examples). Genetic
isolation as adaptive strategy, and thereby sympatric speciation, represents the
most extreme consequence of disruptive selection resulting from discontinuous
variation of the environment.

Several authors point out that in nature the large differences in fitness of phenotypes in different niches required to develop genetic barriers may be rarely present,
and also that the long series of generations necessary to erect these barriers seldom
may be realized (Crossley, 1963; Knight et al., 1956). The consequence then would
be (Mayr, 1963) that it is easy to obtain sympatric speciation by disruptive selection
in a model experiment, but that this hardly ever occurs in nature.

The next, less drastic effect of disruptive selection would be the introduction of
polymorphisms. This will be discussed in the next section and only the most important literature will be cited here. Probably the first experimental evidence of genetic
polymorphism following disruptive selection was contributed by Thoday (1958)
and Thoday and Boam (1959). The chromosomal polymorphism cited above also
belongs to this group. Concerning results from higher plants, the work of Breese
(1964) should be mentioned. This author found a high degree of polymorphism in
Lolium growing in discontinuous environments.

An increase of genetic variance in the presence of disruptive selection is reported
by many authors. Instructive comparisons can then be made with parallel lines kept
under conditions of stabilizing selection (selection of the intermediates). Again
Thoday (1959) found increasing additive genetic variance when disruptive selection
affected the number of sternopleural bristles.

Scharloo (1964) studied the influence of disruptive and stabilizing selection on
the expression of a mutation (see also Scharloo et al., 1967). Disruptive selection
increased and stabilizing selection reduced the genetic variance. The additive component became larger here, too. In the experiments of Gibson and Thoday (1963),
stabilizing selection utilized the additive genetic variance completely but considerable mother effects remained. The lines subjected to disruptive selection diverged
very soon and in an extreme fashion (separating the distributions entirely), although epistasis also played a significant role. Both Prout (1962) and Tigerstedt

(1965), when studying the character rate of development in *Drosophila*, found that the variance of lines subjected to stabilizing selection was reduced whereas that of lines under disruptive selection increased. Other experiments did not produce any effects, such as those with stabilizing and disruptive selection of FALCONER and ROBERTSON (1956) on birth weight in mice, or of FALCONER (1957) on abdominal bristles in *Drosophila*.

Disruptive and stabilizing selection may well lead to a selection and linkage equilibrium, given suitable conditions (see KOJIMA, 1959; CURNOW, 1964; JAIN and ALLARD, 1965; SINGH and LEWONTIN, 1966, and others). This, too, will only be mentioned.

A side effect of increasing genetic variance following disruptive selection is often greater genetic flexibility of the population. This was very plain in the experiments of TIGERSTEDT (1965) and NAMVAR (1971) on disruptive selection for developmental rate with *Drosophila melanogaster*. Samples of 40 flies were moved to large cages to establish new populations. These populations grew fastest when their founders came from lines subjected to disruptive selection. In response to selection such accretion of genetic flexibility should enable natural populations to cope more easily with fluctuating environments. This is probably important for forest tree populations. Genetic variability and flexibility are significant components of the genetic system of a species, and one should expect, therefore, that they have reached a balanced, near-optimum position in a natural population. The section on migration will discuss this further.

To summarize: overwhelming evidence is available to prove that discrete races must exist in nature when environment and the genetic system of a species favor their evolution as an optimal adaptive strategy. The necessary conditions have been well researched in theory and experiments. If the characteristics of such a race are to be explained, a broad spectrum of cumulative and interacting factors must be considered.

For these reasons the chances for development of discrete races differ among forest tree species. Probably such extreme cases will not be found here as STONE (1959) described for the vernal pool mousetail *(Myosurus)* in California. This plant subsists in water holes only during the rainy season, one habitat being isolated from the next. Furthermore, flower structure excludes cross-pollination almost entirely, and seasonal differences in flowering date also exist. The result of such combined effects of the genetic system and the environment is strong homozygosity of the biotypes found by the author and the occurrence of only one to nine biotypes at each water hole. Many of our tree species appear to represent the opposite extreme, that is, large and continuous distribution, high migration rate, etc. However, even here relatively sharp boundaries may develop as the examples *Pinus silvestris* var. *lapponica* and *Picea abies* var. *obovata* indicate.

The explanation of genetic variation patterns therefore requires greater attention to detail. A beginning of this is evident even in the earlier literature, such as in the "pine race studies" of RUBNER (1959).

c) Polymorphisms

Genetic polymorphisms have long been a favorite subject of ecological genetics. We have seen in previous sections that they represent adaptive strategies of populations that have to cope with certain types of heterogeneous environments, or where the properties of their genetic systems, heterozygote advantage, etc., institute polymorphisms. This section attempts to explain some of the well-known genetic polymorphisms of forest trees as adaptive strategies, because the forestry sciences, too, have long struggled with this phenomenon. We are less interested in a classification that allows the grouping of certain observed cases according to origin than in a discussion of the underlying causes, using selected examples. For a general discussion the reader is referred to the literature, particularly to FORD (1964 and earlier).

In plant breeding too, one has tried repeatedly to come to terms with the problem of genetic polymorphisms, i.e., more generally the genetic variability within populations. Experimental results are legion that indicate better buffering of genetically heterogeneous populations than of homogeneous ones in the face of environmental fluctuations. This will only be noted here and reference is made to ALLARD (1961), to cite at least one example. ALLARD's paper reflects the character of genetic polymorphisms as adaptive strategies of populations — and one could mention that here lies a risk of forest tree breeding. FISHER's (1930) prediction that selection will lead to a reduced variation of the characters concerned is valid only for directed selection in a constant environment, and has been repeatedly confirmed for this situation (e.g., ROBERTSON, 1955).

The first example chosen of polymorphism is the color of male and female strobili of conifers. CARLISLE and TEICH (1970) counted the frequencies of individuals in a natural stand of *Pinus sylvestris* in Scotland bearing violet-red, violet-yellow, and yellow male strobili; these were 0.04, 0.39, and 0.57, respectively. Assuming that violet-yellow strobili represented heterozygotes of a biallel locus and the two other classes the two homozygotes, the theoretical frequencies of the three genotypes in the same material were calculated as 0.05, 0.36, and 0.59. Therefore, there is good agreement between observed and expected values. Accordingly, the frequency of the allele for yellow is approximately 0.77 and that of the other allele 0.23. We could regard this as a genetic polymorphism that in some way is maintained by selection. The geographic distribution of the morphs is not known, but the same polymorphism was described in many parts of the species' range and is also commonly found in other conifers in both male and female strobili.

CHING et al. (1965) found different frequencies of the three morphs in different subpopulations of *Pseudotsuga menziesii*. They have also shown that there is only one anthocyanin in *Pseudotsuga* (and other conifers?), as compared with 10–15 in angiosperms. Eleven flavonoids were also discovered but only one of them with red color. RUBNER (1938) found unequal frequencies of the three morphs as indicated by cone color in different stands of *Picea abies*. He anticipated correlations between the color polymorphism of strobili and cones, on the one hand, and other morphological and physiological character on the other, as well as different adaptive values of the morphs. There is unfortunately no indication whether his experiments are based on native populations or mixtures of introduced provenances, which might simulate such correlations. KOZUBOV's (1962) explanation of polymorphism in Scots pine is more likely, that is, that dark- and lightcolored strobili of both sexes ripen at different times depending upon radiation, and therefore the polymorphisms give the population a certain flexibility in flowering time. Hence the polymorphism is then explained as a reaction of a population with concave fitness set in a coarse-grained environment. Differences among subpopulations in frequencies of the three morphs could be temporary (no information is available on geographic gradients), or reflect niche frequencies of the localities in the niche dimension of temperature.

Polymorphisms of flower color as a reaction to radiation are also known from angiosperms. HOVANITZ (1953), for example, describes a blue-white polymorphism of *Hepatica triloba*. Frequencies of the morphs are correlated with tree canopy density above the populations.

In the two preceding sections we have seen that hints can be obtained on the kind of polymorphism and its sustention from correlations to niche dimensions and from the geographic variation pattern. For the case described, such investigations are not yet available. The polymorphism of cone shape of *Picea abies* has been more fully explored in some regions; the overall pattern is still incomplete.

Characters of the cone scale are indicative of morphs in Norway spruce (see Fig. 29). Certain character combinations have been named *acuminata, europaea, obovata*, etc. CHYLARECKI and GIERTYCH (1967), for example, find in Masuria only the forms *europaea* and *acuminata*. Populations of Norway spruce outside the continuous range within this general area predominantly bear cones similar to the *obovata* type. In the Sudete-Carpathian-Beskid mountains to the south, *obovata*-like types are about equally frequent as *acuminata* whereas *europaea* dominates (STASZKIEWICZ, 1967). Similar results are available for other areas but a general outline of the geographic variation pattern of this polymorphism based on the literature cannot be drawn as yet. What appears to be certain is that in the eastern part of the range of *Picea abies* there are exclusively *obovata* forms that have only been observed in high elevations in the west.

Nothing is known regarding the selective forces that give rise to this polymorphism. So far it has only been described or used to derive relationships among populations or discuss problems of paleobotany and descent. Should it turn out that a balanced polymorphism is involved, such as in all known cases of genetic polymorphism, then the preceding attempts would produce extremely doubtful results.

It is conceivable that the selective advantages or disadvantages of the morphs in different environments depend upon the functions of the cones, or of the female strobili whose morphological differences could be correlated with those of the cones. Time of seed release as determined by weather during the period of seed flight, characteristics of the strobili utilizing pollen movement under different conditions, and other causes could be involved. The relevant investigations have yet to be made. One could also imagine that the different cone character combinations making up the few morphs are determined by several gene loci joined together like supergenes. Inheritance of this kind has been reported in several polymorphisms (FORD, 1964, and earlier). It would then be possible to find cone types locally differing from the types often described that are optimal in very specific environments.

With regard to cone form, crown and branch types of Norway spruce have also frequently attracted the attention of foresters and forestry scientists. Part of the reason is that the different types apparently differ in economic value. HOFFMANN (1968) has given a summary of the most important literature. Characteristically, a correlation between cone forms and branch types does not appear to exist.

According to HOFFMANN, the three main types of this polymorphism in *Picea abies* can always be well distinguished. He classifies them as shown in Table 20.

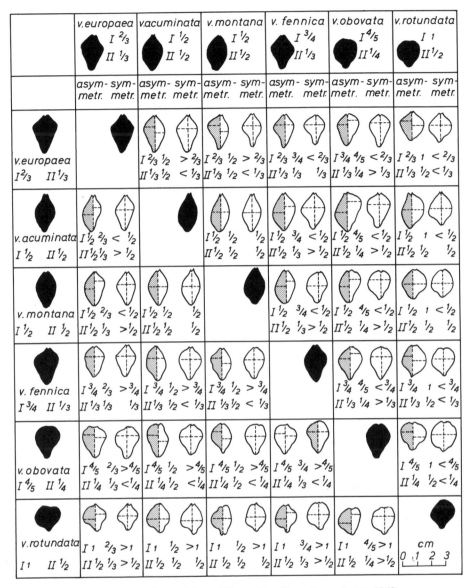

Fig. 29. Cone polymorphism of *Picea abies*. (From Priehäuser, 1962)

Supposedly the differences are most clearly expressed in the middle part of the crown whereas the lower part may always resemble the plate types. Atmospheric damage may influence the crown type and is strongly inherited.

For a long time, consideration has been given to the advantages and disadvantages of the three types in regions subject to ice, snow, or wind damage. According to Hoffmann, the relations shown in Table 21 exist. The low resistance of the brush type to all three influences prompts the question as to how the type maintains itself

Table 20. Main characteristics of Norway spruce types

Branch types	Branches of first order	Branches of second and higher order	Needles	Crown structure
Comb type	Generally horizontal, tip frequently turning upward, relatively long, few in number	Drooping curtainlike, sparsely branched, relatively long	Long, relatively few needles	Conical, large crown base, spreading, open when viewing from below
Brush type	Generally pointed upward or curved, sabre-like, medium-long to long	Horizontal to downward pointing, profusely branched, short frequently overhanging	Long, number of needles still small per cm of shoot	Conical, not as strongly spreading, usually possible to see through from below
Plate type	Horizontal to drooping, relatively short, many branches	Generally branched in the horizontal plane (platy)	Short, larger number of needles per cm of shoot	Paraboloid or cylindrical, narrow densely branched, impossible to see through from below

Table 21

Morphs	Resistance to		
	Snow	Ice	Storm
Comb type	Very high	Low	High
Brush type	Low	Low	Low
Plate type	High	Very high	High

in the population. This is pertinent if different damage frequencies are responsible for unequal frequencies of the types at the various localities, and thereby maintain this polymorphism as earlier authors supposed. The exception is if the brush type represents the heterozygous genotype, though this can hardly be expected since heterozygote inferiority leads to dominance of the one or other allele or "superallele" in such cases (see below).

In spite of this, different frequencies of the morphs may arise as a consequence of these damaging influences. HOFFMANN (1968) cites examples from the Thuringian Forest where the percentage of plate types ranges from 30 to 90 depending upon hoarfrost frequency, and from the French Jura where brush types predominate at elevations of 1000 to 1500 m, from the Harz Mountains, etc. It seems certain, therefore, that the three sources of damage codetermine the frequencies of the three morphs in local populations. Tree breeders have already drawn the appropriate conclusions.

Yet it is questionable whether damage alone maintains this polymorphism. As emphasized also by HOFFMANN: "Typical for comb-spruce is a broad, open crown with regular light channels in contrast to the dense, frequently columnar crown of plate-spruce which resembles the layer-like crown of silver fir." Accordingly, the two extreme types utilize light differently, at least from a certain age on. This is also underlined by arrangement of the branches of second and higher order (drooping in comb spruce, horizontally spread in plate spruce). According to SCHMIDT (1952), the growth rhythms of the two types also differ. Beginning with the third decade of tree age, i.e., the time when crown forms differentiate, growth of individual comb spruces is faster as a result of a larger crown volume and greater proportion of more efficient light needles. Yet on a unit-area basis the production of pure stands of the two types is the same. At this age, therefore, the two extreme types react differently with respect to competition and it is therefore probable that competition contributes to the maintenance of polymorphism, in addition to atmospheric damage. The relevant studies are lacking, but should not be too difficult to carry out. This explanation is supported by observations in other species. For example, *Picea omorica,* which usually develops an open-grown condition in its natural habitat, is monomorphic, resembling the brush or plate type, whereas *P. breweriana,* which always grows in competition and usually occupies the understory, is also monomorphic but similar to the comb type. Both species are specialized; there is homo-selection for the characters serving specialization — in contrast to heteroselection within *Picea abies,* which ranges over a broad spectrum of niches.

A review of the competition in sustaining genetic variation including polymorphisms has been presented by STERN (1969). Therefore, it is sufficient to repeat only the results of some experiments and theoretical investigations that are of significance here. The fitness value of genotypes in plant populations with intraspecific competition is often determined by their own frequencies. Selection is frequency dependent. HALDANE and JAYAKAR (1963) have shown that stable equilibria are then present where gene frequencies may reach a balanced condition. CLARKE and O'DONALD (1964) have contributed several models. FRANKEL (1959) supposes that selection of this type is common even in domesticated plant species, leading to population conditions that he terms relational homeostasis. Equilibria conditioned by competition are here responsible for variation at the gene loci concerned. PFAH-LER (1964) confirms this for the cultivated species in the genus *Avena.* Of interest here also are the experiments of PARSONS and ALLARD (1960). In six populations of lima beans studied over several years, the authors found no interactions with microsite when the plants were grown at wide spacing, i.e., without competition. Closer spacing introduced considerable interactions of that kind. Therefore, competition must have contributed to genetic variation. GREENWOOD (1969) states that very close and very wide spacing may lead to monomorphic populations. With intermediate spacing he obtained the highest degree of polymorphism.

HARDING et al. (1969) found heterozygote advantage at one gene locus of a mainly self-fertilizing plant species with 95% selfing rate. The heterozygote advantage increases with a decrease of heterozygotes, i.e., the less frequently they compete with individuals of the same genotype. Finally, SCHUTZ and URSANIS (1967) stated some of the conditions of competition among genotypes that lead to polymorphisms. Therefore, changing conditions of spacing, probably the normal condition in natural forests, as well as constant spacing and density, allow an equilibrium of competing genotypes. The main causes of the equilibrium appear to be heterozygote advantage, in competition, and/or overcompensation when different genotypes compete. It has been known for some time that in tree species, too, different genotypes react differently at different densities (see, for example, SCHMIDT-VOGT, 1965).

The competitive fitness of genotypes may change, not only at different densities but also at different age. BARBER (1955 and later) found a good correlation between the frequency of an allele and stand age. In this connection the change of trees from one to the other sociological tree class (BUSSE, 1930) is of special interest; this change may be influenced by competition and has been investigated by many authors. Such experiments concerning the change in fitness values with age are of course conducted easily with short-lived plants. To name only one example, FEJER (1967) found genotypes with very different rhythms of development in populations of *Lolium*, which may complement each other in competition. The phenotypic stability of certain genotypes is also determined by the partner in competition. JOHNSSON (1967) reports on progeny of early-flowering individuals of *Betula pubescens* that were superior in growth during the first two decades, but at age 25 the normally flowering neighbors caught up with them. A similar situation was observed in *B. pendula* (STERN, unpublished). Observations in growing stands of other tree species are conflicting. WECK (1958), for example, assumes that the change from one to other sociological classes is an extensive phenomenon. MEYER (1965) reports large changes of this kind in Norway spruce at the age of 30–40 years, particularly in the dominant class. Nearly 50% of the trees left this class and 15% entered it. But ERTELD (1950) finds only a minor change in a 66-year-old stand of Norway spruce. Similar observations were made by HENGST (1962) in beech. For additional results the reader is referred to the literature.

It is obvious, therefore, that the fitness values of genotypes can be modified by competition relative to the population average. Thus, when ADAMS (1970) tested the average competitive fitness of four full-sib families of *Pinus taeda* at the seedling stage in a greenhouse, all possibilities of relative competitive behavior were realized in spite of the small number of families. Following SCHUTZ and BRIM (1967), he termed the mutual influences of two genotypes overcompensation when one of the two genotypes is challenged but the other suffers no loss, or if both promote each other. Comparisons are always made against the yield, or more generally, the expression of the character, in a pure stand. Undercompensation is the competitive compensation of two genotypes if one or both suffer a loss; compensation if one loses or wins; neutrality if none wins or loses. ADAMS' results are shown in Table 22. These results are best given in the author's words:

"It is surprising, that with such a small sample of only four families that the interfamily competitive responses among them are so prevalent and so varied. It is possible that these results are due to having chosen four families which are exceptionally variable in the ability to compete. However, no conscious effort was made to do so and it seems unlikely (especially when the relatedness of the families are considered) that such is the case. Most likely, the results indicate that the ability to compete among families of conifers is highly variable and strongly affects the growth of trees in competition. The results point out several factors that are of

Table 22

Character	Family combination reciprocal pair					
	AB,BA	AC,CA	AD,DA	BC,CB	BD,DB	CD,DC
Height increment	0.0	0.0	+.0	0.0	0. −	+. −
Diameter	0.0	0.0	+.0	0.0	0. −	+. −
Leaf surface area	0.0	0.0	0.0	+.0	+.0	0.0
Weight	0.0	0.0	+. −	0.0	0.0	+. −

Gain is indicated by +, loss by −, and neutrality by 0.

interest to forest geneticists. Since it was clearly demonstrated that variation in competitive ability occurs in seedlings of full sib families of loblolly pine, it would seem logical that such variation would also occur among families of older trees. This must be shown by well designed experiments."

Further conclusions, particularly with respect to progeny tests of forest trees, will not be discussed. For our purposes it is important, however, to note (STERN, 1969) that the naive assumption that "the best genotype" will assert itself in competition is clearly in need of revision. Competition within natural populations, in all stages of the development of populations and individuals, and over all niches in which populations of the species occur, does not lead to the concentration of one genotype and hence to a narrowing of genetic variation. Rather it leads to an equilibrium among a multiplicity of genotypes. Such genetic variety is not reduced when one considers the dependence of competition effects upon environmental variables.

In a quantitative-genetic analysis of a plant population, ALLARD and JAIN (1962) found that the observed genetic variance exceeded the expected quantity $(1 + F) V_A$ (F refers to the inbreeding coefficient of the population, V_A to the additive genetic variance). This is to be expected with heterozygote advantage. In populations of the same species, ALLARD and WORKMAN (1963) found selective values of the heterozygotes that as much as 20–30% higher. Furthermore, there was considerable variation from year to year in the selective values of the homozygotes, leading to a kind of cyclical selection (negative autocorrelation of the environment). This should result in extreme shifts in the direction of selection across the niches (years in this case). Consequently, the optimal solution for the population is large genetic variability and, thereby, genetic flexibility (genetic homeostasis according to LERNER, 1954).

HARDING and ALLARD (1969) extended the analysis of genotype fitness by examining several loci simultaneously. Seven gene loci were included in the study. In 16 of 21 possible comparisons, observed zygote frequencies differed significantly from those expected in simple comparisons. In all 21 cases observed frequencies of the double heterozygotes were above those that were expected.

The result of these two authors cannot be discussed here in detail, but we are drawing upon it to indicate that the many possibilities are still enlarged by epistasis. Furthermore, the genotype's characters of competition should actually be measured in two ways: first, by an active component indicating the role attributable to a certain competing genotype (competitive influence); and second, by a passive component measuring how this genotype reacts to competition from its neighbor (competitive ability). A model applicable to this situation has been given by Stern (1965). Further consequences of the fact that forest trees develop under conditions of competition, like most higher plants, will be discussed in the next chapter.

Thus the polymorphism of crown and branch form in Norway spruce originates at least in two ways: The morphs differ in resistance to atmospheric influences and competitive characters. If this turns out to be true, we cannot expect as distinct a variation pattern as found in other, simpler characters.

The question that remains to be discussed is why the brush type survives, in spite of its inferiority in all three damage situations. A simple answer is that it could be superior in some competitive situations — and this is also the most probable explanation. The assumption of a heterozygote form still requires testing.

Many authors (Fisher, 1930, 1931 was probably the first) have pointed out that in the presence of a polymorphism with two or more extreme morphs, which represent the optimal adaptive strategy, the development of dominance may prevent the rise of inferior intermediates. In essence this means an extension of homozygote selective advantage to the heterozygotes; dominance makes them more similar to one of the two homozygotes (Sheppard, 1960). Clarke and Sheppard (1960) have demonstrated experimentally the existence of dominance inducing modifier systems in *Papilio dardanus* by crossing sympatric and allopatric forms of the same polymorphism. Latter (1964) and O'Donald (1968) contribute models for the evolution of dominance. Finally Parsons and Bodmer (1961) show that in certain conditions, selection may lead to an accretion of heterozygote advantage (overdominance). In contrast, other authors expect that selection would tend to reduce overdominance since this would in turn diminish the genetic load of a population. Hence many physiological possibilities exist that modify gene action, and details of these need not be discussed further. The possibilities range probably from such simple cases like the R-allele of *Primula sinensis*, which lowers the pH value from 6 to 4 and thereby changes the effects of pH-dependent enzymes (Williams, 1964), through many intermediate cases to complex modifier systems.

An extreme case of selection for modifier systems is canalizing selection. This is defined as the fixation (canalization) of developmental pathways to only one possible direction, eventually narrowing the possible phenotypes to a single one. Examples for successful canalizing selection are given, for instance, by Waddington (1960), Rendel (1960), and Waddington and Robertson (1966). Stephens (1945) discusses the importance of such modifier systems for evolution and breeding, using the example of a canalizing gene effect upon the leaf form of *Gossypium*. Only the importance of this type of selection can be indicated here.

A particularly high degree of polymorphisms is evident in pioneer species. An example is *Betula pendula*, which exhibits so many conspicuous polymorphisms that, at times, taxonomists had subdivided it into more than 60 "species." Here again those polymorphisms gained greatest attention that were economically important. As an example we will choose the "brown curly birch." In consequence of irregular cambial function, its wood displays a special pattern that is considered attractive, although this depends upon the whims of fashion. Curly birch is best known in northern Europe where it probably occurs most frequently. Scholz (1960), who has studied its abundance, finds that clumps of various sizes make up different proportions of the birch populations. Vaclav (1963) describes its geographic distribution, though only in relation to general ecology (site requirements). It is known that the heritability of the brown curly pattern is high and may even be inherited through single genes.

The slow growth of curly birch fosters the belief that it is part of the genetic load of the population. However, this is not supported by its frequency in some areas where it meets Ford's definition of polymorphism: its frequency is larger there than could be expected from the assumption of high mutation rates. It is therefore more correct to assume that it possesses a selective advantage in certain environments. For example, such advantages could result from early and prolific flowering, which could confer temporary benefits upon the population. The frequencies of the morphs could change considerably both in time and space, and this has in fact been observed. Unfortunately detailed and long-term investigations are not available.

Additional polymorphisms in forest trees that have been known for a long time and explained in part, are, for example, the color of the seed coat, or the genetically controlled extreme differences in resin production. The former has been interpreted in one case as an adaptation to the color of the ground, i.e., perhaps as a protection against seed-eating animals. Cerepnin (1964) found light-colored seeds of Scots pine more frequently in dry forests and dark-colored seeds on moist soils. The latter is seen in connection with character of resistance in trees. Anonymous (1961) and Gibbs (1968), for instance, found a higher resin content in Scots pine that appeared to be resistant to *Fomes annosus* than in less resistant trees.

Genetic polymorphisms are regulators of the breeding system of populations in many cases. Foremost among these are the genetic mechanisms that prevent self-fertilization, for example pre- and postzygotic incompatibility, different sexes, etc. There are also other polymorphisms that lead to similar results. Levin and Kerster (1967) describe a polymorphism of corolla color in *Phlox*. In areas where the two species *pilosa* and *glaberrima* occur together, the pigmented phase of *pilosa* bore 30% pollen of the other species but the white phase only 12%. The authors conclude that the polymorphisms are maintained by selection for a reduction of gene flow. Conceivably, similar conditions could be found among the many tree species of forests in the tropics and subtropics. In other cases, genetic polymorphisms may be maintained by preferential mating within populations Merrell (1953) has observed this frequently in populations of *Drosophila*. Preferential mating of early- or late-flowering Norway spruces in the same stand as a consequence of the correlation between flushing and flowering date also belongs in this category.

At least reference should be made to the importance of sex-linked polymorphisms in microevolution. This was probably first mentioned by Haldane (1954). An example is given by Stehr (1964) and Karlin (1969) presents a review including the frequent phenomenon of "imprinting" in animals, which may sustain polymorphisms. Its study furthers the understanding of the fauna in ecosystems. This will be considered in the next chapter.

One aspect of polymorphism that has been much discussed in population and ecological genetics is the joint effect of linkage and epistasis. A comprehensive review has been presented by Kojima and Lewontin (1970). Of special interest here is the possible existence of more than one equilibrium point. Consequences of such systems are not easily interpreted. Owen (1952) Moran (1963), Parsons (1963), and others present examples. Moran determines five different equilibrium points, three of which represent stable equilibria, when only two loci and three alleles are involved.

The principal vectors of such systems are the degree of linkage and the kind an strength of epistasis. In a given case the population reaches a certain equilibrium in dependence upon initial frequencies of alleles (here better termed types of gametes because of linkage).

Another interesting case is discussed by LEWONTIN and WHITE (1960) using the example of an inversion polymorphism in *Moraba scurra*. They find the natural populations of this species previously calculated from their experimental data to be always near the saddle of the adaptive surface in the sense of WRIGHT. If this is correct, then the equilibria should be unstable in all cases. In the authors' view, this is conceivable as a result of the combined effect of environmental fluctuation and frequency-dependent selection. ALLARD and WEHRHAHN (1964) point out that this could also be explained by assuming a weak degree of inbreeding for the populations instead of random mating. With inbreeding coefficients between 0.05 and 0.25, for example, the populations are not situated at saddles but at the peaks of the adaptive surface.

Another aspect of the adaptive surface useful in explaining conditions in natural populations is the possibility of observing selection operating not only at the level of individuals within populations, but also as selection between populations. These populations may occupy different adaptive peaks. LEWONTIN (1960, 1965), who developed this concept both in theory and experiment, contrasts such interdeme selection with intrademe selection at the level of the individual. He demonstrates the role of interdeme selection in introducing and sustaining genetic polymorphisms.

As has been shown here and in other sections, a great diversity of possibilities exists that gives rise to genetic polymorphisms. When this is appreciated, the difficult position of modern population biology is clear. Earlier it was possible to work with simple models, e.g., by asking the question as to the stability of a balanced polymorphism (DOBZHANSKY and PAVLOVSKY, 1960), or what conditions are needed for the development and sustention of genetic polymorphisms (DEAKIN, 1968b), or in general to discuss the "theory of balanced polymorphisms" (CAIN and SHEPPARD, 1954), to name only a few examples. In recent times, however, complex models are used more and more and the idea grows that the structures of natural populations are characterized by shifting balance (WRIGHT, 1970) to a greater extent than had been considered possible even a short time ago, and that these structures contain many more elements characteristic of transitory situations (SAMMETA and LEVINS, 1970).

The biochemical polymorphisms mentioned earlier have recently attracted more attention in forest genetics. The isoenzyme polymorphisms will therefore again be considered. Apparently forest trees, too, possess an amazing variability of enzyme systems. For example, BERGMANN (1971) found in the endosperm esterase of a Norway spruce population from Westerhof, Harz Mountains, allele segregation at three loci. There is possibly a fourth locus in this population that may be fixed (see Fig. 30).

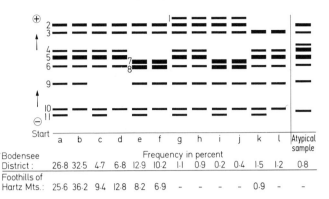

Bodensee District :	a	b	c	d	e	f	g	h	i	j	k	l	Atypical sample
	26·8	32·5	4·7	6·8	12·9	10·2	1·1	0·9	0·2	0·4	1·5	1·2	0·8
Foothills of Hartz Mts.:	25·6	36·2	9·4	12·8	8·2	6·9	–	–	–	–	0·9	–	–

Fig. 30. Esterase pattern of the endosperm from seed of two Norway spruce provenances (Lake Constance region, foothills of Harz Mountains) depicted according to the zymogram. The frequency of individual patterns in both populations is given in percent and is based on 500 seed analyses of each provenance. The esterase bands 1, 2, 3 originate from one gene locus; the combinations (1, 2, 3), (2, 3), and (3 heavy) represent three different alleles (unpublished). The combination of the bands (4, 5, 6) and (7, 8) result from the two alleles (F, S) of the so-called Est-1-Locus (BARTELS, 1971). The bands (10, 11) and (10) represent two alleles of a third esterase locus. The affiliation of band (9) is still uncertain. (From BERGMANN, 1971)

Presumably isoenzyme polymorphisms will follow the same principles as the coarser morphological polymorphisms that have primarily been discussed. For example, frequency-dependent selection in isoenzyme systems was found by YARBROUGH and KOJIMA (1967) at the esterase-6 locus, and by KOJIMA and TOBARI (1969) at the alcohol dehydrogenase locus of *Drosophila melanogaster*. Certain optima were always reached, independent of initial allele frequencies. The effect of individual niche variables upon isoenzyme selection could also be shown. LANDRIGE (1962), for instance, postulated that heterosis in plants and *Drosophila* is largely temperature-dependent because every gene product should have a certain optimal temperature. McWILLIAM and GRIFFING (1965) confirm this for the hybrids from inbred lines of maize. These hybrids were more stable over a broader range of temperature than the inbred lines, which were homozygous at most loci. A parallel example from natural populations is the isoenzyme polymorphism of a fish in the Colorado River (KOEHN, 1968). The frequency of the esterase allele with lower temperature optimum was higher in the upper reaches of the river but that of the allele with higher optimum in the lower part. The heterozygotes, which were particularly frequent at medium levels, produced both polypeptides and therefore were tolerant to a broad range of temperature. The observations of SCHOPF and GOOCH (1971) concerning a frequency cline of two alleles of a leucin-amino-peptidase locus in *Schizoporella unicornis* at the coast of Maine are also explained by temperature dependence of the optima of the two polypeptides.

It is to be hoped that the study of biochemical polymorphisms will contribute new and deeper insights into the mechanisms that institute and maintain genetic variation and with it polymorphisms.

d) Consequences of Gene Flow

The importance of gene flow among populations for clinal and discontinuous variation has been repeatedly mentioned. Beyond this, gene flow is one of the components of the genetic system of populations that influences significantly the optimal adaptation of the population. This has been stressed particularly by LEVINS (1964). LEVIN and KERSTER (1969) and LEVIN et al. (1971) have presented experimental evidence showing that the conclusions of LEVINS are valid at least for higher plants. They found that selfing rate was controlled by population density; with diminishing density there was an increase in the average travel distance of pollen. This will promote gene exchange among colonies of the population in unfavorable years and inferior habitats. Directed transport of pollen (resulting from preferential flight of insects or prevailing wind directions during the period of pollen flight) may have similar effects. Thus the mechanisms of pollen distribution are adjustable and this may bring adaptive advantages. Much earlier, WRIGHT (1931) had directed attention to the fact that certain patterns of geographic distribution of subpopulations create optimal conditions for evolution, whereby isolation and gene flow constitute the most important parameters. Of course the structure of the environment must be considered in each case. In an additional study, WRIGHT (1952) indicated the equilibrium conditions for this case. HALKKA and MIKKOLA (1965) present an excellent example of genetic polymorphism where the genetic variation pattern is balanced by selection, migration, and isolation.

In this study of the effects of gene flow, LEVINS (1964) considers an optimum phenotype, $S_i(t)$, of the ith population at time t. The fitness $W_{ij}(t)$ of the jth phenotype in this population is a function of the squared distance from the optimum:

$$W_{ij}(t) = 1 - [S_i(t) - P_{ij}(t)]^2 .$$

The mean over all phenotypes at time t is then

$$W_{i.}(t) = 1 - [S_i(t) - P_{i.}(t)]^2 - V_P$$

and the mean value over all times t of the expectation value is

$$E(W_{i.}) = 1 - (S - P)^2 - V_S - V_{P_i} - E(V_P) + 2 \operatorname{Cov}(S; P_{i.})$$

where V_P is the variance of the phenotype.

Hence fitness decreases with increasing deviation of the mean phenotype from the optimum, increasing environmental variances, and increasing variances of the mean phenotype.

In a fluctuating environment, greater additive genetic variances imply increasing variances of the phenotypes and mean phenotype. Therefore, additive genetic variance reduces fitness. On the other hand, correlation of the environments of successive generations leads to a greater covariance of the S and $P_{i.}$; the consequence must be that negative variances cancel. The optimal genetic variance is then larger than zero.

The change in mean phenotype resulting from selection (see LEVINS, 1964, for derivation) is

$$\frac{dP}{dt} = (S - P) \sum_i p_i (1 - p_i) (\delta P / \delta p_i)^2$$

where p_i represents the gene frequency at the ith locus. The sum is the total additive genetic variance of the phenotype $V_{A(p)}$ and therefore

$$dP/dt = V_{A(P)}\,(S - P)\,.$$

For simplification it is assumed that $V_{A(P)}$ remains constant. In an environment that is constant in time but variable in space, yet where autocorrelation provides constant environment for each population, and where the proportion m of members is exchanged between populations, the following pair of equations is valid for two populations.

$$dP_1/dt = V_{A(P)}\,[S_1(t) - P_1] + m\,(P_2 - P_1)\,;$$

$$dP_2/dt = V_{A(P)}\,[S_2(t) - P_2] + m\,(P_1 - P_2)$$

with equilibria at

$$\hat{P}_1 = S_1 - m\,(S_1 - S_2)/(2\,m + V_{A(P)})$$

and

$$\hat{P}_2 = S_2 - m\,(S_2 - S_1)/(2\,m + V_{A(P)})\,.$$

Thus the difference between optimal phenotype and the phenotype at the equilibrium is determined by the difference between environment, the migration rate, and the additive genetic variance.

In an environment fluctuating for both populations around the same mean value, and without autocorrelation of successive generations, $P_1(t)$ and $P_2(t)$ are determined by the environments of both populations; the most recent conditions will then have the largest effect. The primary role of migration then is to retard fluctuations of the two phenotypes and thereby to increase the mean fitness of the two populations. If the environments fluctuate in time but in the same manner for both populations, then $P(t)$ is independent of the migration rate m.

As discussed earlier, a response to selection is advantageous for the population only with strong autocorrelation of environments. Therefore, the significance of migration for adaptation is that it allows the population to adjust to consistent differences in the environment while simultaneously putting a damper on local fluctuations. Or, to put it in LEVINS' (1964) words: "... migration is a filter whereby spatial information is used for temporal prediction." Specific optimal migration rates must exist whereby the optimum rate increases with larger variance of environmental variables in time, and decreases with special gradients in the environment. The same author expects the following conditions of migration in typical environments:

1. In regions with little fluctuation of the environment in time, geographic differentiation should be pronounced as migration can be limited without reducing the adaptive values of populations.

2. Stable ecosystems should have an effect similar to climax forests compared with an open field.

3. Along a transect with similar spacial heterogeneity, climax vegetation will be more distinctly zoned than the vegetation of secondary species.

4. A coarse-grained environment can be equated to large fluctuation of environment in time, and hence should lead to a high migration rate.

In forest trees, investigations on optimal rates of gene flow are lacking, and opinions concerning the extent and significance of gene flow differ widely. It seems promising, however, to examine the data on seed dispersal and pollen flight from the viewpoint of optimal gene flow rates.

How much the opinions on, or interpretations of, established facts differ may be seen in the following examples. SLUDER (1969), DORMAN and BARBER (1956), and others conclude that phenological differences among distant populations reduce markedly the probability of gene flow over great distances. On the basis of pollen density measurements several hundred meters above the ground, KOSKI (1970) expected a high probability of fertilization by distant pollen. Indeed, the literature records many instances where a pollen cloud from distant populations arrived in a local population at the time of female flower receptivity and at least participated in pollination (e.g., ANDERSON, 1963). LANNER (1966) discussed the prerequisities of this situation — the presence of "thermal shells," etc. It is recognized, however, that plants derived from pollen of distant populations may be at a disadvantage with respect to fitness (SLUDER, 1969). Thus it is difficult not only to estimate the proportion of seeds and plants arising from distant pollination, but also to evaluate their role in the evolution of the population. For additional, more general forestry literature on this subject see KEAY (1957), OSMASTON (1965), and TAUBER (1967).

Parsons (1963) has reviewed the population genetics literature. He also discussed the consequences of directional and nondirectional migration and the conventional models of analysis. The important question is always that concerned with the evolutionary significance of gene flow. The actual and potential gene flow must be carefully distinguished and treated separately (EPLING, 1947). In many experiments with forest trees this has not been possible and their relevance is mostly restricted to the potential gene flow. Thus EHRLICH and RAVEN (1969) have undertaken the accumulation of experimental evidence for the thesis that gene flow is much less important for the evolution of populations than has been assumed in the past. They argued for the importance of selection both as a stabilizing and diversifying force in the evolution of populations. On the other hand, the examples concerning the formation of local races by selection in Section b of this chapter indicate the consequences for the breeding system very clearly, namely, that selection tends to limit immigration, by reacquisition of self-compatibility or by selection for flowering dates that differ from those of the main population. It is not difficult, however, to arrange our information on forest trees within the framework set by LEVINS (1964). Migration is one of the many factors making up the breeding systems of populations and these factors are being optimized by selection within the system. To that extent one must agree with LANNER (1966), who asks for a new approach to the study of pollen dispersal; but this should not be limited only to pollen because migration by seed is frequently more important (GRANT, 1958).

Forest-tree species with a large range often occupy different ecological niches in different parts of the range. Thus it is not only possible, but highly probable, that the optimal migration rates differ among geographical regions of a species range, and, furthermore, that such differences in the characters determining the migration rate because of selection for different optima.

Observations relevant to these points have been made by MARCET (1951) concerning the pollen-flying capability of different Scots pine provenances, and by

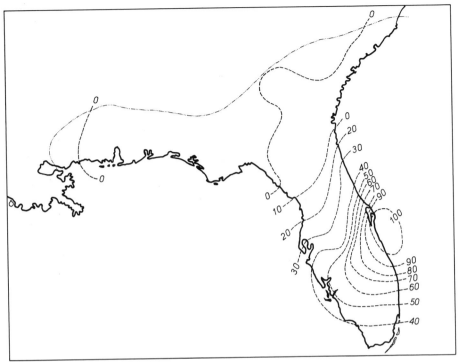

Fig. 31. Percent of trees having high amounts (10% and more) of limonene. (After SQUILLACE, 1971)

ANDERSSON (1955) on variation of pollen characters within provenances that offer good starting points for selection. A similar interpretation is also possible for seed. Here, too, as with all other problems related to the adaptive characters of the breeding system, a clear picture is lacking. Important tasks of forest genetics research are seen in this problem area for the near future. Modern forestry has drastically manipulated the breeding systems of the major species, and further interference through breeding will and must in the future be even more intensive.

Another aspect of migration is best exemplified by the frequency pattern of a gene for resin production analyzed by SQUILLACE (1971) (Fig. 31).

This pattern diverges considerably from those of genes with similar function (Fig. 32a—d). According to SQUILLACE (personal communication), this is probably a new gene that possesses a selective advantage in the population of *Pinus elliottii*. It may have originated through mutation in the center of the species' range, or immigrated from the population of the related *Pinus caribaea* on the Bahama Islands. Migration or mutation first make it possible to substitute new genes with selective advantages for old genes. This is probably one of the principal benefits a population derives from at least moderate migration rates: a means is provided for rapid gene substitution and thereby evolutionary advance at the gene level.

Possibly one may obtain an erroneous impression regarding the process of gene substitution across the species range from the map drawn by SQUILLACE. It has been mentioned repeatedly that, as a rule, the total population of a species consists of

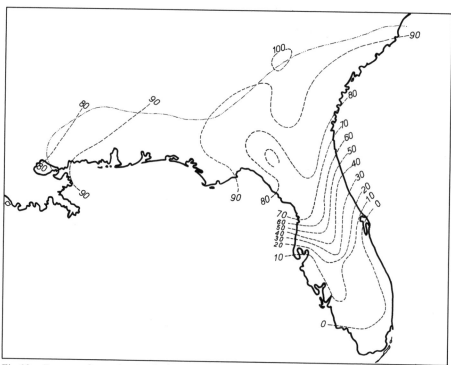

Fig. 32 a. Percent of trees having high amounts (20% and more) of alpha-pinene. (After SQUIL-
LACE, 1971)

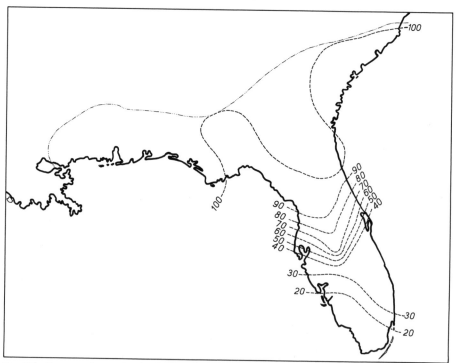

Fig. 32 b. Percent of trees having high amounts (10% and more) of beta-pinene. (After SQUIL-
LACE, 1971)

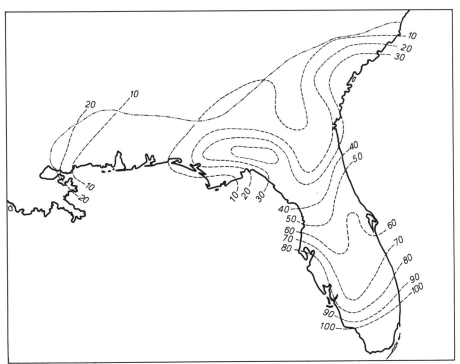

Fig. 32 c. Percent of trees having high amounts (6% and more) of myrcene. (After SQUILLACE, 1971)

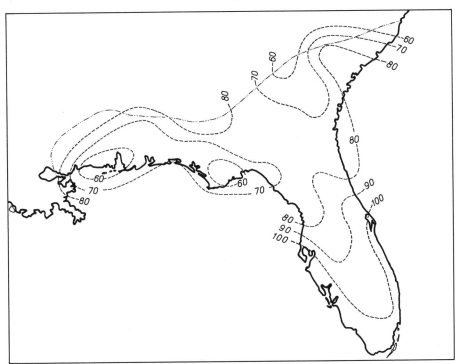

Fig. 32 d. Percent of trees having high amounts (4% and more) of beta-phellandrene. (After SQUILLACE, 1971)

more or less isolated subpopulations. Two phases of gene substitution should therefore be distinguished: first, the substitution within the subpopulation; and second, the movement of the gene into the neighboring or more distant subpopulations, the probability of which is determined by the migration rate. Occasional mass immigration of pollen from populations separated by great distances (ANDERS-SON, 1963, and others) may accelerate the second process considerably. The situation encountered by SQUILLACE, however, does not indicate secondary centers within the range of *Pinus elliottii* from which the gene could have spread, such as would be expected as a result of the presence of significant amounts of distant pollen.

Rather, it appears that recent population genetic models created for the preceding case explain the situation more satisfactorily, such as the steppingstone model of KIMURA and WEISS (1964). The authors assumed that most probably a new gene first enters the next immediate subpopulation where it multiplies and then continues its spread from this stepping stone to more distant subpopulations. This has also implications for the efficiency of selection within and among demes. Details must be obtained from the literature.

BODMER and CAVALLI-SFORZA (1968) present a model for the simultaneous study of the effects of genetic drift and migration in a population divided into subpopulations where mating is primarily within subpopulations. This very general model is probably both applicable and useful for problems in ecological genetics as indicated by the examples from human genetics given by the authors. Furthermore, it should be possible to present the model in an even more general form, thus making it possible to deal with the problem of preferential mating of neighbors in larger subpopulations.

The close relationship of gene flow, and particularly drift, need not be stressed. In nature these two phenomena cannot really be considered separately. However, the methodical study of CROW and MORTON (1954), which is concerned with estimates of the effects of drift, should at least be referred to in this context.

In the past, pollen and seed migration of forest trees could often be estimated only with inadequate techniques. STERN (1972b, c) has tried out two indicator techniques that allow the estimation of both components. After "fertilization" of Scots pine stands with manganese sulfate, the manganese level in tree crowns increased five- to eightfold. It was then possible to distinguish pollen and seed from

Table 23. Percentages of pollen labeled by manganese (MnSO$_4$) in a stand of Scots pine 50 × 50 m in area, constituting a part of a larger stand that is approximately 90 years old. The distance between sample points is 10 m (From STERN, 1972b)

33	38	33	65	40
31	50	44	43	39
42	62	57	40	39
43	52	47	59	35
34	35	27	30	36

Mean values: outer row, 5 m from the edge, 40%; middle row, 15 m from the edge, 50%; center, 57%.

the subpopulations marked in this way from that of all others. Table 23 presents the results from one of these experiments. The labeled pine stand occupied an area of 50 by 50 m and is situated in he center of a large pine stand within the extensive Lüneburg Heath pine region. In the center of the stand the proportion of labeled pollen was 57%, 15 m from the edge only 50%, and 5 m from it approximately 40%. At sample points located 100 m from the stand, the percentage of labeled pollen dropped to 6%. The method makes it easy to obtain a picture of the actual migration rate — with respect to both pollen and seed.

e) Genetic Drift — Accidents of Sampling

The possibilities for random mating of individuals in forest tree populations are often drastically limited in nature. The consequences should be deviations from random distribution of genes between the demes and also within the demes, even when selection is not considered.

In the theory of population genetics this effect of the breeding structure of natural populations has been emphasized particularly by S. WRIGHT. To what degree drift influences the genetic structure of natural populations has been controversial. The fact that drift occurs and plays a role in nature, however, is no longer in doubt. No attempt will be made here to discuss the basic theory and large amount of literature related to drift but instead reference is made to WRIGHT (1970), KIMURA (1964, 1970), COCKERHAM (1969), and CROW and KIMURA (1970). Of course, the phenomenon of drift cannot be viewed independently, since selection, migration, and drift act simultaneously in natural populations and their effects tend to balance each other. The role of drift in this interplay may be either an adaptive or nonadaptive one (VAN VALEN, 1960).

The classical presentation of WRIGHT distinguishes between the island model of isolated, individual subpopulations and the model of isolation by distance. In the first, subpopulations are isolated fully or partially by one means or another from all others of the same species; their finite size leads to accidents of sampling and to changes in gene frequencies. This is expressed either by the variance of gene frequencies over all subpopulations or by the degree of inbreeding of the subpopulations which should increase in each generation by $1/2 N$, where N is the population of individuals taking part in reproduction. The second model is based on the assumption that in continuous populations, such as in plant stands, mating takes place primarily between neighboring individuals. In this way "neighborhoods" arise, the members of which are related. Thus one can imagine that the large continuous populations consist of groups of related individuals that change gradually from one to the other, whereby the relationship of two individuals decreases with distance. Here again the number N, in this case the number of individuals in the neighborhood, is the starting point of considerations. To evaluate the effect of drift, one should therefore first know N. Yet it should be pointed out at this time that in addition to the effects of drift in the original sense of WRIGHT, accidents may influence population structure that can lead to at least temporary changes of gene frequencies, and that must be treated in this connection.

REINIG (1937) has emphasized the importance of advance colonies in settlement or resettlement. These nearly always offer opportunities for the founding of subpopulations with deviating gene frequencies, although usually only for a limited number of generations. A good example is provided by BANNISTER (1965) from the colonization of certain sites in New Zealand by the introduced *Pinus radiata* (Fig. 33 a, b).

Far from the first plantations of *P. radiata* individual trees are found that originated from seed that traveled considerable distance. These trees may become the source of secondary distribution centres. The degree of relationship of the members of the first generation is then at least that of half-sibs.

FRASER (1964) discusses the importance of the founder principle and of drift for populations of colonizing species. The considerable literature on maximum flight distance of seed is usually related to practical forestry problems. For example, as early as 1888 Fliche indicated a maximum flight distance of the seed of *Pinus silvestris*. However much of this literature is not relevant to the problem of distant movement discussed here.

Similar processes are probably active in the regeneration of second-growth forests. Following forest fires, uprooting by wind, and clear-cutting, large areas may

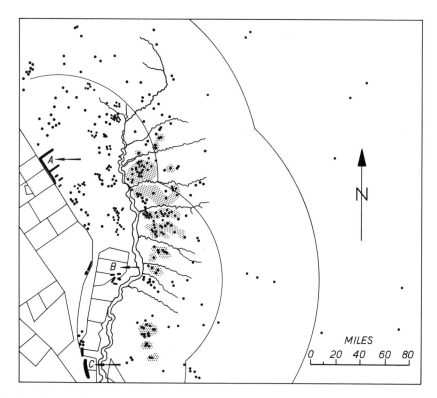

Fig. 33 a. Pattern of spontaneous colonization of *Pinus* radiata, as revealed by aerial photography in 1946. Arable land on left; poor hill country to the right; intersecting arcs show distances of one and two miles from the nearest possible seed source, A, B, and C, which were small plantations of *P. radiata*. Each dot represents a single tree. Fine stipple represents a second generation appearing, probably, as a result of fires. (After BANNISTER, 1964)

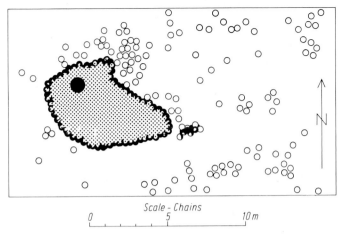

Fig. 33b. Detail of colonization by *Pinus radiata* in Ashley Forest, from aerial photography in 1956. An example of an old, isolated pioneer (solid black circle) that has founded a colony. The stippled area represents a very dense stand of saplings. Open circles represent scattered young trees, mostly from the same mother. (After BANNISTER, 1965)

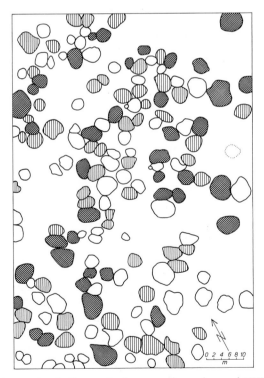

Fig. 34a. The abundance of male flowers in the trees of sample stand XXIII, Tuusula, in June 1957. Abundance is graded as: 1 and 2 (male flowers very scarce or scarce) unshaded, 3 (about average) widely spaced diagonal lines, and 4 (male flowers abundant) narrowly spaced diagonal lines. (After SARVAS, 1962)

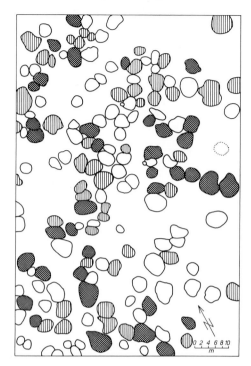

Fig. 34 b. The abundance of male flowers in the trees of sample stand XXIII, Tuusula, in June
1958. Legend same as in Fig. 38 a. (After SARVAS, 1962)

be regenerated by the descendants of a single birch or other species (STERN, 1964). A
particularly interesting example is contributed by JONEBORG (1945). Descendants of
a single tree of *Picea abies*, form *acrocona* had invaded an abandoned field, although *acrocona* (probably a mutation) normally appears to be inferior in fitness
when competing with other trees.

Accidents of sampling of this type should be particularly important for populations at the edge of the species range. This will be treated separately.

To evaluate the effects of drift in the classical sense, the population size should
be known, or, more precisely, the number of members participating in reproduction. Following WRIGHT, this will be designated as the effective population size N_e,
in contrast to the total number of individuals in a population N. Usually N_e is
considerably smaller than N. A typical example is given by SARVAS (1962). Figure 34
depicts male flowering in a stand of Scots pine in three successive years. According
to his results, there were not only considerable differences in the quantity of pollen
produced, but also in each year only a part of the trees participated in pollen
production and these were often different trees.

SCHMIDT (1970) has investigated the production of male and female flowers in a
Scots pine stand of the Lüneburg Heath. He obtained the results shown in Table 24.

The table indicates that approximately 50% of the trees produce 90% of the male
and female flowers. Correlations were: between male and female flowering only,

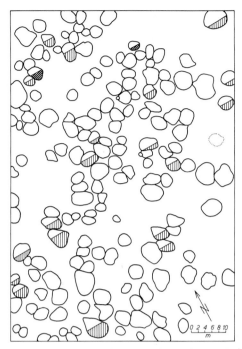

Fig. 34 c. The abundance of male flowers in the trees of sample stand XXIII, Tuusula, in June 1959. Legend same as in Fig. 38 a. (After SARVAS, 1962)

Table 24. Distribution of the quantities of male and female flowers in a Scots pine stand of the Lüneburg Heath region. (From SCHMIDT, 1970)

Trees in order of flower quantity produced (classes)	% of flowers produced a) by classes		b) cumulatively		
	female	male	% trees	female	male
1– 31	35.47	30.10	10	35.47	30.10
32– 62	21.58	18.78	20	57.05	48.88
63– 93	14.44	15.38	30	71.49	64.27
94–124	9.87	11.81	40	81.36	76.08
125–155	7.23	9.80	50	88.59	85.88
156–186	5.29	6.53	60	93.88	92.42
187–216	3.53	4.97	70	97.41	97.39
217–247	1.63	1.94	80	99.04	99.33
248–278	0.74	0.67	90	99.99	99.78
279–309	0.22	0.01	100	100.00	100.00

$+0.312$; between sociological tree class and female flowering, -0.376, and male flowering, -0.542. Results for cone production were similar. Consequently, the effective population size N_e is substantially smaller than the number of individuals in the population N. CROW and MORTON (1955) have estimated the relationship of

N_e/N for various organisms. It ranges from 0.49 to 0.90, indicating that the preceding observations are not unique.

A first estimate of the effective population size here is obtained from the relationship (CROW and KIMURA, 1963):

$$N_e = \frac{4N-2}{V_k+2}.$$

Effective population size obviously varies much from year to year. First, the weak correlation between male and female flowering will be neglected, and separate estimates will be given for the male and female "population." If the correlation were taken into account, the concentration of gamete production on relatively few trees would be even more pronounced. The two estimates for three successive years are

$$N_e \text{ (male)} \quad = 140; 155; 146 \quad \text{or} \quad 45; 50; 47 \% \text{ of } N$$

$$N_e \text{ (female)} = \quad 87; 133; 162 \quad \text{or} \quad 28; 43; 52 \% \text{ of } N.$$

It should be realized that these estimates are probably too high because of the assumption that the sample of mature parents two generations hence represents a randomly drawn sample of 309 pairs of gametes each (one each from the male and female pool). Deviations from this assumption should lead to a further narrowing of N_e.

For the material of SCHMIDT (1970) mentioned above, STERN and GREGORIUS (1972) have calculated the effective population size for three successive years (here, as everywhere, it is the population size effective with respect to inbreeding). They begin with the clarification that incomplete correlation between male and female flowering upsets the ideal monoecious arrangement assumed by all authors. Instead of this correlation they introduce a measure of monoecy M that may assume all values between 0 (dioecy) and 1 (ideal monoecy). They then obtain the following equation for the derivation of N_e, where N'_e supplies the male and N''_e the female effective population component:

$$\frac{1}{N_e} = \frac{1}{4N'_e} + \frac{1}{4N''_e} + \frac{M}{2N_e}$$

The values subsequently estimated for N_e in the same, approximately 100-year-old stand with 309 trees amounted to 139, 169, and 186. The degree of monoecy M was 0.70, 0.82, and 0.78. Numbers that could be obtained from Fig. 34 (a–c) (study of SARVAS) would have indicated even larger annual differences, although Scots pine appears to flower relatively regularly.

The equation for N_e given above is probably most important to control effective population size in tree-breeding programs, e.g., in seed orchards, where the random distribution of clone ramets eliminates distance isolation in natural stands. The equation was developed for this purpose but it also provides information about conditions in natural stands.

Our ideas about genetic drift of this kind in forest tree populations are still inadequate, and this is applicable not only to effective population size. But here it may perhaps be possible to obtain useful estimates at an early date, inasmuch as large numbers of the major forest tree species are subjected to the control of tree breeders in seed orchards (BERGMAN, 1968, and others). Effective population size may change from year to year — often undergoing considerable oscillations. In some species there are pronounced seed years in which all or nearly all mature trees flower and produce fruits; in other years the number of flowering and fruiting trees drops to a few percent, occassionally to zero. If the effective population sizes of previous generations was known, then the mean effective population size could be estimated as a harmonic mean of the N_e over all generations in the sample:

$$\frac{1}{N_e} = \frac{1}{n} \sum_i \frac{1}{N_i} .$$

This approximation is valid if n is relatively small and N_i relatively large. A solution of general validity is given by CROW and KIMURA (1963). It is easily seen that in the preceding equation N_e is influenced particularly by the years when N assumed smallest values, and in which the population passed a bottleneck. This happens especially often in pioneer species, and these should therefore show the most pronounced drift effects.

In climax forests, alternating generations are almost never encountered. Nearly all age classes are represented in one and the same population; the generations overlap. For this case CROW and KIMURA (1963) suggest the equation

$$N_e = \frac{N}{b\tau}$$

where b represents the fraction N_0/N, the birth rate in a given time period, and τ the mean reproductive age. In climax forests a stable age class distribution is often reached. N_0/N is then the mean life expectancy of an individual. We may indicate here that most of the related basic studies were undertaken by S. WRIGHT (1938).

In *Eucalyptus*, BARBER (1958) found selective values of genes for production of flavonoids that changed with age. These genes conferred an adaptive advantage of 5% upon their bearer after removal of the overhead shelter, but the advantage diminshed with age. Thus the problem of effective population size in uneven aged forests is more complex — it also involves the selective values of the genes. TURNER (1970) and CONSTANTINO (1968) have described the resulting consequences.

In many populations estimation of N_e becomes even more complicated in the presence of dormant seeds (EPLING et al., 1960). With an extremely long life period of such seeds, it may easily happen that the parents have long died when a new seedling population develops. In the same species conditions may change much in different parts of the range. One example of this was the adaptation of the cones of conifers where the range included areas that were endangered in various degrees by forest fires. WINSTEAD (1971) has demonstrated different stratification requirements, and thereby dormancy differences, for *Liquidambar styraciflua* from different parts of its range.

The heritabilities of the characters related to flowering and fruiting may also influence N_e. NEI and MURUTA (1966) have shown that high heritabilities of these characters reduce N_e.

Flowering and fruiting are under strong genetic control in trees as in all higher plants. Investigations of these two characters always indicated genetic variation in the intensity of these characters. In *Betula verrucosa*, for example, selection leads to rapid response (STERN, 1961, 1963). Major genes appeared to be involved as in other cases (BEDDOWS, 1962). HEIMBURGER (1958), on the other hand, finds indications that early flowering in poplars is controlled by polygenes. The frequently reported large variation in male and female flowering of clones in seed orchards of many species indicates the common occurrence and adaptive significance of genetic variation in these characters. Thus we may have to accept the idea that variation of flowering and fruiting in stands is largely controlled genetically and acts to reduce effective population size.

Flowering dates, as mentioned earlier, are also subject to genetic control (JOHNSSON, 1968; BERGMANN, 1968, and others). Major genes may again be involved here (see: COOPER, 1959 for *Lolium*, HALLAUER, 1965 for maize, and others). ROWLANDS (1946) describes a case with *Pisum*, in which the flowering date was controlled jointly by a dominant gene for early flowering and a system of polygenes. In other cases polygenes played the principal role. Thus COOPER (1960) found genetic variation of the flowering date in all populations of *Lolium* investigated, but the variation was only discovered with suboptimal to threshold conditions of photoperiod and temperature. Most of the plants were heterozygous and a rapid response to selection was obtained. In this case the apparent uniformity of the populations was not proof of lack of variability: when selection was begun with progeny of a single plant, the variation obtained after three generations was larger than in the initial population. An interesting case is reported by KOZUBOV (1962). He observed that Scots pines with red anthers, which are especially frequent in the northern part of the range of this species, flower one to two days earlier in sunny years because of greater heat absorption. This is probably an adaptive polymorphism in a coarse-grained environment.

In Norway spruce the early flushing forms flower before the late flushing forms, which again may be an adaptive polymorphism, although with probably similar effects from year to year.

Differences in flowering date lead to positive assortative mating. The result upon effective population size should therefore be in opposition to the genetic variability of flowering; the consequence is a larger effective population. GUTIERREZ and SPRAGUE (1959) investigated ten inbred lines of maize in polycross plantations with different marker genes. There were genetic differences in maturity of pollen and stigma, the amount of pollen produced, the duration of the flowering period, and plant size. Even in the first generation after the polycross, distinct deviations from random mating were observed; these in turn led to continued preferential mating and genetic stratification in subsequent generations. On the other hand, in experiments with labeled pollen of six *Pinus sylvestris* in two years, STERN (1972a) observed that the proportion of pollen contributed by these trees to the total pollen quantity remained relatively constant. He concluded that the heterogeneity of microenvironments surrounding the male strobili in crowns of old trees contributed to a leveling of the hereditary differences observed in experiments with young grafts. However, additional experiments are needed here.

These examples of the unknown variables that codetermine effective population size (additional examples are easily found in other sections of this work) denote the weakness of our present appreciation of the effects of drift in forest tree populations. It is to be hoped that an increasing number of studies devoted to breeding systems will soon provide additional and reliable information, although a complete picture is not to be expected.

Not much different is the situation with respect to estimates of neighborhood size in large continuous stands, but additional difficulties are met here. This will best be seen in a concrete example. For a number of years we have been recording the pollen distribution around some Scots pines in stands and obtained more or less regular dispersal of the pollen of an individual tree in all directions and a decreasing density with increasing distance from it. These are the conditions assumed by WRIGHT for his model of isolation by distance and neighborhood size. He describes this dispersal using the equation given by BATEMAN (1947b) by

$$F_D = F_0 e^{-KD}$$

where F_0 is the power of the pollen source and K the decrease of density with increasing distance D. This dispersal function is standardized and the basis of all further calculations. However, when applying it to his Scots pine data, SCHMIDT (1970) found it unsatisfactory; other experimentally derived functions given by several authors were equally unsuitable. He obtained better adjustment of the observed dispersal data by means of the function

$$F_D = F_0 e^{-K\ln D} .$$

There is no fundamental reason why pollen dispersal around a source should follow such a single function, in view of its dependence upon such accidental factors as wind direction, wind speed, turbulence, upward currents, etc., KOSKI (1970) in fact obtained distinct effects of prevailing winds and McELWEE (1970) found at least traces of them in some experiments. Furthermore, one would assume elliptical instead of circular neighborhoods on steep mountain slopes because of the influences of relatively small elevational differences upon flowering dates (STRAND, 1957).

The influence of the chosen dispersal function upon the results, however, is shown by the calculations of SCHMIDT (1970); when using the BATEMAN function the calculated neighborhood ranged from 346 to 30000 trees; with his logarithmic function they ranged from 13 to 77 trees, depending upon the assumptions of dispersal limits beyond which a significant contribution of the pollen source to pollination could no longer be expected. These conditions should be appreciated when considering estimates of neighborhood size for forest trees given in the literature. These range from ten to several hundred trees (for literature see McELWEE 1970). This situation therefore justifies only the assumption that given certain conditions, differentiation into neighborhoods could or should take place; no statement regarding their size or the resulting inbreeding coefficient can be made.

Furthermore, through wind effects pollen dispersal is influenced by ground vegetation (in open stands) or by species mixtures. For example, large stands of pure Scots pine in Scandinavia with their specific mode of regeneration should have a different effect upon the breeding system than occurrence of Scots pine as an admixture to deciduous forests on some sites in central Europe. The related questions are also of interest to natural regeneration. Thus fairly open shelterwood cuts change not only population density but also conditions for pollen and seed dispersal. The consequences for population structure have been pointed out by ZOBEL et al. (1958).

Also, nothing is known regarding the role of seed distribution in determining neighborhood size. We have some information on seed dispersal along the forest boundary or on distant movement but practically none on seed distribution in closed stands. In the meantime, we have initiated studies within stands using manganese-labeled seed (STERN, 1972c), a simple variant of the successful procedure applied to pollen by FENDRIK (1967) and SCHMIDT (1970) (see also KOPECKY, 1966). The problem requires further study (for experiments with seed of Douglas fir, labeled by Sc^{46}, see LAWRENCE and REDISKE, 1962).

Very little is known about the role of drift in the highly complex rain forests of the tropics and subtropics. Their breeding system is often marked by extremely low population density. At times it was assumed, therefore, that drift could be one of the principal causes of the multiplicity of species (FEDOROV, 1966; ASHTON, 1969; regarding the role of reproductive methods of higher plants in race and species formation in general, see BAKER, 1953, 1959; STEBBINS, 1950, and others). Still, there are two tree species the breeding systems which have been relatively well investigated in their natural habitats: *Theobroma cacao* and *Hevea brasiliensis*. The success of breeding with *Hevea* (PURSEGLOVE, 1964, and others), demonstrated by increases in latex production of several hundred percent following the application of simple mass selection is explained with reference to the extremely high degree of inbreeding of the natural populations. The change to a different mating system apparently released strong heterosis effects, such as is often the case in predominantly self-fertilizing species. The same is valid for some species of *Theobroma*. The basis of the high degree of inbreeding is, of course, drift as a consequence of a mating system with low population density. Pollen vectors cannot overcome the often great distances between individual trees or groves of trees to achieve an adequate degree of cross-fertilization. In *Hevea* and *Theobroma*, pollen vectors are small animals.

In species pollinated by birds and bats conditions may be different. The efficiency of these animal species has already been mentioned; their flying ability permits the rapid covering of great distances. Furthermore the flowering habits of these tree species definitely seems to invite cross fertilization in some cases. For these reasons a general evaluation should be postponed until we know more about this subject.

An intermediate position between the large-area stands of the wind-pollinated species in temperate and boreal zones and the tree species of the tropical and subtropical rain forests is occupied by some deciduous species of the temperate zone. *Liriodendron tulipifera* is one that appears to have been most thoroughly investigated. CARPENTER and GUARD (1950) obtained distinct heterosis and higher fertility when crossing trees of different stands (often only a few miles apart) than when crossing within the same stand. *Liriodendron* often grows in groups or small stands. TAFT (1962) observed in his experiments that bees remained mostly within the same group of trees. Thus pollination within the same small groups is substantially more frequent than pollination between groups; drift is favored. Methods and models for the investigation of the mating system of species pollinated by insects, particularly bees, are given by KERSTER and LEVIN (1968) and especially RICHARDSON (1970).

KERSTER and LEVIN (1968) estimated the neighborhood size of *Lithospermum caroliniense*, an insect-pollinated herb. Their investigation included not only pollen transport and selfing rate, but also seed dispersal and the density of flowering plants. From these values they calculated the "genetic density" of the population as a basis of an estimate of neighborhood size (details should be obtained from the original paper). This is given as ranging from 3.9 to 4.6 m in diameter. The consequence is a high degree of inbreeding of the populations and considerable genetic differentiation among colonies.

In another study, LEVIN and KERSTER (1969) investigated the influence of stand density upon the flight of bees and thus pollen distribution. The authors note that bees are strongly oriented toward plant distribution. In dense stands pollen distribution is therefore more limited than in open stands. Thus pollen distribution is clearly dependent upon stand density. As a result local differentiation as a consequence of environmental heterogeneity is expected primarily among dense stands or groups with closely spaced individuals. Density dependence appears to be a means of balancing rates of gene flow. Temporary environmental fluctuations increase the optimal rate of gene flow.

A similar dependence of pollen dispersal upon characteristics of the plant stand has already been discussed in connection with wind pollinators. No other case, however, appears to have been investigated as thoroughly as the one described here, which also demonstrates strongly the need to investigate all components of the breeding system.

Up to this point we have been concerned with the factors that determine the extend of drift in a forest tree population. Insofar as drift is the net result of many factors, the effectivness and joint action of which are difficult to assess in experiments, it would probably be best to assess the effects of drift directly from population structure. This does not diminish the importance of single-factor studies, the results of which are much needed in applied forest genetics (breeding, seed stands, methods of regeneration).

The starting point of studies aimed at estimating drift on the basis of population structure is the idea that drift leads to irregular distribution of genes across subpopulations and neighborhoods. It is easiest to estimate frequencies of certain genes in a population with a given polymorphism, and this has been done in many cases. Such investigations in fact have contributed much to the understanding of the genetic behavior of populations. Well-known examples are the polymorphisms of color and banding pattern of shells of several snail species, industrial melanisms in moths and butterflies, mimicry, and biochemical polymorphisms of blood and serum groups and, in recent times, of protein bodies in general. For a general discussion the reader is referred to the literature, e.g., FORD (1964), and for methods to WRIGHT (1965).

The most significant result of such counts of morph frequencies was that they were determined to a greater extent by selection than by drift. This was quite clear with clinal variation of the frequencies but could also most often be shown in situations similar to drift. Some exceptions were observed as well (e.g., HALKKA, 1964; HALKKA and MIKKOLA, 1965). Often additional experiments were undertaken to clarifiy the role of selection and drift in determining the observed gene frequencies. In forest trees, such experiments are not easily performed.

Suitable polymorphisms are actually not lacking in forest trees. In spite of this, the method has found little application in forest tree populations. Some examples are given by STERN (1964). For example, the distribution of beech with abnormal wood formation (certain grain patterns, e.g., "wimmer beech") in large stands turns out to be strongly clumped, which may indicate drift. However, the method may become more important with increasing simplification of biochemical techniques. Electrophoretic description of isoenzyme patterns for instance should be increasingly used in the future for this purpose. SAKAI and MIYAZAKI (1970), for example, compared the peroxydase pattern in a large population of *Thujopsis*. They found that the patterns decreased in similarity with increasing distance between the compared trees. The inheritance is not known in this case, making it necessary for the authors to estimate the mean degree of inbreeding of the population by using an approximate procedure.

The methods of quantitative genetics also permit estimates of the effects of drift. STERN (1964) estimated drift expressed by WRIGHT's inbreeding coefficient F for two birch species, using the genetic variance within and between inbreeding lines given by WRIGHT as early as 1921. The former variance has the expectation value $(1 - F) V_A$, the latter the value $2 F V_A$. V_A is the additive genetic variance. Of course a restriction is that the effects of drift and selection may be confounded, and hence that drift may be overestimated. Another assumption is that only additive genetic variance exists. In spite of these limitations the results were reasonable. The value estimated for the pioneer species *Betula japonica* was $F = 12\%$; for the climax species *B. maximowicziana*, $F = 5\%$.

In summary, however, it must be stated that forest geneticists have so far made little effort to estimate drift in natural tree populations. The reason could be partly the expense and time consumed and partly the relatively unreliable methods that would have to be used.

Differences in gene frequencies among subpopulations have been of interest to forest tree breeders. They have repeatedly attempted to utilize inbred subpopulations for heterosis breeding. This problem is not well explored in forest trees, and, as one would expect, the conditions are far from uniform. Of course, the complete literature cannot be cited. STAIRS (1967), for example, when crossing seed sources of *Liriodendron tulipifera* from the states of Mississippi and New York, found distinct heterosis and partial dominance of the character frost hardiness. Actually, one should not term this result heterosis if one of the sources grows faster and the other also introduces a character that influences growth but is significant in only one environment, in this case frost hardiness. This is so because the combination of two characters takes place with intermediate or partially dominant inheritance of the character. A large number of additional examples is available, such as the hybrids between northern and southern sources of *Picea abies*.

To cite only one example from maize breeding, which is of special interest here, MOLL et al. (1965) found an increase of heterosis with a certain degree of genetic divergence and then a decline in progenies of crosses from four geographic regions in North and Central America (how genetic divergence was estimated will not be considered here). Here again it may be seen that not only differences in gene frequencies decide but also adaptation to different environments. Adaptation may lead to very different adaptive gene complexes, including different genetic correlations [see Section IV. (g)]. HOFFMANN and SCHEUMANN (1967), for example, found growth superiority in germinant seedlings from low elevations containing anthocyanin, and conversely growth inferiority of the anthocyanin-containing seedlings from high elevations. Such effects of coadaptation may be particularly pronounced in species reproducing vegetatively. Here selection is more for gene combinations and less for single genes. Although evidence from tree species is lacking, experiments with other higher plants confirm a high proportion of nonadditive genetic variance in such populations. AALDERS and CRAIG (1968) and WATKINS and SPANGELO (1968) have shown this for a commercial strawberry that is normally multiplied by clones.

f) Marginal Populations

Populations at the edge of a species' range are exposed to different conditions than those in the center. "Edge" in this sense means a boundary area, where a tree species is limited by some extreme factors such as are found near the tree limit of the far north, in the high mountains, or in arid zones. The range may further be limited by geographic barriers or by competitive superiority of other species where the species could normally thrive without such competition. An instructive example of the different reasons for species limits is given by HALLER (1959). *Pinus ponderosa* occupies a large range controlled at low elevations by excess moisture and at high elevations by temperature. *P. jeffreyi*, on the other hand, grows in a range limited at high elevations by low temperature but at low elevations by competition with *P. ponderosa*. An attempt has been made by VAN BUIJTENEN and STERN (1967) to characterize the special qualities of the niches of marginal forest tree populations as understood here. One should recognize, however, that forest genetics and forest ecology have long been concerned with these populations, which confront forestry with problems of a special nature and difficulty.

High selection intensities characterize the conditions under which such marginal populations exist. Selection in the far north or high evelations is mainly for adaptation to extremely short growing seasons and low winter temperatures, including all accompanying factors such as winter drought, etc. Immigrants from the center of the range will therefore always possess lower fitness, and if they establish themselves in the marginal population, reduce rather than increase its adaptive value. Directional selection thus prevails in marginal populations whereas stabilizing and/or diversifying selection determine evolution in the central populations. The role of competition with other species often diminishes because only a few species are capable of surviving in extreme conditions near the limits of forests and tree growth. Intraspecific competition occurs here at an comparatively low level since one or more environmental factors are at a mininum (temperature, humidity). Stands in such marginal forest areas are therefore often characterized by low density.

Marginal populations are highly specialized in most cases. Their adaptation takes place under the most severe pressure of directional selection, such as has been described by VAN BUIJTENEN (1966) for the "lost pines" in arid regions of Texas. This process, with similar environmental extremes leading to similar stresses upon marginal populations in different geographic areas, often has the consequence of character convergence. Blue coloration of needles or leaves in high elevations or the far north as well as the narrow and spirelike crowns of many species in the same areas

Fig. 35. Forms of Norway spruce with narrow crowns of high-elevation provenance. (Photo Dr. *Krahl-Urban*, Hann.-Münden)

constitute examples. Species of *Fitzroya* in the South American Andes are marked by the same habit as pines and spruces in North America and northern Europe in comparable situations (and the total impression of the flora is the same in spite of a different species composition). Eucalyptus in climatically similar forest boundary areas of Australia bear leaves with the same blue tones as spruce and pines in Europe and North America. A parallel situation exists in marginal populations of the same species that are adapted to similar environmental extremes but in different geographic areas. Figure 35 indicates this for Norway spruces of different provenance. There are differences, but the total impression is the same: spirelike crowns with drooping or short branches.

Marginal populations are often strongly isolated from the main distribution area of a species, as a result of distance or ecology. These populations often exist as typical island populations or populations of small size. Their marginal position and small size makes them highly susceptible to catastrophic events, which are especially frequent in ecologically extreme boundary situations. Destruction of such insular populations and refoundation by few immigrants is therefore common. This, too, has consequences for their genetic constitution. The principal effects to be expected from a harsh, directional selection (the result of extreme environmental conditions) and genetic drift (the consequence of founder effects and small population size) is a drastic reduction of genetic variation within these marginal population, and because of reduced immigration (insular isolation) also larger differences among population means than among subpopulations in the center of the species' range.

The question of reduced variation within populations has been investigated by many authors in diverse organisms. The following studies will serve as examples. CARSON (1959) stated that marginal populations of *Drosophila melanogaster* possess less genetic variation than populations in the central part of the range of this species, the former being subject to the contrasting homoselection (see earlier definition). CARSON and HEDD (1964) found that marginal populations of two nearctic *Drosophila* species *(D. robusta, D. americana)* and a neotropic one *(D. acutilabella)* in Florida were structurally homozygous, and that the central populations to these two species were to a high degree structurally heterozygous. The marginal populations of three other *Drosophila* species *(D. eurotonus, D. nigromelanica, D. willistoni)* were again distinguished by reduced structural polymorphisms. SPERLICH (1964), working with marginal populations of *D. subobscura* in Norway, also found fewer structural polymorphisms. Another instructive example is given by DOBZHANSKY et al., (1963) who investigated a marginal population of *D. pseudobscura* at Bogota, 1500 km distant from the main population of the species. The Bogota population turned out to be related to the Guatemalian one, but much impoverished in polymorphisms, with the smallest genetic load ever recorded in this species. The largest genetic loads (balanced loads?) are apparently carried by the most prosperous populations in the central part of the range.

After studying wing patterns in *Maniola jurtina* of the Scilly Islands, FORD (1958) concluded that where genetic variation of a character is due to polygenes in the small Island populations, it is selection that is solely responsible for genetic diversity among populations. He considers drift and the founder principle to be insignificant (see below).

It has already been mentioned that in the central part of the range selection is usually of the diversifying kind (DOBZHANSKY, 1963). The frequently cited "area effect" of Cain and Currey belongs into this category: the greater amount of polymorphism in shell color and banding pattern of larger *Cepea nemoralis* populations is explained by the niche diversity of larger areas in comparison with smaller ones, although drift should not be neglected here. This difference in polymorphism makes the larger populations that is, larger in area and number — genetically more variable and leads to smaller differences among the means of subpopulations (in connection with a higher rate of gene flow). It is similar to the "Ludwig effect" (LUDWIG, 1950), which, in addition, results from competitional heterogeneity over large areas.

HESLOP-HARRISON (1964), on the other hand, considers the possibility of larger genetic variation at the edge of the range than in the central part, as a result of less interspecific competition. This may be valid for the means of subpopulations, although for different reasons. In the sense of that author, this may apply only to those characters not subject to the extremely strong directional selection in the marginal populations. One should also remember that it is interspecific competition especially that gives rise to a large intraspecific variation. In any case, the majority of experimental results oppose this hypothesis.

Investigations from forest trees are not yet available. We must depend upon observations from experiments started with different objectives in mind. In addition to the character convergence already mentioned, there are indications that marginal populations of tree species are more homogenous within themselves than those in the center of the range. Figure 36 a, b is an example of a Norway spruce marginal population in northern Scandinavia, perhaps the most northerly one in existence.

In northern populations of the same species, LANGNER and STERN (1964) found less genetic variation for flushing date; but ANDERSSON (1965) obtained a surprisingly large genetic variance of the adaptive characters' flowering and fruiting. In northern populations of Scots pine, EICHE (1955) discovered a large genetic load. Perhaps the explanation is a particularly coarse-grained environment for the characters in question (see Section c of this chapter). Here, again, we may have to think anew about the manifold aspects of adaptation to avoid the danger of faulty conclusions. Impressive illustrations for this need are provided by the studies of EICHE (1966) on the adaptation of Scots pine to extremely low temperatures in the far north, or by ANDERSSON (1965, 1966) and SCHMIDT-VOGT (1964) on conditions in marginal populations of conifers in the far north and the high mountains, respectively. New experimental possibilities have been opened by methods of biochemical genetics, particularly in conifers, by the elegant techniques of isoenzyme analysis. A reconsideration of the well-known morphological polymorphisms could also yield interesting results. This applies to the polymorphisms of cone form and crown and branch types in Norway spruce, as well as to others that up to now were understood more as taxonomic aids than as genetic polymorphisms. At this stage, here again most observations point to less genetic variation in marginal populations.

The gene centers of VAVILOV known to plant breeders may also be mentioned here. These are centers of great genetic variability, located probably somewhere in the center of a species' range. Today we would state that they should be centers of diversifying selection; their existence is difficult to understand in any other way. COOK (1961) regards the central populations as the centers of species evolution. A constant stream of genetic variants flows from here to the marginal areas of the range. On the other hand, he also sees marginal populations as focal points of evolution. Although migration tends to smooth differences in the center, isolation and more severe selection at the edge provide better conditions for the evolution of new species. Both theories are actually not contradictory and should lead to the same result: greater variance within populations in the center; and greater among-population variance at the edge, explained not only by drift and isolation but also by founder effects (MAYR, 1963). WOODSON (1964) contributes another example. In butterfly weed he finds orange flowers in the central populations and an increase of the yellow component in all directions — in the southerly direction, a rapid increase, then again an increase of the orange component. From these observations he postulates the evolution of an advanced complex in the center of the range that has not yet spread over the whole range. In the marginal populations conditions are

Fig. 36 a and b. The possibly northernmost population of *Picea abies* at the Menesjoki, south-west of Inari, Finland, 68°33′ °N latitude and 26°08′ E longitude, 460 m a.s.l. Further north there are single trees and smaller groups. (Photo Prof. SCHMIDT-VOGT, Freiburg)

different; they are well isolated and show extreme fluctuations of population size. The author sees in these conditions, and perhaps also in selection, the large differences among marginal populations.

The isolated location of the marginal population and their limitation to certain sites creates special starting conditions for further evolution. FORD (1964) especially has given attention to this situation. Of the many examples in the literature, some of particular relevance will be discussed, PIGGOTT and WALTERS (1954) cite a number of examples concerning soil requirements of marginal populations that are often opposite to those in central populations. The first reason may be simply the lack of competition. This situation opens new evolutionary paths to the marginal populations. In trees the occurrence of *Buxus sempervirens* in Great Britain illustrates the different soil requirements of marginal populations. Additional examples of a similar kind are given by MAJOR and BAMBERG (1963) for populations of plant species from the Sierra Nevada in California, which are outliers of Cordilleran species.

The evolutionary possibilities of marginal populations are further favored by periodic catastrophes. RAVEN (1964) has investigated this for several plant species in the southwestern United States. The trend to arid climate has reduced the range of some formerly more widespread species to pockets. The species are therefore represented by typical marginal populations (small populations, isolated and specialized for certain sites); these also often occupy soils on which the species is not found elsewhere. The periodic catastrophes therefore need to be related to the edaphic endemism of these populations, of which catastrophic selection (see also LEWIS, 1962) is probably one of the causes. This, in turn, may lead to a change of certain gene components, which is a prerequisite for an incisive reorganization of the gene components (coadaptation). From this point of view, too, conditions are especially suitable for rapid evolution of marginal populations, as emphasized by COOK (1961). Additional examples for higher plants are given by STEBBINS (1950).

Another problem of population genetics related to marginal populations is the presence of polyploidy in some species of higher plants (see STEBBINS, 1950 for summary). Apparently there is no general increase in the proportion of polyploids toward the edge of the range. BELL (1964), for instance, obtains no correlation of the degree of polyploidy with longitude, latitude, or elevation for *Eryngium* and three other species of *Umbelliferae*. However, he finds that correlations to ecological factors exist, which give rise locally to isolation, endemism, and speciation as adaptive strategies. This he terms primary correlation. It is only as a result of these primary factors that polyploids attain a large number of survivors in these ecologically marginal areas (secondary correlation). Finally, one may mention here the frequent occurrence of an extremely high phenotypic variation, which often has a genetic background, when a population is moved into an extremely different environment. Foresters, who have experimented with provenances or exotic species, have confirmed this again and again. However, HESLOP-HARRISON (1959) points out that variation of this kind may also be based upon the disturbance of physiological buffering systems and other nongenetic causes. A further discussion is beyond the scope of this book.

To recapitulate, causes of rapid evolution of the marginal population means, respectively, are the invasion of new niches, isolation, founder effects, and possibilities of developing new forms of coadaptation.

g) Linkage

The genetic information of viruses and bacteria is generally united in a single, circular chromosome. This may be seen as a continuous DNA molecule made up of functional units, the genes. Mutation and recombination following crossing over are the only sources of genetic variability; because of immense population sizes they carry viruses and bacteria through the most difficult situations resulting from fluctuating environments. Increasing complexity of the genetic information in the course of evolution led to ever-larger DNA endowments and the creation of special apparatus, i.e., several chromosomes. Their function is the proper transmission of genetic information. The chromosome apparatus also created new dimensions of the space within which the process of evolution takes place. It is now not only a (gene-controlled) distributor but also coordinator of genetic information; and in addition it offers new facilities for storing genetic variability. Also by having at least two homologous chromosomes in the higher organisms, it brings new forms of gene interactions into play, such as dominance and epistasis. To evaluate correctly the role of linkage in the evolution of forest tree populations, we should know more than has been given in the preceding sections. Reference has been made to the possibility of tight linkage, which establishes supergenes or complex loci, i.e., the basis of many polymorphisms; and to the assembly of alleles on whole chromosomes of some populations as a consequence of simultaneous selection at many gene loci. Here, too, some additions will be made. Chromosome polymorphisms in *Drosophila* have turned out to be such supergene-like arrangements. A new inversion affecting a certain number of gene loci will lead to a structural polymorphism if an allele combination arises in the specific chromosome section as a whole that leads to properties of polymorphism (meeting the conditions mentioned in previous sections). In this case the allele endowment of the inversion will be protected against recombination since no crossover is possible within the inversion. The only additional prerequisite is a selective advantage that compensates for gamete loss resulting from the disturbed meiosis of heterozygotes. The investigations of SAYLOR and SMITH (1966) have already been mentioned; some of their results also contain hints regarding the relative frequency of such polymorphisms in *Pinus*.

Knowledge gained from studies of chromosome polymorphisms has contributed evidence for the theory that optimal linkage relationships exist (see index of recombination, Chapter III).

The phenomenon of simultaneous allele endowment at several or many loci, which is found in demes subject to complex conditions of selection or along parallel geographic gradients of several ecological variables, has been termed genetic coherence. The frequent occurrence of character coherence in interracial hybrids has been noted especially by CLAUSEN (1959). He concluded that it reinforces the genetic flexibility of species in a geographically heterogeneous environment and supports the maintenance of the allele combinations typical of races in zones of contact. A good example of this in trees is given by HALL (1952): *Juniperus Ashei* and *J. virginiana* form an extensive hybrid swarm in Oklahoma; the species can be well distinguished in several morphological characters that are also expressed in their typical form in the hybrids; in the hybrid swarm, segregation of the characters typical of the species is correlated.

CLAUSEN (1959, see also 1951) further indicates that there should be an equilibrium of coherence and variation under these circumstances; we might say, a linkage equilibrium that is codetermined by selection and one that should change with a modification of the environment that shifts the selection pressure that maintained this linkage equilibrium. In tree species only very few investigations have been made; one example is given by STERN (1964). He found a correlation between initiation and cessation of growth in *Betula maximowicziana* but none in *B. japonica*. In Japan, where the material originated, the range of the two species overlaps on a large area. The former species occupies a position near the climax but the latter is a typical pioneer. The result thus makes sense if one assumes that it is to the advantage of a pioneer species to offer a broad genetic variation in each generation whereas a climax species is more specialized. Based on the assumption of two loci with two alleles each, there would be only three possible genotypes with strong linkage, but nine with independent segregation.

FORD (1964) offers another example based upon *Pinus radiata*. Its range consists of several effectively isolated areas. The genetic differences observed between the genetically independent geographic subgroups of the species are interpreted as selection effects by the author. Considerable genetic variation also exists within populations, except in the population Cambria. The breeding system of the species and a large index of recombination result in a lack of character combinations, except those for development. Apparently under the special ecological conditions of this species, there is relatively little genetic coherence.

Regarding CLAUSEN'S second conclusion that coherence supports the maintenance of allele combinations typical of races in zones of contact, thereby making the population genetically more flexible, no experimental evidence exists in forest trees. This consequence of genetic coherence is, however, a plausible one, provided there are two distinct niches and sufficient autocorrelation exists between the environments of successive generations (see Sections a and b of this chapter). Finally, optimal linkage relations should also exist along ecoclines that are determined by selection and migration.

We shall return now to the problem of the genetic effectivness of distant pollen. The incorporation of genes or gene complexes from similar subpopulations along short distances of a cline or a system of clines is a consequence of genetically effective migration by means of pollen or seed.

The differences among pollen or seed mean gene endowments of the population and immigrants increases with distance along the cline. Provided the cline or system of clines can be interpreted as the result of the interplay of selection and gene flow, the selective disadvantage of the immigrants must increase with distance along the cline. This effect is still reinforced by genetic coherence. Here, too, certain optima exist in nature that codetermine evolution, and which are unknown as yet. STERN (1970) has suggested the study of haploid conifer endosperms for a number of reasons, one of these being that estimates of linkage parameters should be possible with proper planning. These parameters have remained practically unexplored but should be known, at least to some extent, to check many of the conclusions found in the forest genetic literature. These conclusions, e.g., concerning the role of distant pollen in population structure, have been developed on the basis of insufficient information. The approach of SQUILLACE (1971) in dealing with a new gene(?) substitution (see Section IV. 3d) is an example among many others of an attempt to find new solutions to old problems.

One problem related to linkage, which has been the concern of forest genetics since its early period, is the correlation of characters observed on young and old trees. This correlation is the basis of all so-called early tests (STERN, 1961). Here the linkage correlations are actually undesirable because with early tests the intention is to utilize the genetic correlations, i.e., correlations resulting from pleiotropy. Both types of correlations, those from linkage and those from pleiotropy, exist side by side in our forest tree populations. Linkage correlations could be expected in large populations with different conditions of selection in different parts of their range. Earlier authors occasionally calculated such character correlations from provenance means, but these are typical only for the specific sample and include all effects of parallel selection. Examples were given in previous sections of this chapter.

Pleiotropy is actually often the cause of pronounced correlations between different characters. PUGSLEY (1965), to cite only one relevant example, found that a single factor explained drastic differences between two wheat varieties in several characters responding to day length. Such genetic correlations (from pleiotropy) are probably more often responsible for the reaction of other characters than those directly considered in a selection program.

This will be illustrated by another example. In the tetraploid, pseudogamous species *Rubus nitidioides*, HASKALL (1959) selected two plants, each with dense spines and sparse spines plus a very-early-flowering one. Although selection was continued in the progeny over several generations, additional characters were changed. The presence of dense spines was closely correlated with late flowering, but the selection process also affected the size of plants and fruits, leaf shape, and color, as well as general fertility. The new leaf shapes that appeared were typical of those of some closely related species. In this experiment, as in many others with similar results, it is usually not possible to separate the roles of pleiotropy and linkage.

Linkage may often lead only to "hitchhiking effects" in the carrying along of those genes that are closely linked to those directly involved in selection. It may then often be a matter of accident as to which effects are correlated with selection. Hitchhiking effects also play a role in experiments with marker genes; not isolated genes, but whole chromosomes at first, and later smaller chromosome sections, determine the results. Such experiments also have been conducted occasionally by forest geneticists.

A typical example for parallel selection at linked loci was given by DADAY (1964). In *Medicago sativa*, close linkage exists for the characters' cold resistance and capability of growing at low temperatures. However, the author has reason to believe that the two characters are basically independent. Nevertheless, the two selection principles nearly always lead to parallel, unidirectional selection for both characters in nature.

This may be different in other cases. GRANT (1967), for instance, investigated the linkage of genes controlling morphology of reproductive organs with those responsible for seedling vitality. These genes turned out to be strongly linked, and the author interpreted this as a means of protecting proven combinations of both characters against breakup by accidental crossing, i.e., he saw linkage as playing a role similar to the genetic coherence of CLAUSEN (1959). Presumably different optima exist for correlations from linkage in different populations. This, in fact, appears to be so, for ALLAN et al. (1968) found that in wheat correlations exist between two character complexes, yet these correlations are typical of populations, and vary from one to the next.

The importance of the characters determining fertility — especially those of the generative apparatus — has often been underestimated by earlier writers in forest genetics. As is to be expected, many examples demonstrate their priority in selection, by far exceeding that of the characters determining vegetative processes. JAIN and MARSHALL (1967) exemplified this in a study that included the whole life cycle of barley.

The development of certain linkage relations is not only a consequence of parallel, directed selection leading to specific gene endowments at the loci of the same linkage groups following simple niche differences, or of linkage at such loci that respond to different factors of selection in the presence of complex niche differences. THODAY (1960) has noted and given experimental evidence that disruptive and stabilizing selection also influence linkage relations. For example, if disruptive selection is applied over two niches, where greater fitness is provided in one niche by the dominant alleles of two linked loci and in the other by the two recessives, then the coupling phase will be stabilized. Conversely, if stabilizing selection occurs for an intermediate niche, whereby the genotypes Aa/Bb possess greatest fitness, then the repulsion phase is favored. We could say here that linkage changes the form of the fitness set.

It may further be noted here that linkage relations, which originated under the influence of selection and are often remarkably stable, are very important both for evolution and the way it is controlled by man, namely, breeding. In their monographic treatment of linkage, BODMER and PARSONS (1962) have emphasized this especially. The breakup of existing linkage relations should lead to new genetic multiformity in many cases, as many experiments on linkage and its role in the maintainance of genetic variability (often hidden variability) have indicated. For example, SARSON and MURTHY (1968) applied disruptive selection for flowering date in one population of *Brassica campestre* over five generations, by selecting the earliest and latest plants in each generation and crossing them. The result was a population with considerably greater genetic variation, including yield attributes that offered better prospects for selection than the initial one. The authors subscribed to THODAY's assumption that the effect of disruptive selection is breakup of linkage correlations that were maintained in the original by uncontrolled ("natural") selection and then broken, leading to greater genetic variation. Using an example from wild and cultivated forms of *Sorghum*, which in Africa often grow close together, DOGGETT and MAJISU (1968) demonstrated that disruptive selection, and with it also selection for certain linkage relations (as are to be expected following selection of this kind), influenced evolution of *Sorghum* in these areas in a significant manner. In their view the evolution of maize may have been influenced in a similar way.

Reference has already been made to the importance of stabilizing selection in the evolution of forest tree populations. This has opposite effects. LEWONTIN (1964) investigated the interactions of linkage and selection in optimum models. Seven different situations were assumed. The principal result is that selection under optimum conditions for many generations may lead to semistable equilibria.

Thus the combined effects of directional, disruptive, and stabilizing selection in our forest tree populations probably instituted a certain linkage position somewhere near the optimum of each population. However, these optima are determined not only by selection and linkage, but also by the kind of gene effects — by dominance and epistasis. SCHNELL (1963) has presented a general model. The main impression gained from it is that covariances become very complex between relatives (including between parents and progeny) when linkage is involved and that it becomes very difficult to predict the response to selection in such systems. For this reason simpler models will be discussed here, although even these convey an impression of the difficulties to be expected. Frequently, solutions of even these simpler models are possible only by means of Monte-Carlo simulations.

LEWONTIN and KOJIMA (1960), for example, considered the joint effects of linkage and epistasis upon the evolutionary dynamics of a complex polymorphism. Even in this simple system they observed several adaptive peaks; i.e., conditions of a higher mean fitness of the population at a certain level of linkage and gene frequency, which then decreased in all directions (see also CORMACK, 1969). Population equilibria at these peaks were stable but the maximum possible rate of approach of a population to a peak was not realized as a result of selection pressure. The equilibrium position in the absence of epistasis was not influenced by linkage, which merely had a delaying effect. Even if epistasis is present, only very close linkage will influence the position of the equilibrium or equilibria. Very close linkage may give rise to permanent disequilibria or influence the gene frequencies typical of adaptive optima. With intermediate frequencies, equilibria were often only possible with very close linkage. Gene frequencies were also responsible for the rate of approach per generation toward the equilibrium.

The interplay of linkage and epistasis is comprehensively reviewed by KOJIMA and LEWOTIN (1970). Two general results of their discussion are of interest here. First, the statement that multilocus systems with linkage have not reached their equilibrium even after many generations, and, further, that the development of fitness (with equal conditions of selection) is not necessarily uniform but may consist of several ups and downs. In the authors' view many natural populations may never reach an equilibrium because of small but consequential changes. Second, the authors support TURNER'S (1967) conjecture that selection may tend to achieve optimal recombination. In essence this is also the conclusion of LEVINS (1965) concerning the development of linkage in optimal genetic systems. The optimal degree of linkage (or, vice versa, the optimal recombination rate) is influenced by autocorrelation of the environments of successive generations and the relative size of gene effects. For example, without any autocorrelation every response to selection is a disadvantage; with certain linkage relations, deviations from equilibrium then reduce the influence of gene effects at the loci concerned upon the fitness value. With high autocorrelation, however, a response to selection benefits the population. Therefore, with relatively small gene effects (relative with respect to size of niche differences), a maximum recombination is optimal since it guarantees the most rapid response to selection. On the other hand, relatively large gene effects may generate too much genetic variance, which can then be reduced by linkage. Here a limitation of recombination by linkage is beneficial. This theory apparently also covers all arguments advanced by CLAUSEN (1959).

h) Introgression

In some closely related tree species with overlapping ranges, variation patterns are often codetermined by gene flow across species boundaries. Following ANDERSON (1949), this process is termed introgressive hybridization or introgression if it results not only in the occasional occurrence of spontaneous hybrids but also the incorporation of genes and gene complexes of one species into the gene pool of the other species (see also BAKER, 1951; STEBBINS, 1950). Species hybrids have long been of special interest to forest genetics. Work with them is not only illuminating with respect to the taxonomy of the species involved, but may lead to promising heterotic hybrids — e.g., in the larches, pines, and poplars — and plays a role in many breeding programs. A few typical cases are considered in order to discuss the most important problems and results.

The prerequisite for the origin of species hybrids is contact of the species involved. Species with overlapping ranges usually occupy different ecological niches, either in the same or neighboring habitat. An overlap of species ranges is often limited to certain regions because the difference in species ecology is not only expressed by occupation of different ecological niches in the same region, but also especially by different adaptive values in different geographic regions. Because of species ecology, therefore, favorable conditions for the crossing of species usually exist only in certain areas of the ranges. A second prerequisite is an overlapping of flowering dates. This may also be limited to certain portions of the common area. Finally, the species involved must be cross-compatible.

Frequently all of these conditions are met, but nevertheless there is no gene flow in either direction. Thus hybrids between the pine species in the southeastern United States are occasionally found in all overlapping zones but have no bearing on evolution of the species (LITTLE et al., 1967, and others). There may be ecological or genetic reasons for this (an appraisal of the role of species hybrids in *Pinus* is given by MIROV, 1967). Genetic reasons for the reproductive failure of hybrids could be different coadapted gene complexes of the parental species or sterility as a consequence of the different structure of homologous chromosomes as discovered by SAYLOR and SMITH (1966) in species hybrids of the genus *Pinus*. There are numerous examples of the presence of different, coadapted gene complexes in forest trees and resulting disturbances in vegetative development. Thus MOFFET (1965) found chlorosis in the progeny of the cross between *Acacia decurrens* and *A. molissima*. Clearly segregated and intermediate forms appeared in the F_2 generation. Langner (oral communication) repeatedly found pronounced hybrid weakness as early as in the F_1 generation is species hybrids involving *Picea abies, P. omorica, P. sitchensis,* and *P. jezoensis*. The hybrids of *Betula pendula* and *B. pubescens* investigated by STERN (1963) also exhibited distinct hybrid weakness in the F_1 generation. In the cross *B. pubescens* × *B. cordifolia* the alleles of two loci were responsible for hybrid weakness; the loci were not linked (STERN, 1960). Occasionally hybrid weakness does not appear prior to the F_2 generation (F_2 breakdown), as observed by LANGNER (oral communication) in the hybrid *Larix decidua* × *L. leptolepis* on heath sites in northwestern Germany. Hybrid weakness and hybrid sterility may prevent introgression as effectively as geographic barriers or cross-incompatibility.

STEBBINS (1958) presented a comprehensive review of inviability, weakness, and sterility of hybrids. These physiological conditions appear not only in interspecific hybrids but, as may be noted here, also in hybrids between races of the same species. Thus STERN (1960) observed distinct hybrid weakness in progenies of crosses between Scandinavian and central European provenances of *Betula pendula*. On the other hand, species are known where these conditions do not occur, such as in some widely distributed species of the F_2 generation in *Drosophila*, although hybrid weakness is not rare in other species of this genus. In plants, maize constitutes a prominent example (POLLACK et al., 1957).

Genetic incompatibility barriers may have come about as by-products of allopatric speciation. The two species may be isolated for a long time and evolve along different paths while acquiring qualities that prevent crossing when they again come into contact, either in nature or during experiments (see KITAGAVA, 1967, and others). The crossability pattern of the species within a genus then permits inferences concerning their degree of relationship. Such crossability patterns have been worked out for several tree general — by WRIGHT (1955) for genus *Picea*, by CLAUSEN (1970) for *Betula*, and by MIROV (1967) for *Pinus*. The physiological incompatibility barriers between two species may be very different (see MIROV, 1967; KRUGMAN, 1970; CHIRA and BERTA, 1965 for conditions in *Pinus*).

However, cross-incompatibility between species may also be the direct consequence of natural selection for incompatibility. Since KOPPMAN's (1950) first experiment on this subject with *Drosophila pseudoobscura* and *D. persimilis*, it has been known that such selection may lead to a sharp limitation of the gene flow between the species. Similar experiments have been made on many organisms. A recent experiment with maize (PATERNIANI, 1969) confirms this result for a higher plant species.

This course is selected only where there is secondary contact between two species and the hybrids are at a selective disadvantage: individuals crossing exclusively within the same population produce more progeny than those crossing in both directions and thereby use up some of their gametes for the production of inviable hybrids. The cross-incompatibility that arises in this way (i.e., in angiosperms usually incompatibility of pollen and stigma), and is often related to the self-incompatibility system, is usually superimposed on the cross-incompatibility resulting as a by-product of evolution. Therefore, for these and for other reasons, crossability patterns are not always reliable guides to the actual relationships. Thus in one experiment (unpublished) with the European species *Betula pendula* and *B. pubescens*, and the Japanese species *B. japonica* and *B. ermannii*, we have found cross-incompatibility of *B. pendula* and *B. japonica* when crossed with *B. pubescens* on the one hand and *B. ermannii* on the other. *B. pendula* crosses well with *B. japonica*, and *B. pubescens* with *B. ermannii*. However, *B. ermannii* is the most distantly related species; it is tetraploid and belongs to another section than the three other species.

Incompatibility barriers between sympatric species may have come into being only in parts of the common range of two species but may (still?) be absent in other parts. The two European oaks *Quercus robur* and *Q. petraea* probably constitute examples. DENGLER (1941) obtained 4% germinable seed from the cross *robur* × *petraea* but only 1% from *petraea* × *robur*. Crosses within the species yielded an average of 50%. In nature the flowering times of the two species overlap, and thus there is no limitation of gene flow in this respect. On the other hand, SALISBURY in

England (1919, 1940) found hybrids frequently on soils intermediate in quality between the typical soils of the species.

Simple environmental variables may also influence the degree of introgression. VAN VALEN (1963), for example, found that in population cages containing both *Drosophila persimilis* and *D. pseudo-obscura*, no hybrids were obtained at 16% relative humidity but at 25% many hybrids. Thus the populations were strongly isolated in the first environment, but not at all in the second. Such results unfortunately are not available for forest trees, although the following situation illustrates the same principle. BENSON et al. (1967) investigated introgression of *Quercus douglasii* and *Q. turbinella*, the ranges of which overlap in California over a distance of 250 km. In the center of the overlap area there are not only hybrids but also hybrid swarms. One of these hybrid swarms was studied more closely. On northeast slopes it consisted of hybrid forms closely related to *Q. douglasii*, and on southeast slopes the hybrids were close to *Q. turbinella*. The authors infer that evolutionary sorting of the hybrid types takes place as a result of differences in the direction of selection in these contrasting environments. Here, too, introgression depends upon environment. Ecological specialization of the species is often a serious barrier to introgression. In the case of *Q. robur* and *Q. petraea* (SALISBURY, 1919, 1940), the former occupied heavy and calcareous soils and the latter the lighter and more acid soils. The hybrids were found in the zone of transition between both soil types. Selection for mutual exclusion is here responsible for the separate distribution of the two species. Hybrids apparently excel in competitive ability on the intermediate soils. MULLER (1952) described similar situations for a series of oak species: *Q. mohriana* occurs on calcareous soils or calcareous soil beneath the surface horizon, *Havardi* on sandy soils, and the hybrids on intermediate soils; *Grisea* is restricted to soils from igneous rocks and the hybrids with *Mohriana* are found in the transitional areas; *Stellata* occupies soils with sandy loam or gravelly, loamy soils and the few hybrids with *Mohriana* occur on intermediate soils; *incana* prefers deep sands and *marilandica* loamy sand, the hybrids intermediate contact areas; *margaretta* is also restricted to deep sands and its hybrids with *stellata* to the transitional soils of this species as indicated above. Similar conditions may be typical for genera other than the oaks. Ecological restrictions are also enforced by climate. Thus *havardi* and *stellata* are partially isolated by precipitation and hybrids are found in areas with intermediate precipitation. Results similar to those reported by DENGLER and SALISBURY for *Q. petraea* and *robur* were found by MULLER for *Q. emoryi* and *gravesii*, which overlap on a broad front but form hybrids only in a single location. Similar conditions apply to *Q. hypoleucoides* and *gravesii*: the incompatibility barrier has not yet been developed everywhere.

Again in oaks, STEBBINS et al. (1947) studied conditions controlling the occurrence of hybrid swarms in contact areas of *Q. marilandia* and *Illicifolia*. They found them on disturbed habitats. In stands growing on soils typical of either species, hybrids are not found, or only very rarely. COOPERRIDER (1957) confirmed a similar situation for *Q. marilandica* and *velutina* in Iowa. Apparently the extreme ecological isolation of these two oak species prevents introgression from contributing to the evolution of populations as long as the environment remains the same. The opening of new niches may create new conditions in every case. The role of introgression in shaping the genetic variation pattern of the species should be seen in the same light.

In birches introgression may be complicated by different types of polyploidy of the species involved. Dugle (1966) described the relationships among *Betula fontinalis*, *B. glandulosa*, *B. glandulifera*, *B. resinifera*, and *B. papyrifera* in western Canada. She found seven putative hybrids, and in some cases distinct introgression. Introgression proceeded twice in the direction of higher ploidy, once toward lower ploidy, and once in both directions.

The birches constitute a genus of particular difficulty for the taxonomist, not the least because of widespread polymorphism and morphological similarity of sympatric species. For these reasons, the earlier literature often described individual morphs and hybrids as new species. Thus the so-called blue birches (*B. caerulaea* and *B. caerulea grandis* of the old literature) turned out to be a hybrid swarm of *B. papyrifera* and *B. populifolia* Brayshaw, 1966), and similarly another example of a hybrid is *B. andrewsii* (Froiland, 1952). In the birches, too, there are indications that hybrids possess a selective advantage only on certain sites. Thus Clausen (1959) found hybrids between *B. papyrifera* and *B. pumila* predominantly in disturbed habitats. Elsewhere they were found with different frequencies throughout the overlapping range, except in Canada, where different flowering periods appear to isolate them.

The nordic birches have long been of special interest. Elikington (1968), for instance, compares variable populations of shrub birches *(B. pubescens* and *B. nana)* in Iceland with populations of *B. pubescens* in Great Britain. The Icelandic populations were found to be strongly influenced by introgression between the two species, which proceeds via triploid hybrids. This, of course, substantially impairs fertility, but more in the types close to *B. pubescens* than in those resembling *B. nana*. In Finland, Vaarama and Valanne (1970) obtained similar results (Fig. 37).

Fig. 37. *Betula tortuosa* (trees in the background), *B. nana* (small shrubs of knee height at left) and their hybrids (at right) in Finland. (Photo Vaarama and Valanne, 1971)

One would expect that the polyploid series in the genus *Betula* would consist of alloploids in most cases, and that the evolutionary potential of this genus is codetermined in large measure by introgression of species with different chromosome levels. This may apply even more to the willows.

That interspecific hybridization may constitute a basis for speciation has been supposed for many species in several genera of forest trees, and, in fact, it constitutes a good possibility for rapid evolution (STEBBINS, 1950, and others). For example, *Pinus washoensis* occupies a zone between *P. ponderosa* (lower elevations) and *P. Jeffreyi* (higher elevations). It still crosses easily with *P. ponderosa* but not with *P. Jeffreyi*. HALLER (1959) has postulated that it arose from the introgression of *Jeffreyi* into *ponderosa*. Further examples of a similar kind in the genus *Pinus* have been assembled by MIROV (1967). STUTZ and THOMAS (1964) contributed a particularly instructive example of speciation from the introgression of *Cowania* and *Purshia* in Utah. Hybrids are found particularly along slopes with changing local climate where seasonal isolation, the main barrier to gene flow, is eliminated. Gene complexes of one species may possess selective advantages of such magnitude in the population of the other species that they advance from the hybrid zones by as much as 200 km into the range of the other species. *Purshia glandulosa* is described by the authors as a stabilized hybrid between *P. tridendata* and *Cowania*. Theoretically it would be possible for a new species to arise after introgression in a single genera-

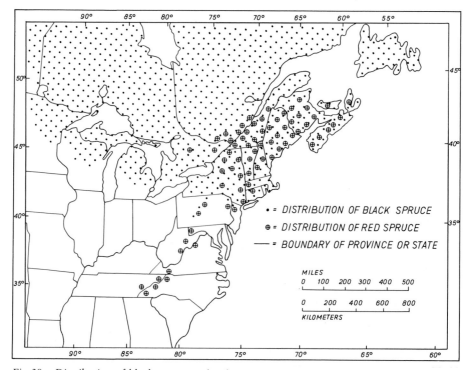

Fig. 38a. Distribution of black spruce and red spruce in eastern North America. (From MOR-GENSTERN and FARRAR, 1964)

tion. RAO and DE BACH (1969), for instance, after crossing different species of
Aphytis, obtained hybrids that were fully fertile and partially isolated from both
parents. In one case they were completely isolated. Of greatest interest is the hybrid
family; its males were isolated from females of both parental species. Such cases are
probably rare, yet they indicate the possible past and future role of introgression in
evolution.

   An interesting case of the influence of introgression upon the genetic variation
pattern of a genus was investigated by MELVILLE (1939). *Ulmus* in Great Britain
exhibits a distinct nothocline (gradient of different stages of introgression) when
the pure and transitional forms are assessed quantitatively using biometric methods
(see Fig. 40).

   In concluding this discussion of introgression, we shall examine the results of a
thorough study by MORGENSTERN and FARRAR (1964). It covers all significant genetic
and ecological aspects of the introgression of *Picea mariana* and *P. rubens*, and
reflects the type of problem encountered in the study of forest trees.

   The two species are easily distinguished on the basis of typical characters. Their
ranges overlap in eastern Canada and the northeastern United States (Fig. 38a). The
overlapping area includes almost the entire range of *Picea rubens*. Here the species
occur mostly in adjacent habitats; each possesses greater fitness on certain sites. *P.
rubens* is very tolerant of shade in its juvenile stage and therefore grows in pure

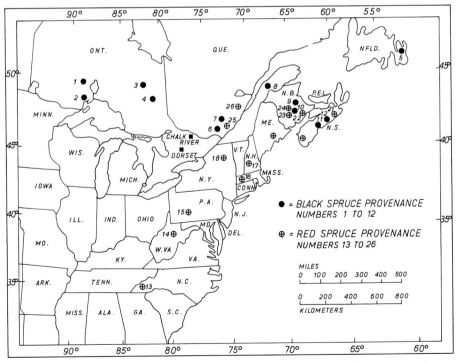

Fig. 38b. Map showing the origin of the provenances that were grown at the Petawawa Forest
Experiment Station near Chalk River, Ont. (From MORGENSTERN and FARRAR, 1964)

stands as well as in unevenly aged mixed stands with tolerant desiduous species or
in mixture with nontolerant deciduous and coniferous species, but *P. mariana*
mainly occupies sites where it is subject to little competition by other tree species. In
the southern part of its range, *P. mariana* is restricted to swamps and bogs that
almost eliminate competition and where reproduction by layering is possible. In the
Great Lakes region, *P. mariana* occupies dry plains together with *Pinus banksiana*,
and regeneration generally follows fires to which it is adapted as a "fire species,"
having serotinous cones. Large pure stands often exist here.

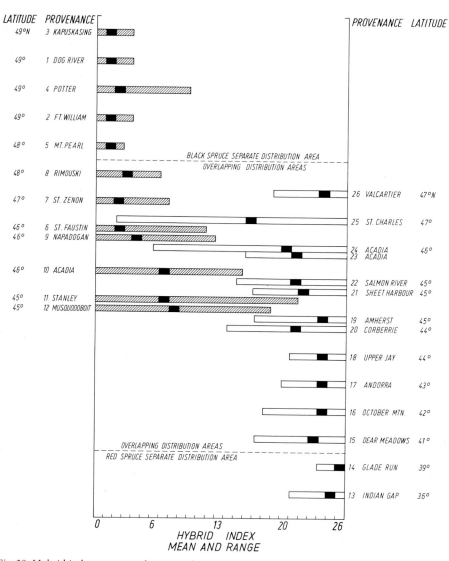

Fig. 39. Hybrid index means and ranges of populations of red spruce and black spruce. Ranges
overlap in the Maritime Provinces and Quebec but means remain apart

No genetic barriers exist between the species. Seasonal isolation appears to be insignificant also. Geographic isolation of course exists for the larger portion of the population of *P. mariana* and the population of *P. rubens*. Hybrids, hybrid swarms, and often small populations made up exclusively of hybrids are therefore not rare in the sympatric range of these species.

Ecological isolation differs markedly throughout the common range. In the Acadian region, for example, it is weak; in the Great Lakes region strong. Twenty-six provenances were available to the authors for analysis and their distribution across the species ranges is shown in Fig. 38 b.

A hybrid index (ANDERSON, 1949) was developed that assumed the value of 0 for pure *P. mariana* and 26 for pure *P. rubens*. The result is shown in Fig. 39: provenances from areas where *P. mariana* or *P. rubens* occurred alone were characterized by mean hybrid indices near 0 and 26, respectively, and relatively small variances; provenances from the area where the species occur sympatrically showed large variances and much deviation from these ideal values. The degree of ecological isolation in the overlapping distribution areas codetermined the degree of introgression decisively. Provenances from New England or New York State, for example, where ecological isolation is strong, indicated less introgression than provenances from the Maritime Provinces where ecological isolation is only weakly developed.

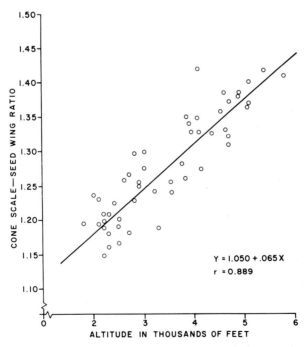

$$Y = 1.050 + .065 X$$
$$r = 0.889$$

Fig. 40. Relationship between altitude and cone scale morphology in the white-Engelmann spruce complex in British Columbia. Each point represents 100 cones. Pure forms of white spruce and Engelmann spruce are distinct taxonomically and occupy quite distinct niches at low (white spruce) and high elevations (Engelmann spruce). The intermediate zone is occupied by a hybrid swarm. (From ROCHE, 1971)

In the first area the means of the hybrid index ranged from 0 to 2 for *P. mariana* and 24 to 26 for *P. rubens*; in the second the means were approximately 8 and 20, respectively. Of special interest was the situation in Quebec where there was marked introgression into *P. rubens* but little into *P. mariana* because *P. mariana* is here more closely restricted to swamps and bogs than in the Maritime Provinces. We have already seen that the net result of introgression is often codetermined by environment, as in the case referred to above of the oak hybrid swarm codetermined by slope position.

MORGENSTERN and FARRAR emphasized that their investigations did not indicate clearly whether the degree of introgression was greater in disturbed habitats. They even found populations with strong introgression in very stable habitats and considered it possible that introgression opened new habitats to the species in certain cases.

The example of introgression given by ROCHE (1969, 1971) for *Picea glauca* and *P. engelmannii* is also particularly illustrative. The former species occupies lower elevations; the latter is a typical high-elevation specialist. In a comparison of cone form, a nothocline is obtained between typical *engelmannii* forms in the higher elevations and *P. glauca* in the lower elevations (Fig. 40).

V. Forest Ecosystems

The great variety of forest types found even in a restricted area has posed problems to forest botanists since early stages in the development of forest science. The dependence of species composition upon soil, climate, topography (microclimate), management, and the possibility of describing site on the basis of vegetation and sometimes to measure its productivity all offered numerous opportunities to forestry science and practice to engage in vigorous debate. The discussion of the idea of the "perpetual forest" (Dauerwald) is one that will remain in memory. Complex systems such as the forest ecosystem were well suited to the application of the dialectic method to an object of natural science, and it was apparently unavoidable that the endeavors of forestry sciences in coping with such complexity were hamstrung by ideology under the influence of dialectic holism. The most prominent model of ideological thought, the so-called "forest organism," has exerted a bad influence until the most recent times and found supporters, especially in central Europe. This of course was not an autonomous development of forest science; rather, holisms had taken hold throughout phytology (to name only the closest science). TANSLEY's (1935) essay on this problem is still worth reading today. Even as late as 1967, LEWONTIN was forced to argue against research-paralyzing models of this kind, indicating that the debate with supporters of the dialectic method still continues.

Viewing a biological association as a system has at least two advantages. First, this makes it possible to apply the modern methods of systems theory to our object, thereby introducing new research categories. Second, this provokes thought along evolutionary lines. Every biological unit is the result of an evolutionary process, or, better, represents a stage within this process. This of course also applies to the ecosystem. Considering the increasingly higher levels of organization, nucleotide, DNA molecule, gene, chromosome, genome, population, and ecosystem, the elements of coevolution will be introduced as determining evolutionary factors at the level of the ecosystem. The evolution of every population in a certain ecosystem, be it microorganism, plant, or animal, is significantly influenced by presence and evolution of other species of organisms in the same system. It would be presumptuous to claim that we are in a position to measure and understand such a highly complex system, or even only its major relationships. Here we probably find the difference of decisive importance for practice and experimental research between the ecosystems approach and the theory of the forest organism.

Probably the first attempt to describe highly complex situations in genetics by means of such a model is represented by WADDINGTON's (1957) book "The Strategy of the Genes." Models from the theory of games have also been proposed and applied by other authors for the same purpose, e.g., by LEWONTIN (1961), SLOBODKIN (1964), and others. In these models populations act as players that can adopt different strategies. Our fitness value W or the Malthusian parameter m suggests that the aim of the population in such games is maximum profit. However, the common strategy to minimize the expectation of maximum loss is also conceivable.

Attention has been called to the special process of information gathering. KI-MURA (1961) was probably one of the first to do so. This process results from DNA as the basis of all life and the properties of Darwinian evolution. Natural selection accumulates information on the one hand and exchanges less useful for more useful information on the other. One of the problems of optimization of a genetic system is to store and transform information gained by means of a primitive trial-and-error procedure, and this over a long series of generations and in an extremely fluctuating environment. In another type of model, SCHMALHAUSEN (1960 a, b) conceives evolution as a cybernetic regulator. The biogeocoenose is the regulator of the population. The system should be controlled by feedback mechanisms, and this information-retrieval function is performed by individuals. In this conception the ecosystem (the biogeocoenose as applied here cannot be anything else) clearly becomes the regulating unit for the evolution of all participating populations.

Nearly all of these models reflect the difficulty of finding generally applicable definitions for the selection coefficients of genes, fitness values of individuals, and adaptive values of populations. However, within the framework of such a system, suitable *ad hoc* definitions may be found. DARLINGTON's (1939) genetic system, together with LEVINS' (1961) concept of the population or species as an adaptive system, and the possibilities of searching for and finding optima at least in subsystems (to mention only some of the problems) probably indicate substantial progress in methodology. It is also probable that these definitions will support endeavors "toward a predictive theory of evolution" (SLOBODKIN, 1967), but of course it is not certain that they will result in the achievement of this goal.

In relation to the subject discussed here, it is possible only to become familiar with the general ideas of recent developments in this field, although it would be tempting to pursue them further. For these reasons, we shall choose one of the more recent and sufficiently general models for the evolution of ecosystems. This has the advantage of accounting for all phases of interest to us in a fairly logical manner.

1. A General Model of Evolution Including Evolution of Ecosystems

WARBURTON (1967) describes evolution using a model of a guessing game and develops the concept of a Darwinian guessing game for this special purpose. His premise is that storage and transfer of genetic information constitute one of the major problems of biology. This is not information in the same sense as defined in general information theory; there the fundamental problem is to reproduce at one point with stated accuracy the message selected at another point. The importance of the message is immaterial but it is necessary to select it from a set of possible messages. The system of transferring messages must be such that it functions for all possible messages of this set. In contrast to this conventional definition of the term information, genetic information is always "information about something."

The origin of this information is the interplay of the genetic system with the environment. Adaptive genetic information may therefore be interpreted as information about the environment of the species concerned. To every adaptation corresponds a related phase of the environment (this coincidence of niche and phenotype has been emphasized in the preceding chapters). It is conceivable that at some future date a geneticist may be able to read and interpret DNA sequences much as present biologists interpret adaptations. In the final analysis, the source of information leading to a certain adaptation is the complex of niche dimensions that calls for this adaptation.If this deduction is correct, then one could say that environment is the source of all genetic information. Such relativity of the genetic information makes it necessary to define it in a way that differs from the abstract definition applied by SHANNON's information theory. This definition should still be oriented sufficiently toward general information theory to facilitate the application of its mathematical procedures: Let x be a variable that may assume one of N_x possible values, and S_x be a set of possible values of x. Every x_i has the probability p_i associated with it. In accordance with SHANNON's definition the entropy of S_x is given by the relation

$$H_x = -\Sigma p_{i_1} \log p_i .$$

Let y be a second variable that is sufficiently correlated with x to allow reading y-values from x; y then transmits

$$I_x = H_x - H'_x$$

units of information about x (e.g., binary digits if logarithms have the base 2 in the preceding equation). We may state that x is the object of the message and y a message about x.

In the majority of cases, it will be impossible to calculate the entropy of the information set. Therefore, the messages subsequently considered will consist of symbols that designate an answer to a multiple-choice question. All possible symbols that may be given in a message must be written in the form of a list after the questions have been asked. Following this one obtains probabilities for the appearance of certain symbols in the message; these simplified conditions then permit the estimation of the entropy of all possible messages. With this technique it is possible to adopt other important components of conventional information theory such as equivocation and redundancy.

We shall now consider the guessing game itself. Participants in the game are one or more players and one referee. To win in T experiments, the player must come up with one "correct answer." There are N' correct answers, which together form the set S'; all the player knows at the start of the game is that these answers are contained in the set S of N elements, which form the set of all possible answers. To designate the correct answer, the player must accumulate $\log (N/N')$ units of information. It is the referee's task to supply this information to the player on the basis of the player's guesses. There are r possible answers to every guess. Therefore, the referee can supply a maximum of r units of information for a correct answer per experiment. In the more interesting variants of the guessing game [N large, $T \log r$ not much larger than $\log (N/N')$], there are no chances for the player to win the game by accident. He must then choose a certain strategy to increase his chances to win, i.e., to force the referee to supply him with as much information as possible.

The characteristics of the guessing game are as follows:

1. The player divides the set of possibilities into more than r subsets, giving one element of each subset to the referee. Every element is a guess and the simultaneous presentation of a group of guesses is an experiment.

2. The referee announces whether one of the guesses was better than other guesses. A referee of this kind may be termed selector, since he arranges the guesses in order from the worst to the best according to a system unknown to the player. Another possible type of referee in such a trial-and-error game is that of a comparator who measures each guess by comparison with the correct answer — by classifying it as too large, too small, etc. A comparator is easily set up whereas a selector consists of a highly complex mechanism.

3. The player assumes that his best guess is contained in his best subset of guesses. He therefore will divide this subset further in his next experiment. This procedure is continued until all T possibilities are exhausted or the correct answer is found. To facilitate repeated division of the subset, there must be a possibility for a hierarchical classification of the subset. In every experiment the player must decide which of his subsets can be divided further. This requires data or information storage. Complete storage of all information obtained in the game is not necessary. Rather, there will always be a strategy allowing the choice of the subset to be divided further and to classify it within the hierarchy. As a result it is enough to store the last best guess and the number of experiments carried out to date.

Of special interest is the situation when the player of a trial-and-error game with selector can store only the last best guess. This is the unique Darwinian guessing game, which is defined as a trial-and-error game with one selector where the player can store only his last best guess (or a number of best guesses in the last experiments).

We shall now develop the analogy between the process of evolution as controlled by selection and the preceding model of information theory. The player is represented by a population of a sexually reproducing species. Every individual is a guess. The environment functions as referee as described above, by deciding whether a guess possesses greater fitness than others or not. The set of possibilities is given by the set of all individuals of this species offered since its origin. The entropy of this set cannot, of course, be defined in the sense of conventional information theory (see above). This is immaterial here, however, because answers of the selector represent answers to multiple-choice questions of which the information content can be estimated without knowledge of this entropy. A correct answer is an organism that is as well adapted as any organisms that we find will survive permanently in any niche of a complex ecosystem. At any stage of the evolution of this organism a population in this specific ecosystem (and in neighboring ecosystems) consists of the "best guesses" of previous experiments and progenies of this population. The "player" therefore possesses no information about experiments that failed or their total number. This last statement implies that early history of the population determines the result as much as selection in recent historical times and at the present time. Previous experiments decide the character of the niche in which the organism resides, i.e., perhaps which position it takes in the food chain, and which opportunities are open to it in future experiments.

If one wishes to discuss the most interesting problems of a Darwinian game of this kind this is best done under the headings given by Warburton:

1. *The Guesses Must be Expressed by Numbers.* In this model of the Darwinian guessing game this means that the player must formulate guesses that represent subsets of the same magnitude. Consequently, the criteria used to form the hier-

archical classification must guarantee subsets approximately equal in size. A hierarchy based on subdivision criteria that can be used in every case without changing the relative size of subsets is a permutable hierarchy. In the Darwinian game the player is unaware of the level of classification and thus also of previous subdivision criteria. For this reason he should classify the set of possibilities in a permutable hierarchy. In this classification every element of the set is indicated by a sequence of digits, given by alternative symbols of a small alphabet. Every letter represents a different condition of this digit. The choice of one of these letters for the ith position in the sequence, for example, then does not change the alphabetic order used to fill the subsequent positions. In this way it is possible to represent every element of the set in digital fashion.

Obviously, the genetic material of an organism corresponds to this digital description. Whether we think of the sequence of nucleotides in the DNA chain or the sequence of genes in the chromosome is immaterial. The set of all possible genomes could also be given in the same permutable manner. It appears to be so that a digital arrangement of this kind represents one of the necessary conditions for evolution by selection.

2. Equivocation in the Answers of the Referee. The reason for this may be indecision of the referee or interactions. For the first case, suitable models may be found when considering several components of the genetic system simultaneously. In the second case the genetic background is involved; selection operates at the phenotypic level and the referee is never aware of the sequences of symbols offered to him.

3. Evolution of a Strategy. The decisions of the player in a Darwinian game concern all components of the genetic system. He must know how many guesses he will make in each case, how large the proportions of true replications or mutations, respectively, should be, and so on. If he would have an idea of the magnitude of equivocation on the part of the referee, and would know the real position of the last guess in the sequence of possibilities represented on the scale of selection, and also the future rules applied by the referee (plus other conditions not discussed here), then he could come to an intelligent conclusion regarding his strategy. However, he is without this information and therefore chooses his strategy more or less blindly. Yet the genetic system offers possibilities to develop adaptive strategies in a trial-and-error procedure, but the theoretical possibilities available have not been studied and remain vague for the sole reason that processing of information for this purpose has not been given much theoretical attention.

Special problems are created in games where several players compete. It would then be conceivable that the referee eliminates from time to time players making the poorest guesses. This then requires looking at the whole population, for example, in a certain generation, as one guess. Each of the two (or more) species of organisms competing in the same ecosystem at any phase then represent players in a second-order Darwinian game. Every player would repeatedly play new strategies (compositions of the population). The biggest prize, i.e., complete elimination of the competitor, would simultaneously create new problems.

4. Games with Several Correct Solutions. The referee could make the rule, for example, that different guesses would be judged equally under certain conditions. In this way models of the game can be developed; the games then have certain final values that are predictable but not the course of the game itself; or the games could be without predictable final values.

We may recall here that very similar abiotic environments have given rise to a very dissimilar fauna and flora although these may be equally well adapted. Neither the course nor the result of the evolutionary process could be predicted here. Then the question immediately arises, as to whether such a result, which is more or less final, could exist at all. For these reasons, instead of considering the result, it is preferable to consider the stage of the ecolutionary process or Darwinian guessing game that characterizes the game, and in which the player takes part.

A set of correct answers in a second-order Darwinian game represents a set of organisms that are able to coexist. We know the conditions for this, i.e., the conditions for the development of ecosystems, with the same superficiality as the conditions experienced by the player, namely, the genetic system. At the same time we know that the evolution of ecosystems is the inevitable consequence of Darwinian evolution; and therefore the biology of ecosystems must be seen under genetic aspects just as the evolution of single organisms is to be seen under ecological aspects.

It is clear that this section could not present more than a rapid survey of the subject. This may be adequate, however, to indicate to the reader the necessity of applying unifying, if only descriptive, models to such highly complex situations.

2. Two Main Axioms of Coevolution in Ecosystems

The initial assumption that forests are systems in the sense of TANSLEY (1935) requires some remarks on the general nature of systems to make a connection with general systems theory. This account cannot give all necessary details and will be limited to some notes. First of all, a system is a set of elements, and also consists of a second set of axioms about these elements. For the time being, axioms are defined as unproven statements. The set of elements in an ecosystem then is perhaps the set of species in the system; the axioms concern assumptions about the species, such as position in the food chain, evolutionary past and future possibilities, dependence upon other species, and so on. A knowledge of the most important qualitative and quantitative relationships would make it possible to predict future development of the system.

ODUM (1969) attempts to explain the strategy of ecosystem development using examples from ecological successions. He names 24 ecosystem attributes, which are distributed to six complexes. Every attribute characterizes the evolution of the ecosystem in a certain way (Table 25).

The reader will recognize at once that significant components of ecosystem development depend upon evolution and coevolution of the participating species. The number of species, for example, which is small in initial stages and increases with progressive development of the system, is codetermined by regulation of competition during coevolution; closely correlated with this is evolution of niche specialization (attribute 12 in Table 25) and adaptation of the developmental cycles of organisms to the special requirements of their niche (attribute 14). As specialization develops, entropy and information content are simultaneously determined (attributes 23 and 24), and so on.

Table 25. A tabular model of ecological succession: trends to be expected in the development of ecosystems. (After ODUM, 1969)

Ecosystem attributes	Developmental stages	Mature stages
Community energetics		
1. Gross production/community respiration (*P/R* ratio)	Greater or less than one	Approaches one
2. Gross production/standing crop biomass (*P/B* ratio)	High	Low
3. Biomass supported/unit energy flow (*B/E* ratio)	Low	High
4. Net community production (yield)	High	Low
5. Food chains	Linear predominantly grazing	Weblike, predominantly detritus
Community structure		
6. Total organic matter	Small	Large
7. Inorganic nutrients	Extrabiotic	Intrabiotic
8. Species diversity–variety component	Low	High
9. Species diversity–equitability component	Low	High
10. Biochemical diversity	Low	High
11. Stratification and spatial heterogeneity (pattern diversity)	Poorly organized	Well-organized
12. Niche specialization	Broad	Narrow
13. Size of organism	Small	Large
14. Life cycles	Short, simple	Long, complex
Nutrient cycling		
15. Mineral cycles	Open	Closed
16. Nutrient exchange rate, between organisms and environment	Rapid	Slow
17. Role of detritus in nutrient regeneration	Unimportant	Important
Selection pressure		
18. Growth form	For rapid growth ("*r*-selection")	For feedback control ("*K*-selection")
19. Production	Quantity	Quality
Overall homeostasis		
20. Internal symbiosis	Undeveloped	Developed
21. Nutrient conservation	Poor	Good
22. Stability (resistance to external perturbations)	Poor	Good
23. Entropy	High	Low
24. Information	Low	High

In choosing only two axioms out of many that characterize an ecosystem, we should be well aware of the arbitrary nature of this choice and the resulting incompleteness of the discussion. The aim of this discussion is therefore to show by example which models may be helpful to investigate a complex situation such as this, and to test these models critically. That we choose two axioms of special interest to the geneticist is a consequence of the title of this book but does not imply an emphasis of some axioms over others. Some of these have been discussed in the first volume of the series on Ecological Studies.

a) Competing Species

Consider two species with overlapping niches, i.e., two species the individuals of which compete in a certain phase of their life cycle. In a Gause-Volterra model, the growth of each of these is represented by a differential equation:

$$\frac{dN_1}{dt} = r_1 N_1 \left(\frac{K_1 - N_1 - \alpha\, N_2}{K_1} \right),$$

$$\frac{dN_2}{dt} = r_2 N_2 \left(\frac{K_2 - N_2 - \beta\, N_1}{K_2} \right).$$

Here N_1 and N_2 indicate the population sizes of the first and second species; K_1 and K_2 the carrying capacity of the two species, i.e., the population sizes possible in the limited area if one of the two species were absent; r_1 and r_2 the intrinsic rate of increase; α and β the mutual influences, or, more accurately, the reduction in the increase of N_1 (or N_2) caused by presence of one individual of the other species.

Since competition of the two species depends upon the probability of contact between individuals, each α_{ij} should be measured by

$$\alpha_{ij} = \sum_h p_{ih} p_{jh} / \sum_h p_{ih}^2 .$$

Here the index h stands for the hth environment, and p_i and p_j are probabilities for the appearance of individuals of the ith and jth species, recpectively. The α_{ij} apparently may be interpreted as a regression coefficient that measures the dependence of environmental utilization of one species upon the other species. A regression coefficient of this kind seems quite sensible.

The fraction $1/\Sigma\, p_{ih}^2 = B_i$ is used as a measure of width of the niche by some authors (e.g., LEVINS, 1968). It is seen at once that α_{ij} assumes the value of one with equal niche width of both species, and becomes smaller than one when i has a wider niche but j is more specialized.

For an ecosystem with more than two competing species one could write a set of equations of the form

$$\frac{dN_i}{dt} = r_i N_i \left(\frac{K_i - N_i - \sum\limits_{\substack{j \\ j \neq i}} \alpha_{ij} N_j}{K_i} \right).$$

For a community at equilibrium, the following relationship should be valid for all values of i (only in these conditions is $dN_i/dt = 0$)

$$K_i = N_i + \sum_{\substack{j \\ j \neq i}} \alpha_{ij} N_j.$$

If we introduce the column vector N of the N_i

$$N = \begin{matrix} N_1 \\ N_2 \\ . \\ N_i \\ . \end{matrix}$$

and write as a matrix

$$AN = K$$

then

$$A \equiv \begin{pmatrix} 1 & \alpha_{12} & \alpha_{12} & . & . \\ \alpha_{21} & 1 & \alpha_{22} & . & . \\ \alpha_{31} & \alpha_{32} & 1 & . & . \\ . & . & . & . & . \end{pmatrix}$$

is the community matrix the α_{ij} elements of which constitute the complete set of competition coefficients.

Thus the statement that an ecosystem is in equilibrium condition can now be expressed precisely: the equilibrium of the participating species is adequately represented by the preceding matrix (an effect additional to equilibrium will be discussed in the next section). Therefore, evolution toward the equilibrium, the climax as understood, for example, by plant ecologists, has the ultimate aim of fulfilling these conditions.

It is likely that the coefficients of the community matrix can never be estimated completely and with sufficient precision, at least not for complex communities that would be of greatest interest, and for which the more ambitious estimates concerning stability of ecosystems applied by plant ecologists should be valid. As frequently experienced earlier, the problem encountered here is to analyse and describe a highly complex system; in doing so we wish to express the relationships within the system in as complete a manner as possible. This, however, is possible in principle also without an accurate knowledge of the individual relationships. The formal discussion of the above relationships is already helpful at least in arriving at more precise questions and to make some general statements concerning the properties of such systems.

Next, we ask the question as to how many competing species may exist in an ecosystem that is at equilibrium. For example, this question is relevant to attempts by forest ecologists who wish to increase the stability of ecosystems by introducing new, often exotic, species. First, it is evident that in a one-dimensional niche, i.e., where competition occurs in only one niche dimension, every one of the competing species may possess only very few positive α_{ij}. As niche dimensions increase, the variance of the α_{ij} can be reduced. Therefore, very likely a larger number of species may be expected if competition takes place at several niche dimensions.

From the community matrix it may also be derived that its determinant must be positive at equilibrium. Beyond this, the symmetric determinant must also be positive, i.e.,

$$\alpha_{ij}^* = \alpha_{ji}^* = (\alpha_{ij} + \alpha_{ji})/2 \,.$$

Finally, the same requirement must be met by every subdeterminant arising from the elimination of a row and the corresponding column. Therefore, the means of these determinants can be examined with respect to the statistical distribution of the α values (e.g., see LEVINS, 1968). For example, if one derives the recurrence relation of the expectation D_n of the determinant of rank n (n rows, n columns), then the result is given by the pair of equations

$$D_n = D_{n-1} - (n-1)\,\bar{\alpha}^2\,T_{n-1} - (n-1)\,\mathrm{cov}\,(\alpha_{ij}\alpha_{ji})\,D_{n-2},$$

$$T_n = D_{n-1} - (n-1)\,\bar{\alpha}\,T_{n-1}$$

with initial values of D_0, $D_1 = 1$. T_n is here the determinant of the community matrix in which the first column is replaced by 1. Since covariances of the competition terms reduce the value of the determinant D_n, one expects to find fewer species in a community with niches of approximately equal width than in another community with more dimensions and niches of unequal width. Accordingly, it is to be expected that in an old community, where evolution has long continued in the same direction, the variance of α decreases and the number of species increases (see also MACARTHUR and WILSON, 1967 regarding the evolution of biological communities on islands). On the basis of different yet principally similar considerations, WILLIAMSON (1957, 1958) arrived at analogous results. He described negative density-dependent factors, which influence death rate of populations, as control factors. He concluded that the number of possible species in a community depends in part upon the number of such control factors. On the other hand, these control factors could also give rise to genetic polymorphisms (see section on polymorphisms). Without comitting too large an error, one can probably equate control factors with niche dimensions.

In forestry, it is perhaps of interest to ponder on the question of what happens in a community if one eliminates selectively one or the other species. This process is not rare. Conversely, one may inquire what happens if one introduces new species or when the forester influences the matrix of competition factors or suspends them entirely by regulating competition by thinning or other silvicultural measures (artificial regeneration, etc.). Obviously, the evolution of participating species will then move in a new direction, because the stability of the community matrix in an ecosystem at equilibrium is ultimately guaranteed by continual selection, and this is the result of a guessing game where previously "right" answers are now perhaps "wrong." This will be further discussed in the last chapter.

The coefficients of the community matrix are only valid for a certain environment. This may vary seasonally, which is especially important for organisms with a short life span. In a tropical habitat LEVINS (1968) found that in spite of large seasonal fluctuations of population size the community matrix reflected the equilibrium conditions at least approximately at any time of the year. He emphasized that this result could not necessarily be expected. In long-lived organisms, especially trees, zoned variation of soil and local climate should materially influence the spacial structure of the community. Many examples from plant ecology are known. ELLENBERG (1956) has made a survey of this field. One could account for this situation by compiling sets of matrices for a larger area, which would then provide a better picture of the conditions. Correlations of the coefficients of the same species in different environments are also of interest in this connection.

FISHER (1958) considered competition of two species in the light of game theory models advanced by VON NEUMANN and MORGENSTERN. He concluded that the best adaptive strategy for both competitors is to play a randomly mixed strategy, e.g., to offer different morphs. The study of WILLIAMSON (1958) mentioned earlier had arrived at a similar result. Already DARWIN had come to similar conclusions. LEVIN (1971) investigated this problem on the basis of special examples.

He assumed that every genotype h of the ith species possesses an intrinsic rate of increase r_{ih}, a specific carrying capacity K_{ih}, and an associated group of competition coefficients. This group must account for all possible genotype x species interactions, He further defined the competition coefficient β_{ijhk} as the effect of a single representative of the genotype k, species j, when it narrows the resource base of genotype h of the species i (β_{ijhk} is larger than or equal to 0 for all i, j, h, k). If $i = j$ there is intraspecific competition. The coefficients β_{iihh} must assume the value of one, but not the $\beta_{iihk}(h \neq \mathrm{V}k)$.

Furthermore, when n_{ih} is the number of individuals of the hth genotype of the ith species (there are g_i genotypes in this species), then the change of population size with time is obtained from

$$\frac{dN_i}{dt} = \sum_{h=1}^{g_i} \frac{dn_{ih}}{dt} = r_{ih} n_{ih} \left(1 - \sum_{j=1}^{2} \sum_{k=1}^{g_i} \frac{\beta_{ijhk} n_{jk}}{K_{ih}}\right).$$

The model assumes two asexual species. It is an extension of that given by MACARTHUR and LEVINS (1967) and introduces genotype categories.

What is involved here is the extreme case of selection, termed competitive exclosure, ecological exclosure, mutual exclosure, or similarly. RESCIGNO and RICHARDSON (1965) have shown that suitable models can also be derived for more than two species (see also LEVINS, 1968 and others). They arrive at the well-known example that the number of species cannot be larger than the number of niches. One of the problems that arise here is to investigate whether this result is valid if there are polymorphisms in all, or at least some of the participating species. AYALA (1970) has already shown that with frequency-dependent selective mechanisms, stable equilibria are also possible when the number of niches is smaller than the number of species. This problem is the second of those studies by LEVIN (1972) that are of interest to use here.

The Malthusian fitness of the genotype h of the species i is obtained from

$$M_{ih} = r_{ih} \left(1 - \sum_{j} \sum_{k} \frac{\beta_{ijhk} n_{jk}}{K_{ih}}\right).$$

This means that at any time the fitness of the genotype is determined by its intrinsic rate of increase (r_{ih}), its demands upon the environment (K_{ih}), its competitive attributes, and the total number of individuals in the two competing populations. The preceding relationship is too complex for a general statement of the effects and relations among the factors involved; but for certain restricted situations it may be given in the customary manner.

For a first restriction, it is assumed that genotypes differ only in the sizes r_{ih}. The fitness function is then given by

$$M_{ih} = r_{ih} \left(\frac{1 - \sum_j \sum_k \beta_{ij.k} n_{jk}}{K_{i.}} \right)$$

since all genotypes h possess equal carrying capacity. Consequently, the change of genotype frequencies with time is

$$\frac{dp_{ih}}{dt} = p_{ih} (r_{ih} - \bar{r}_{i.}) \left(1 - \sum_j \sum_k \frac{\beta_{ij.k} n_{jk}}{K_{i.}} \right).$$

In this equation $\bar{r}_{i.} = \sum_h p_{ih} r_{ih}$. Parentheses on the right-hand side enclose a density function that is equal for all genotypes. The product of this function with its intrinsic rate of increase measures its Malthusian fitness. Consequently, there is no change in fitness order of the genotypes as long as the density function remains positive. Only the frequencies of the genotypes change, depending upon the size of the competing populations. If the density function becomes negative, the relations of the fitness values are reversed. Genotypes with a lower rate of increase decrease less rapidly and therefore are better adapted — which is an unrealistic statement having only formal significance.

If the genotypes differ only in values of K_{ih}, then the Malthusian fitness or change of genotype frequencies with time is given by

$$M_{ih} = r_{i.} \left(\frac{1 - \sum_j \sum_k \beta_{ij.k} n_{jk}}{K_{ih}} \right)$$

$$\frac{dp_{ih}}{dt} = p_{ih} r_{i.} \left(\sum_h \frac{p_{ih}}{K_{ih}} - \frac{1}{K_{ih}} \right) \left(\sum_j \sum_k \beta_{ij.k} n_{jk} \right).$$

Here the density function is positive and equal for all genotypes of the species i, and therefore the rank order in fitness of the genotypes is a function of the saturation level. Selection proceeds steadily toward the genotype with highest value of K.

Finally, when the genotypes differ only in the group of competition coefficients, the corresponding functions are

$$M_{ih} = r_{i.} \left(1 - \sum_j \sum_k \frac{\beta_{ijhk} n_{jk}}{K_{i.}} \right)$$

$$\frac{dp_{ih}}{dt} = p_{ih} \frac{r_{i.}}{K_{i.}} \left(\sum_h p_{ih} \sum_j \sum_k \beta_{ijhk} n_{jk} - \sum_j \sum_k \beta_{ijhk} n_{jk} \right).$$

The relative fitness of the genotypes is here determined by their competitive reaction to the opposite group. At every point in time the genotype with the lowest sum of the $\beta_{ijhk} n_{jk}$ is the fittest. However, the order of fitness values as well as the magnitude of changes in gene frequency is determined by frequencies and densities of the competitors (see also TURNER, 1970 for these and other dependencies of the fitness values upon effects not yet accounted for in classical selection theory).

On the basis of the preceding relationship it is now possible to examine the conditions of two competing species simultaneously. It becomes apparent that selection alone for the size of r cannot create a stable coexistence although it may lead to other interesting results. If the genotypes within the species vary only with respect to the values of K, the rate of change in population size for the species i is given by the relationship

$$\frac{dN_i}{dt} = \bar{r}_{i.} N_{i.} \left(1 - \sum_j \frac{\beta_{ij..} N_j}{\bar{K}'_i}\right)$$

where $\bar{K}'_i$ is the harmonic mean

$$\frac{1}{\bar{K}'_i} = \sum_h \frac{n_{ih}}{K_{ih}}.$$

Selection on a variable K should make $\bar{K}'_i$ larger. K therefore can influence the qualitative result of selection (determination of the surviving species) by changing the relations between the competition coefficients and the ratios of K values of the two species. For example, if at any time the parameters typical for elimination of the first species indicate the situation $\beta_{12} > \bar{K}'_1/\bar{K}_2$ and $\beta_{21} < \bar{K}'_2/\bar{K}'_1$, then selection on K could change these in a way that would create conditions for the elimination of the second species. A condition of stable coexistence is also possible, namely, with $\beta_{12} < \bar{K}'_1/\bar{K}'_2$ and $\beta_{21} < \bar{K}'_2/\bar{K}'_1$. However, because these conditions depend upon the specific reciprocal relations between the sizes of K, it is not possible that they are fulfilled by unstable "inner" equilibria of the two species. The final result is always fixation of genotypes with a maximum value of K_{ih}. Therefore, the conditions required for such a stable competition equilibrium can also be described by $\beta_{12} < K_{1\,max}/K_{2\,max}$ and $\beta_{21} < K_{2\,max}/K_{1\,max}$.

At this point readers are reminded that the sequence of indexing is: first, the species considered; second, the species competing with it; third, the genotype considered; and fourth, the genotype competing with it.

If only the competition coefficients vary, the change of population size in two asexual species is given by

$$\frac{dN_i}{dt} = r_i N_i \left(1 - \sum_h p_{ih} \sum_j \sum_k \frac{\beta_{ijhk} n_{jk}}{K_{i.}}\right).$$

In this case the direction of selection depends upon the frequencies of the different genotypes in each of the participating species. The resulting possibilities are too complex for a simple prediction; the frequent opinion found in forestry literature that in each of the competing species there is a genotype that is not susceptible to competition but able to assert itself (unidirectional selection), is based upon a special case. This could be described by the assumption that β_{12hh} and β_{21hh} are smaller than K_1/K_2 or K_2/K_1, respectively. These are the only conditions allowing stable coexistence of the species without inner polymorphisms.

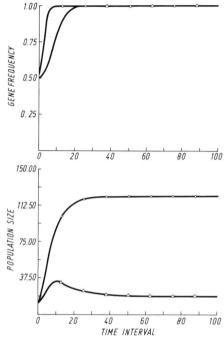

Fig. 41. Selection operating on the variability in saturation level. (After LEVIN, 1971)

$$K_{11} = 80, K_{12} = 20, K_{21} = 130, K_{22} = 80$$
$$\beta_{1111} = \beta_{1122} = \beta_{2211} = \beta_{2222} = 1.0$$
$$\beta_{1211} = \beta_{1212} = \beta_{1221} = \beta_{1222} = 0.5$$
$$\beta_{2111} = \beta_{2112} = \beta_{2121} = \beta_{2122} = 0.4$$

Gene frequencies and population sizes for two competing species

Table 26. Values of the parameters used in runs testing the effects of unidirectional selection on the competitive interaction terms. The first genotype of each species is least sensitive to competitive inhibition. (After LEVIN, 1971)

Fig. No.	Species (i)	Genotype (h)	r_{ih}	K_{ih}	Competitive Coefficients β_{1111}	β_{1112}	. . .			β_{2233}
42a	1	1	0.5	100	1.0	1.0	1.0	0.7	0.5	0.3
		2	0.5	100	1.0	1.0	1.2	1.4	1.2	1.0
		3	0.5	100	1.2	1.0	1.0	0.8	1.6	1.8
	2	1	0.5	100	0.25	0.5	0.7	1.0	1.0	1.0
		2	0.5	100	0.8	1.0	1.6	1.1	1.0	1.2
		3	0.5	100	1.4	1.8	1.2	1.3	1.1	1.0
42b	1	1	0.5	100	1.0	1.0	1.0	0.7	0.5	0.3
		2	0.5	100	1.0	1.0	1.2	1.4	1.2	1.0
		3	0.5	100	1.2	1.0	1.0	0.8	1.6	1.8
	2	1	0.5	100	1.25	1.5	1.7	1.0	1.0	1.0
		2	0.5	100	1.8	2.0	2.6	1.1	1.0	1.2
		3	0.5	100	2.4	2.8	2.2	1.3	1.1	1.0

This will be demonstrated by two examples (from LEVIN, 1971). Table 26 gives two different sets of parameters for two competing species with three genotypes each.

The results of the first set are given in Fig. 42 a. There is coexistence here because a certain relationship exists between the size of the minimum interspecific competition coefficients and the ratio of carrying capacities. The result of the second set is found in Fig. 42 b. Species 2 is here eliminated.

Furthermore, it can be shown that if both competition coefficients and saturation levels are variable, the direction of selection depends upon the frequencies of genotypes in each of the two species. A general prediction again is not possible.

It is also interesting to specifiy the conditions where selection leads to oscillations of population size (reversal of dominance, etc.). Some such cases are specified in Table 27. The conditions to achieve this are restrictive. To obtain more than one reversal in numerical dominance, there must be genetic variability of the β_{ijhk} in at least one of the two species, and also an inverse relationship of interspecific to intraspecific competitive behavior of the genotypes. To achieve

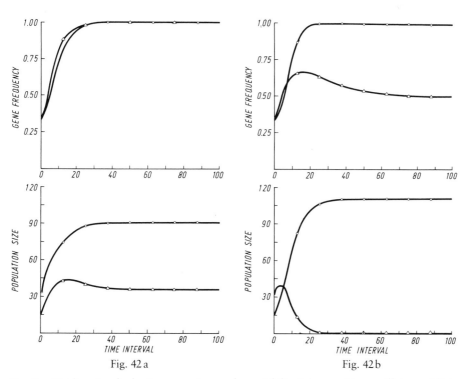

Fig. 42 a   Fig. 42 b

Fig. 42 a. Unidirectional selection operating on the variability in competition coefficients. Minimally sensitive genotypes specifying coexistence. Parameter values from Table 26. Gene frequencies and population sizes for two competing species

Fig. 42 b. Unidirectional selection operating on the variability in competition coefficients. Minimally sensitive genotypes specifying the elimination of species 2. Parameter values from Table 26. Gene frequencies and population sizes for two competing species

Table 27. Values of the competition parameters used in the runs made for the examination of the conditions for continuous reversals in numerical dominance. (From LEVIN, 1971)

Figure	Species (i)	Genotypes (h)	r_{ih}	K_{ih}	β_{1111}	β_{1112}	...	β_{2222}	K_j	K_{ih}
43a	1	1	0.5	100	1.0	0.75	1.0	1.0	0.875	100
	1	2	0.5	100	1.25	1.0	0.75	0.75		
	2	1	0.5	100	1.0	1.0	1.0	0.75	0.875	100
	2	2	0.5	100	0.75	0.75	1.25	1.0		
43b	1	1	0.5	100	1.0	0.5	1.25	1.25	1.0	100
	1	2	0.5	100	1.5	1.0	0.75	0.75		
	2	1	0.5	100	1.25	1.25	1.0	0.5	1.0	100
	2	2	0.5	100	0.75	0.75	1.5	1.0		
43c	1	1	0.5	100	1.0	0.5	1.0	1.0	0.875	100
	1	2	0.5	100	1.5	1.0	0.75	0.75		
	2	1	0.5	100	1.0	1.0	1.0	0.5	0.875	100
	2	2	0.5	100	0.75	0.75	1.5	1.0		
43d	1	1	0.5	100	1.0	0.5	2.0	2.0	1.75	100
	1	2	0.5	100	1.5	1.0	1.5	1.5		
	2	1	0.5	100	2.0	2.0	1.0	0.5	1.75	100
	2	2	0.5	100	1.5	1.5	1.5	1.0		
43e	1	1	0.5	100	1.0	0.75	1.0	1.0	0.875	100
	1	2	0.5	100	1.25	1.0	0.75	0.75		
	2	1	0.5	100	0.5	0.5	1.0	0.875	0.4375	100
	2	2	0.5	100	0.375	0.375	1.125	1.0		

continued reversal of dominance while coexistence is maintained, additional conditions must be met (here for the case of two species with two genotypes each):

1. The fitnesses of the genotypes in both species must be distributed symmetrically around the gene frequencies of 1/2.

2. The mean competition coefficients of both species must be smaller to or equal to their ratios of saturation levels:

$$\bar{\beta}_{12} \leq \bar{K}_1/\bar{K}_2 \quad \text{and} \quad \bar{\beta}_{21} \leq \bar{K}_2/\bar{K}_1$$

where

$$\beta_{ij} = \sum_h \sum_k \frac{\beta_{ijhk}}{4} \quad (\text{for } i \neq k)$$

and

$$\bar{K}_i = 1/4 \sum_h \sum_K K_{ih}\beta_{ijhk} .$$

3. The two species must influence each other in the same way, i.e.,

$$\bar{K}_1/\bar{K}_2 - \bar{\beta}_{12} = \bar{K}_2/\bar{K}_1 - \bar{\beta}_{21} .$$

In Fig. 43a the mean values of β are smaller than the ratios of K (figures of the set 43a in Table 27). In Fig. 43b they are equal to the ratios of K. This results in a decreasing amplitude in the first case and constant amplitude in the second. Fig. 43c shows the case of asymmetrical fitness of the genotypes; continuous oscillations do not occur. If the mean values of the competition coefficients exceed the ratios of K, there is no reversal of dominance (Fig. 43 d), and this also applies with unequal competition effects of the species, although permanent coexistence is possible even here.

The reason for the detailed consideration of the preceding results is to show to what extent the coevolution of competing species in complex ecosystems can be explained when using the conventional Gause-Volterra equations as the basic model. We will leave the question undecided whether the model leads to satisfactory, or the best, explanations possible at this time. Possible differences in competitive behavior of the genotypes with age or developmental stage of the individuals were not considered. Possibilities of alliances noted by FISHER (1958), which may play a role particularly in highly complex systems, were also excluded. To name a special case, root grafting in trees has often been considered as a means of regulating competition. It appears that with great population density all individual trees of a species may be linked to each other by root connections (BORMANN and GRAHAM, 1959; DE BYLE, 1964; KUNTZ and RIKER, 1955; LA RUE, 1934; SAUNIER and WAGLE, 1965; SCHULTZ and WOODS, 1967; YLI-VAKKURI, 1953, and others). The opposite

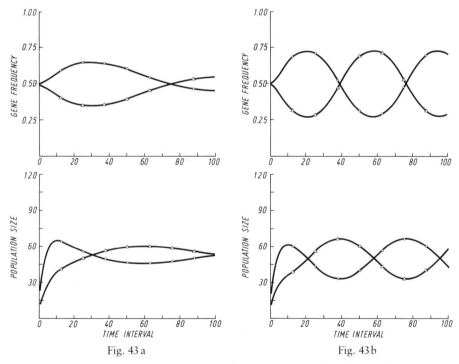

Fig. 43 a   Fig. 43 b

Fig. 43 a–e. Selection leading to continuous oscillations in numerical dominance. Gene frequencies and population sizes for two competing species. a Mean competition coefficients less than ratios of saturation levels. Parameter values from Table 27. (After LEVIN, 1971). b Mean competition coefficients equal to the rates of saturation levels. Parameter values in Table 27. c Asymmetric fitness relationships. Parameter values in Table 27. d Mean values of competition coefficients greater than the ratios of saturation levels. Parameter values in Table 27. e Species unequal in their competitive effects. Parameters in Table 27

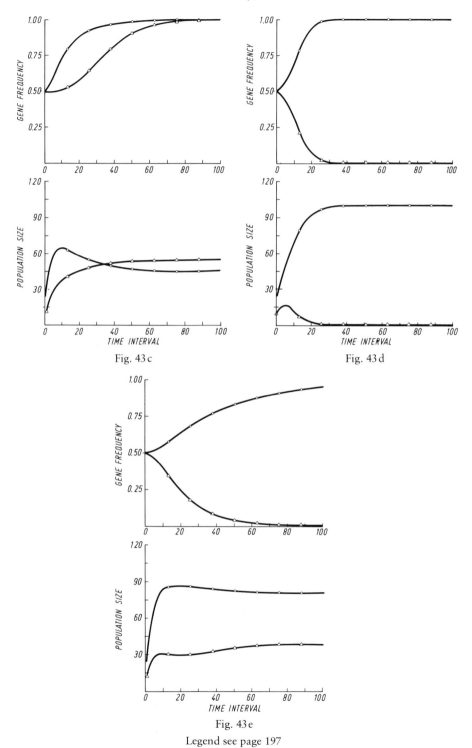

Fig. 43 c

Fig. 43 d

Fig. 43 e

Legend see page 197

effect possibly could be achieved by the soil poisoning of a tropical rain forest tree species in Queensland, a case that was mentioned earlier (see LEVIN, 1969 for theoretical consideration). Similar conditions exist in *Pithecellobium* and *Entrelobium* of Costa Rica (HATHEWAY and BAKER, 1970). These species cannot regenerate in the shade of older trees of the same species. Seeds and young seedlings are found frequently, but trees larger than 1 m only in openings, and thickets only on large open areas, which reflects a behavior typical of pioneer species also found in the tropical rain forest. Another factor that was not considered is the importance of mutual enemies for population equilibria such as were investigated by SLOBODKIN (1961) for animal populations.

More progress could perhaps be made by using other models, and thus at least a few will be noted. PIMENTEL (1965) considered competition and other factors in ecosystems from the viewpoint of cybernetics and pointed out the possibilities of genetic feedback mechanisms. MACARTHUR (1970) surmised that competition equations should be considered as first elements of a Taylor series. The trend of these proposals is to lead to more complex models, which may be more realistic for individual cases, but, because of their complexity, are difficult to handle in an experiment. Spatial distribution must also be regarded in the evaluation of competition and equilibrium as PIMENTEL et al. (1965) have shown in an example with two competing fly species. In doing so, it is not necessary that spacial or areal parts represent different niches.

For the purposes at hand, there is no need for a detailed consideration of these more complex models, as even the relatively simple Gause-Volterra model indicates most of the important difficulties experienced. One of these difficulties is the determination of the fitness of genotypes, or the definition of the fitness value, respectively (WADDINGTON, 1967; TURNER, 1970; KOJIMA, 1971, and others). This is correlated with, but not identical to, the question regarding the definition of the adaptive value of populations that we have encountered repeatedly. Our relatively simple model indicates at least some of the problems that arise. Furthermore, the role of relative frequencies of the competing species can be elucidated; i.e., the significance of initial frequencies of both species for the result of evolution in the ecosystem or, with genetic variability, for the community matrix of the equilibrium community (e.g., NARISE, 1965). Similarly, the model provides at least some answers to the question regarding the origin of species frequencies (e.g., PRESTON, 1962a, b). The distribution of species frequencies is often surprisingly similar in different ecosystems. This observation, which has led to the recognition of the phenomenon of dominance of certain species, has been illuminated anew by MCNAUGHTON and WOLF (1970). In contrast to its definition in phytology, dominance is here understood as a control function that some very frequent species exert over all others. The authors conclude that the problem is best dealt with by using an evolutionary theory, as is attempted in this book. Of central importance in this theory are such concepts as niche and adaptation, although the authors emphasize the unsatisfactory understanding of these concepts and others that are derived from them (e.g., width of a niche; for a definition of the niche and its understanding, see also LEVINS, 1968b).

For forest trees it may appear to be meaningless to apply the value of K in our model. The maximum number of individuals that may be reached per unit area and in the reproductive stage is probably not important. Longevity and a large crown serve the same purpose of assuring presence and reproduction of the species. The concept of maximum security of the population probably carries much more weight than the maximum number of individuals (for the importance of longevity in this sense see, for example, CANNON, 1966). The significance of canopy stratification has already been noted. This is especially pronounced in old communities and those with many species. MARSHALL (1966) has pointed out that in the formal model of the niche every species possesses an axis for territorial requirement, which, at the same time, assumes zero values for other species competing for the same territory. Strata provide a means of enlarging the territory. The mechanism resembles that in bird species where members disregard the division of the same territory by other species such that competition for territory may become strictly intraspecific.

Related to the concept of dominating species is that of the regulator species. This is understood as a species that exerts important regulating functions within an ecosystem and thereby assists in maintaining its stability (DARNELL, 1970, and others). This concept is difficult to apply in higher plants although analogies exist. Thus some species of higher plants may serve as basic food material and thereby assume key positions in the ecosystem. However, the related questions are quantitative rather than qualitative.

Forestry is a field of endeavor characterized by long-range planning, and relative inflexibility in comparison with other fields of economic activity, while fulfilling ecological functions and preserving the landscape. Consequently, questions regarding the balance in ecosystems and how to maintain it in spite of drastic interference by management have long played a large role. This is where the dialectic school found its principal training ground, and it is here that the need for such teaching models as the forest organism could be easily "proven." However, it was not only the dialectic school that has argued (in its characteristic manner) these problems. Certain types of forest ecosystems resisted the methods of classical forestry and required the application of certain procedures (not philosophical principles). This is also valid for countries outside of central Europe. Examples are given in HOEKSTRA's (1961) remarks concerning the management of rich, highly complex deciduous forests with many species at the edge of the Appalachian Mountains. An example, again one of many, for a view that considers forest genetics in categories of ecosystems is provided by the work of SPURR (1960). The author discusses the risk of management in different forest ecosystems, the significance of the evolutionary stage when evaluating them, and the possibilities available to the forest geneticist to guide this evolution, for example, by introducing exotics into communities with few species, etc.

b) Coevolution of Host–Parasite Systems

In addition to the types of coevolution in ecosystems already discussed — pollen and seed dispersal by animals and competition — the coevolution of host–parasite systems is also of interest for an understanding of conditions in ecosystems. An attempt will be made to present a rapid survey of the subject. Other types of coevolution will not be discussed, interesting though they may be in the context of forest ecosystems, such as the coevolution of tree species and micorrhizae. The reader will recognize additional possibilities without difficulty when examining the classification of mutual influences between two organisms in an ecosystem such as has been developed by STAKMAN and HARRAR (1957).

Species A

Species B   + +   + 0   + −
                      00   0 −
                      − −

The authors designate the six possibilities as follows:

+ +   synergism
+ 0   metabiosis
+ −   competition
00   neutrality
0 −   antagonism
− −   mutually negative relationship

It is likely that host–parasite relationships always belong in the category + −, which was here somewhat unsuitably designated as competition; there are, in fact, many difficulties in designated conditions in ecosystems by means of such simple categories. We thus are well advised to define the conditions of host–parasite systems in a more precise manner.

Let us assume that species A represents the source of food or one of the principal sources of food of species B. A may then produce food for B in different ways, which can also have different consequences for its own niche. In one way B may be indispensable for the breakdown of certain products of A, for instance, for the decomposition of its leaves, which, when not accomplished, could destroy the livelihood of A. However, the "saprophyte" B may become a threat to A, such as is known for fungi that normally live on dead tree branches that rot and drop to the ground, probably to the advantage of the tree species, and then in special circumstances become "parasites" by attacking living tissue of the host species, causing the death of these individuals. An example is the fungus *Valsa nivea*, a harmless saprophyte on European aspen, which becomes a dangerous parasite on the hybrid *Populus tremula* × *tremuloides* in certain areas. Similar examples have become known from micorrhiza fungi. Therefore, there is no sharp and clear distinction between parasites and saprophytes. Environment as well as the physiological condition of the "host" determine the relations of organisms A and B in many, if not most, cases. We shall therefore accept the rule that in the simplified host–parasite systems such as considered here, the relationship + − will always be valid, i.e., species A will be damaged in the presence of B. The degree of damage in turn depends upon the environment, which may influence the frequency of B. Weather conditions of certain years may lead to an increase in the frequency of B so that it poses a threat to A, whereas in other years B acts as a harmless parasite on weak population members. Fluctuating density of the population of a species D, i.e., of a "hyperparasite" that controls B, could have the same effect. These conditions have been described by many authors in greater detail than is possible here — by ANDREWARTHA and BIRCH (1954), or from a forestry viewpoint by SCHWERDTFEGER (1963, 1968). For our purposes the discussion can be limited to simple cases.

Parasitic behavior of a species often presupposes extremely specialized adaptations of the parasite to the peculiarities of a single or only a few hosts (see Section II. 1. 2). The adaptive phase to the host and consequently of the parasite also is determined by a single or very few genes. For a comparable situation, FLOR (1954) developed a gene-for-gene hypothesis, i.e., a model where for every "resistance gene" of the host there is a winning "virulence gene" of the parasite. This model has greatly stimulated the development of ideas regarding coevolution in host–parasite systems by providing a simple basis upon which theory can be built in several directions. For example, cases are known where, with complementary action of the alleles of a heterozygous locus for resistance, both biotypes of the parasite are dominant (BRÜCKNER, 1967, and others); or where epistasis occurs among alleles of different loci that determine resistance (compare WILLIAMS, 1964; WRIGHT, 1958, and others).

In view of all the population genetic studies dealing with complicated equilibria that are determined by intra- and interspecific competition, heterozygote advantage, and other factors, it is appropiate to examine the conditions determining equilibria in host–parasite systems, which are determined by R genes in the host and V genes in the parasite. MODE (1958, 1960, 1961) was probably the first to define special cases where certain sets of R genes and V genes may give rise to host–parasite equilibria. In several investigations PERSSON (1959, 1966, 1967) found evidence for the presence in nature of at least some of such cases that met MODE's theoretical requirements. In species with large ranges a component of space appears. Because the migration rates of the R and V genes are limited, conditions may vary in different parts of the range (PIMENTEL, et al., 1963). However, where the host is a tree species, all these models for host–parasite systems may have limited value, as foresters have pointed out (HATTEMER, 1967). Here the extreme difference in generation length of host and parasite must be taken into account.

It is possible that the parasite produces and tries 100 generations during a single generation of the host. Our first question is: Are there major genes for resistance in tree species? The answer is clearly in the affirmative. SOEGARD (1966), for example, reported on resistance of two *Thuja* species to a virulent fungus attacking during the seedling stage. In the generations following species crosses he clearly obtained a Mendelian segregation pattern. Furthermore, even the statement that supergenes could have been or are likely to be involved, does not weaken the result since this does not change the evidence materially. KINLOCH et al. (1970) provided another example from the five-needled *Pinus lambertina* in the western United States. The species is extremely susceptible to the fungus *Cronartium ribicola* (which was introduced from the Old World) like all five-needled pines in the New World except those in areas where environment is unsuitable to the disease (VAN ANSCHL et al., 1961, and others).

The second question then is whether major genes can be the cause of an equilibrium in a host–parasite system. This question is particularly important in situations where the parasite enjoys clear advantages on the basis of its genetic system, such as *Cronartium ribicola* against the five-needled pines, extremly rapid sequence of generations, and infection by means of haploid spore populations where every mutation and recombination is offered directly and mass-produced vegetatively without the shelter of dominance. In the Old World, well-functioning genetic equilibria appear to exist between *Cronartium* on the one hand and *Pinus cembra, P. peuce,* and *P. griffithii* on the other. A good survey of the problems with this disease is given in the symposium edited by BINGHAM (1972). It is improbable that this system can lead to an equilibrium here on the basis of interactions between sets of R genes and V genes, even if a large number of R and V genes is involved, and even if there is a spacial component. Selection is admittedly extremely effective in parasites (see ROANE et al., 1960 for an example) and could lead to the loss of certain V genes in a few years, yet the special devices of the genetic system of parasitic fungi and other microorganisms for the development and mass reproduction of new genetic variants should make up for this.

PIMENTEL (1968) describes the interplay of the genetic systems of host and parasite as a genetic feedback system. Most of the models mentioned in the previous section could be applied on the basis of this theory. This certainly is one of the more general models for the description of the mutual influence between the two systems. Using it, one could attempt to derive the conditions that are necessary to obtain fairly stable equilibria in host–parasite systems where trees act as hosts, including particularly the avoidance of large fluctuations in host populations. This attempt has in fact been made several times. Several examples have already been cited; they belong to this group because ultimately all information from the past is genetic feedback. ROSENZWEIG and MACARTHUR (1963), for instance, gave examples of the stability conditions for predator–prey systems. OKABE and HASHIGICHI (1968) tried to determine conditions for optimal mixtures of multiple-line varieties with the aid of the theory of games. In forestry this problem exists primarily, but not only, in populations reproduced by clones. The main result of all of these investigations seems to be that in simple gene-for-gene systems and major genes of the R and V type, stability is hardly possible if the host is a tree. In contrast, polygenic inheritance on both sides should provide suitable conditions for stable systems of this type. In some circumstances these should be accompanied by physiological and biochemical threshold values for susceptibility, which are subject to both genetic and environmental effects. For example, the lower terpenes, the genetic variability of which was discussed in relation to another problem, could well constitute a qualitative basis for a resistance system in the conifers. In addition to the findings in *Pinus taeda* by SQUILLACE (1971) cited above, the results of other authors confirm the existence of genetic polymorphisms of this type, e.g., HANOVER (1966) and WILKINSON et al. (1971), who, like SQUILLACE, find frequency clines in *P. strobus.*

To cite only one result pointing in this direction: MÜLDER (1953, 1955) compared the intensities of disease attack and disease frequencies in *Pinus sylvestris* by *Peridermium pini* with that in *P. strobus* by *Cronartium ribicola.* In *P. sylvestris* there was no attack at seven years of age in the 5000 trees of a stand; at the age of 17, 0.4% of 2000 trees bore one infection court each. At age 33 of 600 trees, 5% were attacked with a maximum of four infection courts; at age 40, 25% were diseased and there were as many as 54 infection courts on an individual tree. By means of graphic presentations the author supported the hypothesis that one lot of trees was still above the threshold of susceptibility, but the other, diseased lot differed gradually in susceptibility as indicated by the number of infections per tree, which had already been observed by HAACK (1914). Haack confirmed his assumption of gradual differences in resistance by making artificial inoculations on infected, less infected, and healthy trees. On the other hand, MÜLDER's (1955) study with *P. strobus* was concerned with a very young host–parasite system. Infections leading to mortality were often recorded, even in seedlings, and infection percentages were higher in all age classes than in *P. sylvestris* with respect to *Peridermium pini*. Unfortunately, no data are available for comparison of the old host–parasite systems of the Old World with the five-needled pines as host and *Cronartium ribicola* as parasite. Presumably the parasite is only as harmful to the host population as *Peridermium pini* to *P. sylvestris*, or even less.

Forest genetics therefore must interpret "old" and "young" host–parasite systems differently, particularly with respect to chances of breeding for resistance. Selection of resistant forms in old systems could even be detrimental under certain conditions by disturbing the equilibrium, but selection of resistant forms in young systems may make it possible to reach an equilibrium between host and parasite or may accelerate evolution in this direction. STERN (1969) has reviewed the considerations regarding this problem.

Forms of the host that are resistant against a new parasite (in many cases this may be an introduced parasite) are often so rare that they must be counted with the group of preadapted genotypes of the host population. In some cases they may be carriers of rare single genes; it is not unusual that these types are at a selective disadvantage before the parasite appears. The parasite itself may be quite generally considered as introducing a new selection principle. Unfortunately, relevant examples from forest trees and their parasites are unknown to us and we therefore depend upon results from analogous conditions in other organisms. FENNER (1964), for instance, reported on the coevolution of the rabbit and the *Myxoma* virus, which severely decimates the host population when it enters a new territory, a process that may be repeated in the same territory. Apparently every new wave of attack leaves some resistant forms but these are at a selective disadvantage in normal conditions, i.e., when the virus is absent. An equilibrium is then only very gradually developed, partly through accumulation of the resistant forms and modification of the effects of the resistance genes of the host, and in part possibly also through selection of less dangerous forms of the parasite. HUWALD (1965) described the evolution of resistance in *Tetranychus urticae* against tetraethyl pyrophosphate. In such cases of resistance against an abiotic agent, conditions are simplified since only one side reacts to selection. He found that the frequency of resistance genes, accumulated by continuous selection, diminishes rapidly when selection ceases. Therefore, *R* genes are not neutral with respect to other selection principles. For example, the TEP-resistant forms are not as resistant as other forms to shortage of nutrients. The previous history of the population also plays a role — selection for TEP resistance on one or several earlier occasions.

Examples for successful preadaptation against abiotic agents are also known from forest trees. HOFFMANN and BERGMANN (1966) report on differential resistance of Scots pine clones against a herbicide (Dalapon) and VOGL et al. (1968) on variation in resistance of *Larix leptolepis* provenances to industrial fumes.

Here again the importance of genetic variability for the functioning of host–parasite systems becomes apparent. Investigations of the variability of parasites on forest trees are still in their infancy, nevertheless, there is no doubt that such variability exists, including variability in virulence. For example, HOLMES (1965) crossed two compatible races of *Ceratocystis ulmi*, the causal agent of Dutch elm disease. There was considerable genetic variation in the progeny of the fungus, not only in morphological character but also in virulence against two clones of the host, one of which was considered resistant and the other susceptible (against previously tested lines of the parasite). Some progenies from this biparental cross of the parasite attacked both clones, others only the susceptible one. On the other hand, progenies of a cross of two randomly selected biotypes from Holland were more homogeneous. Characteristically, there were also pronounced interactions of virulence between test sites and clones of the host. Therefore, the coevolution of host and parasite in forest trees is without doubt influenced by the genetic variability on both sides. Consequently, disturbances of the established equilibrium may be as catastrophic as is known from the history of agriculture and horticulture. Sudden outbreaks of diseases in clonally propagated trees, such as poplars, provide evidence; there are also examples from shrubs, formerly members of forest ecosystems, that have recently been subjected to agricultural cultivation. GOHEEN (1953) describes the history of highbush-blueberry, which is found in large numbers in its natural habitat but is practically free of serious diseases. The prevailing idea was that this shrub was not subject to any diseases and this impression persisted as long as cultivation was restricted to small and isolated fields. However, modern forms of agricultural utilization, characterized by clones (of genetically uniform material) and large fields, have completely refuted this assumption. Several of the "harmless" parasites have become serious pests.

Since little is known about genetic variation in harmful animals of the forest, we shall indicate the gaps in our knowledge by at least referring to some examples from the literature on enemies of crop plants that have been more closely investigated. We begin with geographic variation. It is known from many species that insects are locally adapted; *Drosophila* has been the standard object of the related experiments. One of the first authors in this field, SPAGER (1938), found evidence for the existence of locally adapted "races" of the moth *Cheimatobia brumata*. These races are adapted especially to the specific weather cycles and extremes of such specific localities. We shall leave undecided here the actual value of general formulas for developmental cycles of harmful insects, in view of local adaptation of the genetic mechanisms that control these cycles.

The first indication that genetic variation influences fluctuations of the size of insect populations probably came from FORD and FORD (1930). They investigated a colony of *Melitea aurinia* in Cumberland over a period ranging from 1881 to 1935. The species was frequent until 1897, but then decreased markedly. Between 1917 and 1919 it was one of the rare insect species of that region. Subsequently its frequency expanded while its genetic variation within the population also rose. As the population reached steady size, it became more uniform and was distributed around a new mean phenotype. The authors explain the increase of genetic variation during the growth phase as the result of reduced intraspecific competition. Whether this assumption is correct or fully explains the observations is an open question.

Interesting, in any case, is the fact that the mean genotype was changed after the phase of population growth that could be interpreted as a change in direction of selection. In a vole species, TAMARIN and KREBS (1969) discovered a different direc-

tion of selection in the growth phase of the population than in the phase of population decline as indicated by a transferring-locus with identifiable isoalleles. GERSHEN-
SON (1945) investigated a melanism in hamsters and found that the frequency of the
melanic form depended upon enemy frequency and its own population density
(intraspecific competition?). Be that as it may, HUXLEY's (1956) supposition that an
understanding of population dynamics requires insight into the genetic architecture
of populations appears to be basically correct. In recent times especially FRANZ
(1964) and CARSON (1967) have emphasized this. The latter author in particular has
considered the population genetic background of fluctuations in population size.

He assumed that periods of exceptional population growth occur in many
populations and are conditioned mostly by environment. Since every population is
subject to stabilizing selection, which is at least partly discontinued during such
phases, a maximum of population size must coincide with a maximum of genetic
variation. This is to be expected in species with a high recombination index that are
capable of generating genetic variants in large numbers within a short time. What
takes place here is, in principle, a revaluation of genotypic fitness although elements
of the genetic load are still eliminated. Population growth may also start new
colonies beyond the existing range. New evolutionary centers may be created that
lead to new forms of adaptation. Based on his own results and those from the
literature, the author also provided an interesting discussion on how the population
balances its (optimal?) size as it reaches a certain genetic constitution. As strongly
inbred fly populations became adapted, they reached a constant size with little
oscillation. After the introduction of a single new fly the population size increased
rapidly, reaching a peak after approximately nine generations. Subsequently, there
were oscillations around an equilibrium point situated slightly below the previous
one. The author also gives examples of genetic equilibria in populations with
seasonal cycles in population size.

In natural populations of harmful forest insects such experiments are difficult.
However, they have seldom been tried. They are also complicated through the
presence of the frequent hyperparasites. PIMENTEL and AL-HAFIDH (1965) studied the
coevolution of a housefly population and a population of its predator *Nasonia
viripennis*. Pupae of the housefly were exposed to females of the predator wasp for
a period of 24 hours each, and the host population reproduced only from the pupae
that had remained free of attack. This resulted in a more resistant housefly population and a population of the predator where the number of potential progenies per
female had been reduced from the original 135 to 30.

The adaptive value of genotypes of a harmful insect is often determined by
utility of colors and patterns for camouflage, including the extreme case of mimicry. KETTLEWELL (1956) has investigated such characters in the typical forms of the
(harmless) butterfly *Cleora rependata* found in the vestigeal native pine woods of
Scotland. He found that 10% belonged to the form *nigrina*, which closely resembles
the melanic form *nigra* of industrial areas, although there is no industry near the
woods where the study was made. *Nigrina* is more easily visible when at rest on a
pine stem but not as much endangered while in flight the stage when insects are
predominantly caught by birds. BROWER and BROWER (1962) reported on a study of
Papilio glaucus, one of the few (although insignificant) insects attacking *Liriodendron tulipifera*. This butterfly is a member of a mimicry complex imitating *Battus*

philenor. B. philenor is avoided by enemies because of repulsive substances. The frequency of mimic forms of *P. glauca* indicates a distinct and clinal dependence upon the frequency of the model *(Battus).*

In the previous example of the form *nigrina* of *Cleora rependata*, the relative safety of the butterflies during flight depended in part upon the speed and pattern of flight, i.e., behavior characters. These are subject to strong genetic control and include tendencies of animals to separate from the group or to congregate during specific activities, to prefer or avoid certain localities according to definite criteria, etc. Behavioral genetics has so far been little used to explain population genetic phenomena and research is fast beginning; no attempt will be made here to correct the situation. Two additional examples may be sufficient to indicate the significance of behavior for an explanation of conditions in ecosystems. DEL SOLAR (1968) found that eggs are predominantly laid in clusters by several species of *Drosophila*: a female prefers a spot where other females have previously deposited theirs. As a result, many suitable spots are never used. Selection in three populations for clustered and isolated egg deposition was successful in all three cases. The importance for forest ecosystems of spot selection may be illustrated by the study of HATTEMER et al. (1969). Females of *Diprion pini* chose spots on pines where eggs were laid depending upon the tree species next to potential hosts. If there were non-host five-needled pines near one of these pines, the pine tree was avoided. In the discussion of conditions for biological equilibria in forest ecosystems, which is now centuries old, this phenomenon has received little attention, although the role of harmful animal agents has been recognized. Behavioral characters are definitely influenced by genetics, and an immense genetic variation must exist within populations.

Characters of behavior are also involved in host preference and may even lead to the formation of new parasite species. CLUTTERBRUCK and BEARDMORE (1961) added peppermint or juniper oil to the food medium of *Drosophila melanogaster*. After very few generations females preferred the food source on which they had been raised. In nature similar examples were found by SINGER (1971) for the butterfly *Euphydras editha* and by BUSH (1969) for *Diptera*, genus *Rhagoleties*. The last example is especially interesting. Pairing and mating on the host plant of the larvae produce a direct correlation of host and partner selection. Newly developed races on certain hosts or even sibling species may be explained in this way (sympatric speciation). An important prerequisite for the functioning of host–parasite systems is often an adjusted distribution of the two partners (ATSATT 1965, and others). If the parasite expects an uncertain environment for the next generation, the answer of the population may be a genetic polymorphism. For instance, ATSATT and STRONG (1970) found that annual grassland hemiparasites are confronted by an environment changing from year to year, resulting from modification of the frequency and dominance of host plants. In this way individuals of the hemiparasite *Orthocarpus* may occur by chance on host plants of different species, genera, and families. The genetic variability of the hemiparasites therefore reflects mainly interactions with its own genotypes and with those of the hosts; this is only partially cushioned by autotrophy. Reactions to different hosts may vary considerably among the hemiparasites. *Perentucellia* benefited from all hosts; *Bellardia* benefited only from one and even suffered losses in combination with three others. The beginning of specialization becomes very obvious here.

Coevolution of host–parasite systems is surely one of the most important phases in the evolution of ecosystems. The major problems could be discussed only with reference to some selected examples, but it may have become clear that while the general problems are much the same, there is a multitude of biochemical, physiological, and ecological variations of the basic theme. Detailed research is therefore the basis of the development of knowledge, especially here, exemplified perhaps by the symposium edited by Fuchs and van Andel (1968) on the fundamental aspects of the coevolution of parasite and host.

3. Three Principal Types of Forest Ecosystems

Forest ecosystems may be classified according to different points of view. Principles of such a classification could be found in the cited general literature or the special forestry study of Galoux. Contributions to the compendium on ecosystem models as a basis of resource management (edited by van Dyne 1969) indicate the possible differences in emphasis of ecosystem characteristics. A similar volume with more attention to methods has been edited by Watt (1966). The biological side of the problem perhaps has been best described by Kershaw (1964) and an excellent survey of the formal basis is given by the contributions to the symposium on systems theory and biology that was edited by Mesarovic (1968). For an introduction to mathematical and statistical methods that are used in the description of parts of ecosystems the reader is referred to Pielou (1969).

Although it is possible to classify forest ecosystems more logically, we shall not attempt this here. Instead, we shall select for discussion three principal types of forest ecosystems that have played a large role in forestry literature and perhaps indicate the different genetic backgrounds that different ecosystems provide for the adaptation of tree species. Our basis will be the four major ecoclines of ecologists (who do not define "ecoclines" in the same way as geneticists), as these are shown by Whittaker (1970) (Fig. 44). These ecoclines refer to ecological gradients along which there is a systematic change of the principal variables of first of all, the abiotic environment.

The ecocline depicted in Fig. 44a follows a geographic gradient of water supply. The example chosen is a transect from the Appalachian Mountains in the southern United States to the Sonora Desert at the border of Mexico. The extremes consist of the mesophytic forests in the east and desert vegetation in the west. An ecocline determined by water supply in tropical South America has a tropical rain forest as one of its extremes and, again, desert vegetation with its special adaptations as the other.

Another ecocline, mainly defined by temperature, characterizes the change in vegetation from the foothills to the peaks of mountains. Our example represents such an ecocline in a tropical mountain area (Fig. 44c). The south–north ecocline in forests with sufficient moisture but diminishing temperature and length of the vegetative period is similar in principle (Fig. 44d). Additional examples for ecoclines could be found.

A comparison of the conditions along such ecoclines indicates that some general conclusions can be drawn from the systematic decrease in humidity, temperature, and other factors. For simplicity, Whittaker (1970) will again be cited:

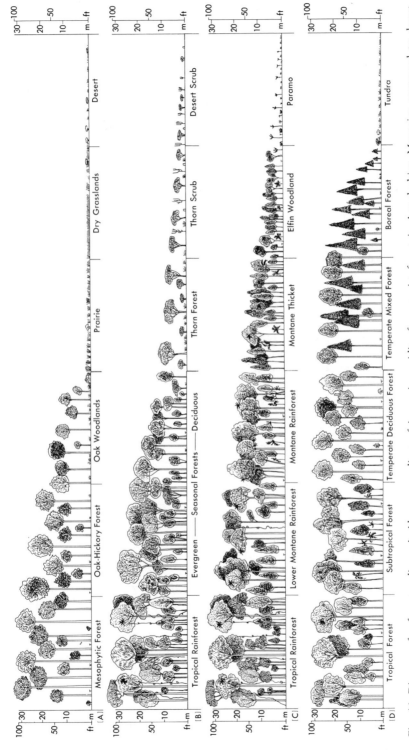

Fig. 44. Diagrams for four ecoclines. A Along a gradient of increasing aridity from moist forest in Appalachian Moutains westward to desert in the southern United States. B Along a gradient of increasing aridity from rainforest to desert in South America. C Along an elevation gradient up tropical mountains in South America. D Along a latitudinal gradient. (After WHITTAKER, 1970)

"1. Along a gradient from a 'favorable' environment to an 'extreme' environment there is normally a decrease in the productivity and massiveness of communities. The decrease in amount of organic matter per unit area is expressed in decrease of height of dominant organisms and percentage of the ground surface covered. Thus on land a climatic ecocline may lead from a high forest with a dry weight biomass exceeding 40 kg/m^2, canopy tree height of 40 m, and coverage (counting overlap of different strata of trees, shrubs, and herbs) well above 100 per cent, to a desert with a biomass less than 1 kg/m^2, plant height below 1 m, and plant coverage less than 10 per cent.

2. Related to these are gradients in physiognomic complexity. Toward increasingly unfavorable environments there is a stepping-down of community structure and a reduction of stratal differentiation with generally smaller numbers of growth-forms arranged in fewer and lower strata.

3. Trends in diversity of structure are broadly paralleled by those in diversity of species. In general (but with exceptions as regards both particular ecoclines and particular groups of organisms), alpha and beta species-diversities decrease from favorable to extreme environments, whether the latter are extremes of drought, or of cold, or of adverse soil chemistry, or (for the sea coast) of tidal exposure.

4. Each growth-form has its characteristic place of maximum importance along the ecoclines-rosette trees in some tropical forests, semishrubs in desert and adjacent semiarid communities, and so on. (Some growth-forms, for example, grasses and grasslike plants, may have more than one area of importance along the major ecoclines.) A growth form, like a species, has dual aspects of adaptation to niche and habitat. Growth forms are both broad-niche categories among plants and broadly significant expressions of plant adaptation to physical environment.

5. The last observation implies that the same growth forms may be dominant in similar environments in widely different parts of the world. Because of this fact, along with the relationships in points 1 and 2, similar environments on different continents tend to have communities of similar physiognomy. This adaptive convergence at the level of the community is one of the major generalizations about the geography of life."

In selecting the three types — tropical rain forests, forests of the temperate and boreal zone, and forests in northernmost regions of the boreal zone (the subarctic) — we are following essentially the south–north ecocline of ecologists. In particular, three ecological-genetic problems will be examined: the great complexity in species composition and stand structure of the tropical (and subtropical) rain forest; the phenomenon of succession, which is most significant for the tree species of the temperate zone and adjacent part of the boreal climatic zone; and the problem of adaptation to an extreme environment that requires persistent recolonization and fitness in the extremely fluctuating environment of the subarctic zone.

a) Forest Ecosystems in the Humid Tropics and Subtropics

The wealth of species in humid tropical and subtropical forests has always fascinated botanists — and not only taxonomists but also ecologists, geneticists, and students of evolution. Much research remains to be done in these forests, not the least with respect to their rational economic utilization.

The first attempt to treat the problems of the tropical rain forest in mono-graphic fashion was made by RICHARDS (1952). Among the regional monographs, that are particularly suitable for our purposes is the study of ASHTON (1964) on species-rich Dipterocarp forests in Malaysia. Figure 45 was taken from this work to illustrate the wealth of species and stratification of such complex forest types.

Other regional monographs include the ecological investigations of the rain forests in southern Nigeria by JONES (1955, 1956). The number of publications has increased in recent times in response to the growing interest in utilization of the forests in the humid tropical and subtropical areas.

The wealth of species that characterizes such ecosystems literally provokes an evolutionary explanation. Initially it was difficult to comprehend the existence of so many competing species in the same habitat. The first illuminating answer to this question was probably given by DOBZHANSKY (1950), when he referred to the great variety of niches found in tropical rain forests of the Amazonas. A contribution on the evolution of tropical rain forests is also found in the symposium on evolution in the tropics edited by HUXLEY et al. (1954). This contribution (CORNER 1954) largely confirms DOBZHANSKY'S assumption. The variety of niches probably depends to a significant degree upon crown stratification, which is more pronounced here than in any other forest type. MARSHALL (1960) has pointed out that stratification allots living space to many different species and species groups (compare Fig. 45).

This still leaves the question open as to how it is possible that many, often closely related, species compete in the same stratum (see Fig. 45). This problem (the occurrence of related species in the same area and same community) has been discussed in relation to another subject, reference is made to the older study of DIVER (1940), and that of WILSON (1964), which is of special interest here. Unless sympatric speciation is assumed, every newly developed species must have gone through a colonizing stage before finally occupying its range. The author indicated the conditions that allow an immigrant species to assert itself (in some circum-stances this may be a close relative of the present species with which it may cross) by using a theoretical model given by BOSSERT (1963). This problem will be discussed later.

Until recently, the prevailing opinion apparently was that sympatric speciation is the only possible explanation for the multitude of species in the humid tropical and subtropical forests. The work of FEDOROV (1966) is typical for this view of the special features of evolution in the humid tropics and subtropics. This view, which is still widely held today, definitely involves some correct premises.

FEDOROV assumes that in the temperature and boreal zones closely related species usually grow apart from each other (e.g., parapatrically). Upon close examination, ecological isolation of the species is always found. The situation is different in the humid tropics and subtropics, especially in southeast Asia, but also in Africa and South America. Furthermore, the tree species in these forests belong to genera that are restricted to tropical forests. Thus, in the genus *Shorea* (all subsequent genera are limited to the tropics) there are 167 species; in *Hopes*, about 100; in *Dipterocar-pus*, about 80; in *Vatica*, 87; in *Balanucarpus*, 20; in *Anisoptera*, 14; and in one Dipterocarp genus endemic to Ceylon, 12 species. He points out that in other tropical tree genera there are even more species; in *Ocotea*, for example, there are 697. Within families and genera, the species usually can be grouped along evolu-tionary lines.

The author states that it is incorrect to assume a distribution of the multiplicity of species — e.g., in the genus *Shorea* to different strata. Therefore, the possibility that the ancestors of two present species were previously isolated ecologically is excluded. A large majority of the species in *Shorea*, for example, is found in the uppermost stratum. Thus we must assume that species are isolated from each other by special mechanisms, and that each of them occupies a special niche. The following types of distributions are found in four genera of the Dipterocarp family on the Malay Peninsula:

1. One group of species, particularly endemic species, is found only in few localities.

2. A large group of species, including endemic as well as widely distributed species, occurs in small groups.

3. Some species are characteristic for high elevations or low elevations in mountains. However, many species are distributed in forests of the whole peninsula; these never occur in larger numbers and their population density seldom exceeds one or two individuals per acre.

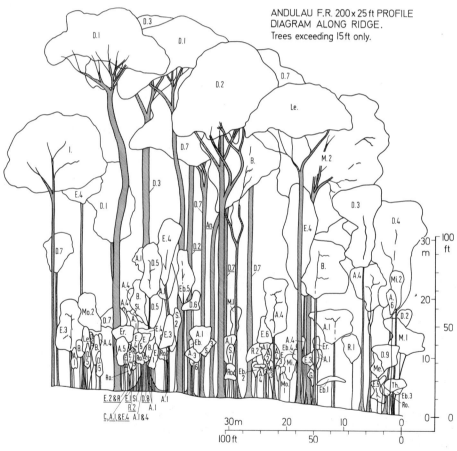

Fig. 45. Legend see opposite page

Fig. 45. Andulau F. R. profile diagram along ridge. (After ASHTON, 1966)

Anacardiaceae
A.   1. *Mangifera havilandii*
          (11 individuals)
A.   2. *Mangifera sp.* (1)
A.   3. *Melanorrhoea torquata* (1)
A.   4. *Parishia? insignis* (8)
A.   5. *Semicarpus rufovelutinus* (1)
A.   6. *Swintonia schwenkii* (1)

Anonaceae
An.   *Polyalthia sumatrana* (1)

Burseraceae
B.   *Dacryodes expansa* (6)

Celastraceae
C.   *Lophopetalum subobovatum* (1)

Dipterocarpaceae
D.   1. *Cotylelobium melanoxylon* (13)
D.   2. *Dipterocarpus globosus* (3)
D.   3. *Dryobalanops aromatica* (2)
D.   4. *Shorea acuta* (1)
D.   5. *S. dolichocarpa* (5)
D.   6. *S. geniculata* (1)
D.   7. *S. multiflora* (5)
D.   8. *S. ovata* (1)
D.   9. *Vatica micrantha* (1)

Ebenaceae
Eb.   1. *Diospyros buxifolia* (1)
Eb.   2. *D. ferruginea* (1)
Eb.   3. *D. hermaphroditica* (1)
Eb.   4. *D. bantamensis* (1)
Eb.   5. *D. toposoides* (1)
Eb.   6. *D. indet.* (1)

Ixonanthaceae
Er.   *Ixonanthes reticulata* Jack (1)

Euphorbiaceae
E.   1. *Agrostistachys leptostachya* (1)
E.   2. *Aporosa elmeri* (1)
E.   3. *Cleistanthus winkleri* (4)
E.   4. *Koilodepas sp.* (8)
E.   5. *Mallotus griffithianus* (2)
E.   6. *Pimeleodendron griffithanum* (2)

Guttiferae
G.   *Garcinia parvifolia* (1)

Icacinaceae
I.   *Platea fuliginea* (1)

Lecythidaceae
L.   *Barringtonia sp.* (2)

Melastomaceae
Me.   *Memecylon sp.* (1)

Moraceae
Mo.   1. *Artocarpus odoratissimus* (1)
Mo.   2. *Prainea frutescens* (1)

Myristicaceae
Mi.   1. *Knema cinerea* v. *patentinervia* (1)
Mi.   2. *Knema kunstleri* (1)
Mi.   3. *Myristica lowii* (1)

Myrtaceae
M.   1. *Eugenia sp.* (2)
M.   2. *Whiteodendron moultonanium* (1)

Olacaceae
O.   *Gonocaryum sp.* (1)

Rosaceae
Ro.   *Parastemon urophyllum* (4)

Rubiaceae
R.   1. *Canthium sp.* (1)
R.   2. *Gardenia tubifera* (1)
R.   3. *Randia jambosoides* (1)

Sapotaceae
S.   1. *Madhuca crassipes* (1)
S.   2. *Payena sp.* (1)
S.   3. *Payena lucida* (3)

Simarubaceae
Si.   *Eurycoma longifolia* (1)

Ternstoemiaceae
Te.   *Adinandra cordifolia* (1)

Thymeliaceae
Th.   *Gonystylus velutinus* (1)

This distribution does not permit the conclusion that there is partial geographic or ecological isolation. FEDOROV explained that the outstanding characteristic of the tropical rain forest is low population density of the species, and consequently low effective population size. A greater frequency of one species and larger effective population sizes are really found only in the lower strata. In the upper stratum even the dominant species occur only in small groups. The tropical rain forest also lacks seasonal variation. The result is that rhythms of flowering and fruiting are irregular. Many species flower rarely, perhaps once every five to ten years. Few seasonal ties exist. Not only is it true that closely related species flower during the same period, but this also applies to neighboring individuals of the same species. One could assume from this that self-fertilization is favored.

FEDOROV points out that this situation should encourage drift to an extreme degree. Elimination of the genetic load from mutations, increasing homozygozity, and accumulation of favorable mutations in certain localities are to be expected. Such conditions and automatic processes favor rapid evolution and speciation. New species could arise continually (and, one could add, sympatrically) without elimination or suppression of existing species. Thus a tropical rain forest would be "flooded" by a multiplicity of small and new populations that could be seen as a series of closely related species that arose at the same locality where they now exist. They may easily possess different characters that may assume high adaptive value in the future.

FEDOROV asks whether the series of related species in the tropical rain forest could be interpreted as a single, nonintegrated species, i.e. a series of species in *statu nascendi*. This is contrary to the idea that speciation is particularly favored in boundary areas, e.g., at the edge of mountains where speciation takes place in optimal environments.

He also points out that competition in the tropical rain forest is almost exclusively interspecific. Furthermore, he refers to character convergence of different species in the same stratum, which has been mentioned here in connection with another subject. At this point we must question how speciation functions in species other than trees in the same area. The literature contains a large number of examples of beginning or advanced stages of speciation, especially in the humid tropics and subtropics. Those from the genus *Drosophila* are most prominent, especially the species *D. paulistorum* and *D. willistoni*. To give only one example (a reasonably complete literature cannot be given here): In genetic experiments DOBZHANSKY and SPASSKY (1959) found that the species *D. paulistorum* can in fact be interpreted as a "cluster of species in *statu nascendi*." But here, as well as in *D. willistonii*, the initial stages of speciation were found especially at the edges of species ranges. This opposes FEDOROV's ideas, but the degree of inbreeding in *Drosophila* is probably not as high as in tropical tree species. Admittedly not enough is known. Investigations of the biology of flowers, fruits, and regeneration in tropical forests, such as those of PHILIPS (1926) and of other authors already mentioned, are still much too rare to draw general conclusions at this stage. Of course the pollination system can result in effective isolation, as GRANT (1949) has shown, and it stands to reason that the conditions for such isolation are more favorable for many tree species in the humid tropics and subtropics than elsewhere. However, we do not know whether this is enough to set in motion the process of continuing speciation that FEDOROV has in mind; in other words, we do not know whether certain incisive conditions are fulfilled that constitute essential prerequisites (in addition to the literature cited here, see CROSBY, 1970; MAYNARD SMITH, 1962; MOREE, 1953, and others).

FEDOROV's ideas have not gone unchallenged. They almost invite opposition inasmuch as they implay that drift and related processes are the principal cause of evolution in the humid tropics, in contrast to all other zones. ASHTON (1969), already cited in conjunction with a related topic earlier, developed his opposing views on the basis of genetic-ecological studies in the same regions where FEDOROV worked.

ASHTON reexamined, first of all, statements in the literature on population density and distribution of tree species. He concluded that pureley visual counts must lead to the wrong conclusions. Reasons for errors are historical causes for the distribution of a species such as catastrophic events and subsequent difficulties in the reestablishment of a certain species, arbitrary exclusion of the largest part of the population of a species by setting a lower diameter limit of the trees included in counts, zoned heterogeneity of the soil, and inadequate size of sample plots. Factors that play a role in species distribution are size of recent openings and their position in relation to the next seed tree, existing regenerations, and the ability of species to be present (waiting seedlings, seed dispersal). Accordingly, clumped distributions are frequent in species with irregular fruiting and limited possibilities for seed dispersal; they are rarer or less pronounced in species with frequent fruting and efficient devices for distant dispersal. In species growing in clumps there are opportunities for cross-fertilization within and between clumps. But if the species in fact do occur as single trees, an efficient seed-dispersal mechanism could provide for a continuous gene exchange. Therefore, one would expect that the best conditions for rapid evolution are not provided in species that grow individually and distribute their seeds efficiently, but in species that occur in clumped populations. (Comparisons of pollination vectors unfortunately are lacking here, it is conceivable that birds and bats could achieve an efficient gene exchange between clumps.)

ASHTON has discussed the conditions for selection and interspecific hybridization, and concluded hat in the tropical rain forest hybrids between closely related species are extremely rare and probably eliminated by competition. As in the temperate zones, they should be expected in disturbed habitas where the specialized climax species colonize with difficulty or not at all. In contrast to FEDOROV, he expected a variety of niches in the tropical rain forest and even within the same stratum, particularly when thinking of the multitude of forms and stages that may decide the result of competition. New techniques make it possible to demonstrate ecological diversity in the tropical rain forest that looks so surprisingly uniform at first sight, and to correlate such diversity with adaptive characters. In doing so, it should be kept in mind that the adaptive significance of a character is not immediately apparent and can only be determined with the aid of such correlations. FEDOROV's claim that in the tropical rain forest interspecific competition dominates, and intraspecific competition occurs only as an exception, is also questioned by ASHTON. (Authors note: This assumption was one of the pillars of LYSSENKO's ideology, although not specifically with respect to the tropical rain forest.) On the contrary, because mechanisms for seed dispersal are relatively inefficient in many Dipterocarp species, competition on the first developmental stages is primarily intraspecific. The same also applies to species in other families. When this is investigated, it should be noted that character convergence makes seedlings and saplings of different species morphologically indistinguishable. During the later developmental stages of the trees, interspecific competition becomes more and more important.

If this is actually so, greatest diversity should be expected among the species of the upper stratum. Every mature tree in this stratum has passed through the largest number of phases with different biological pressure. Observations in Southeast Asia confirm this prediction.

Only the rare and the competing species of a complex ecosystem would develop in conditions of interspecific competition and accumulate optimal genotpyes while selection for mutual avoidance takes place (perhaps better termed mutual exclusion). Thus, the success or failure of a seedling is determined by competition and accident (sample of the competitors, etc.): "Each individual rain-forest tree occupies a succession of microhabitats during its life span and it is impossible to determine whether the species comes finally to occupy a habitat at maturity arrived there through a process of selection through a complex series of competitive hurdles alone, or in part by chance. We may safely assume that both are involved and that the relative importance of each will vary; selection though will always play a crucial part." (ASHTON, 1969.)

Pollen and seed distribution are of decisive importance in the initiation and maintenance of species diversity in the tropical rain forest, as FEDOROV points out. They determine the effective population sizes of species, as well as drift and related effects. Here, again, ASHTON's results differ. Pollen distribution is determined mainly by land forms (as a factor governing continuity of species occurrence), flight distance, and periodicity of flowering. Figure 46 (from ASHTON, 1969) indicates, for example, that in a rain forest in Sarawak with a multitude of species, 26% of the 722 tree species exceeding one foot in circumference were dioecious, i.e., depended upon cross-fertilization. In comparison, dioecy is only found in 5% of all the flowering plants in Great Britain. An additional 14% of the 711 species have unisexual, protandrous, or proterogynous flowers. Surprisingly, most of the dioecious species were found in the lower strata, where spacial isolation among the individuals of a species is most pronounced. Therefore, a tendency for cross-fertilization apparently exists even in the tropical rain forest. In spite of this, of course, earlier statements concerning differences in effective population size between temperate forests and those in the humid tropics and subtropics may remain valid.

Pollination by wind is rare in tree species of the humid tropics and subtropics. Only one of 760 tree species on 100 acres in Brunei was a wind pollinator. Characteristically, it grew on mountains tops. Wind pollination is more frequent only on extreme sites, such as on river banks, in heath areas, and on mountains tops; here it may even dominate (*Casuarina, Actomeles, Agathis, Quercus,* and others). Conditions for wind pollination here are more favorable than in the interior of the densely closed, species-rich forest types. The efficiency of other pollen vectors (insects, birds, bats) was discussed earlier. These replace wind, which is the dominant pollen vector of temperate and boreal zones. We still know too little about them to develop a comprehensive view. In any case, birds and bats are exceptionally well equipped for distant pollen transport.

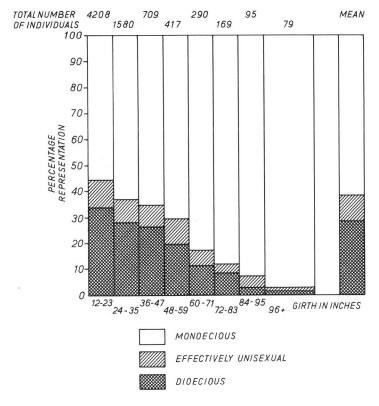

Fig. 46. Size distribution of monoecious and dioecious trees, and effectively unisexual trees, in mixed dipterocarp forest at Bt. Raya, Sarawak. (After ASHTON, 1969)

ASHTON explained that seasonality of flowering, or at least periodicity (e.g., periods of 4½ months in *Ficus sumatrana*), is the rule also in tropical rain forests, in contrast to the statements of earlier authors. The controlling factors are unknown. There is also simultaneous flowering of individuals of the same species in groups or over larger areas. The contradictory observations stem mostly from artificial plantings in botanical gardens or similar areas — away from the natural habitats and normal control factors. In addition, in Sarawak the flowering periods of *Dipterocarpaceae* differed among sites, and even the closely related species in the same habitat flowered at different times, effectively preventing interspecific crosses. (The importance of a continuous supply of pollen and fruits over the whole year to specialized distributors has already been indicated.)

The dispersal of seed is, as a rule, a more efficient population genetic mechanism for migration than the dispersal of pollen. Here again the distribution of light, wind-distributed seed in forests of the moist tropics and subtropics is a rare, if not exceptional, phenomenon. Distribution of heavy fruits with large food reserves predominates, allowing seedling establishment in shade. Distribution by animals is frequent. Probably a distribution pattern is common that has an effect similar to that of the pollination pattern: most of the seeds drop near the parental groups; a few are transported farther.

The wealth in species of tropical rain forests is not only the result of environmental stability, which allows extreme specialization in narrow niches, but also of the age of the tropical ecosystems: every evolutionary process takes time. The more time that is available in a constant milieu, the greater is the degree of specialization achieved and number of species that evolve.

ASHTON summarized the results of his studies concerning the evolution of Dipterocarp forests in Southeast Asia as follows:

"1. Highly specilized adaptation to their biotic and physical environment, notwithstanding that a single microhabitat may be filled by one of several alternative species in part by chance.

2. Limited efficiency of fruit dispersal imposing contagious distribution of individuals within their habitats and inability to cross all but the narrowest dispersal barriers.

3. An unspecialized pollination system in which autogamy is usual, but outcrossing between individuals of a clump, and to a lesser but significant extent, between clumps of a population, frequent enough to allow gene exchange throughout populations in a continuous habitat.

4. Allopatric differentiation between populations in response to differential selective pressures.

5. Rarity of hybrid populations.

6. Great morphological constancy within taxa even when with a widespread but disjunct distribution; rarity of clinal variation.

7. Long life-cycles combined with low number, but high numerical constancy, of chromosomes.

The complexity of the rain forest ecosystems has been explained in terms of:

(a) The seasonal and geological stability of the climate which has led to selection for mutual avoidance, and, through increased specialization, to increasingly narrow ecological amplitudes, leading to complex integrated ecosystems of high productive efficiency. As the complexity increases the number of biotic niches into which evolution can take place increases but they become increasingly narrow.

(b) Their great age."

According to this view, the evolution of ecosystems in the humid tropics and subtropics in principle is determined by the same factors as in all other regions of the earth, except that the constant and favorable environment controls these factors and their interplay in a different manner.

The previous graphic presentation of the north–south ecocline with increasing species diversity (assuming favorable habitats for plant growth) toward the south has prompted many explanations, only a few of which have been included in the preceding discussion. PIANKA (1966) considers six of these theories, which may be summarized and briefly discussed as follows:

1. The theory of time. This assumes that the number of species in a community increases with age. Compared with the ecosystems of the humid tropics and subtropics, those of the temperate and boreal zones are impoverished mainly as a result of glaciations but also because of the extremely short-term fluctuations of climate. Either the existing suitable species have not had sufficient time to migrate into an

ecosystem with few species (ecological time theory), or the short time since the last incisive change of environment has not permitted the evolution of many specialized species. Because of the similarity of models, reference is made to the book of MacArthur and Wilson (1967) concerning conditions on islands.

2. The theory of spacial heterogeneity. It is assumed that environmental heterogeneity increases in a southerly direction. This includes both macrodiversity resulting from topography and similar factors and microdiversity within communities created by biotic and abiotic niches.

3. The theory of competition. This is especially advocated and discussed in detail by Williams (1964). Here it is assumed that selection controlled by the physical environment is the prime factor in the temperate and boreal zones, but that biological competition is most important in the humid tropics and subtropics. This is the reason why there is more specialization for narrow niches and, as a result, a greater variety of species. The constancy of the abiotic environment, both in the short and long term, has been emphasized in connection with this theory. In northern latitudes, catastrophic events affecting entire ecosystems have led to increased fertility, an acceleration of development toward the reproductive stage, and the neglect of competitive relations (selection for high innate capacity of reproduction).

4. The theory of enemies and parasites. The uniform abiotic environment in the humid tropics also favors the evolution of specialized pests and parasites. Dangers of this kind are sufficient to keep the host population at low density, which also implies reduced competition within and between competing species. A low degree of competition favors coexistence, which, in turn, leads to the choice of other hosts and the evolution of new parasitic organisms.

5. The theory of climatic stability. This was newly formulated by Klopfer (1959). It emphasizes the importance of a stable climate for the evolution of advanced specialization whereby constancy of food supply is also included. The consequence is specialization for narrow niches (see above) and hence a multiplicity of species.

6. The theory of productivity. This was formulated in a modern way by Conell and Orias (1964). The authors assume that a larger production of organic substance leads — ceteris paribus — to a greater species diversity.

There are valid points in all of these theories. Certainly the interplay of factors and the predominance of one or the other is responsible for the evolution and establishment of new species. It is, however, difficult to examine this in individual cases. It would be even more difficult to determine how selection is set into motion and how it creates such complicated self-regulatory systems as the toxic effects of a species upon its own seedlings or those of other species. This is selection for low values of K, i.e., low carrying capacity. This kind of selection probably requires a threat by animal agents and diseases while population density is too high, and is successful only when the breeding system of the species concerned is suitably adjusted and the parasite allows changes of its own genetic constitution.

This admittedly superficial discussion of the highly complex ecosystems in the humid tropics and subtropics, which can be easily exparded on the basis of the cited literature, sets the stage for a discussion of the evolution of ecosystems at the other end of the ecocline — in the extreme areas of the boreal zone.

b) Forest Ecosystems in the Subarctic

Just as the studies of FEDOROV (1966) and ASHTON (1969) formed a basis for the discussion of the major problems of forest ecosystems in the humid tropics and subtropics, we can refer to a monograph by DUNBAR (1968) with regard to ecosystems in subarctic areas. Although this work is not particularly geared to problems of forest ecosystems, its main arguments should be valid for the forests of northern marginal areas. The discussion therefore will be based primarily upon this monograph and reference is made in advance to the ample literature cited therein.

The recent history of the fauna and flora in the subarctic, the boreal, and the presently temperate zones of the northern hemisphere is determined by repeated glaciation, and in general by climatic changes over relatively short periods. "Short" in this sense implies a scale relative to the generation interval of forest trees. For example, beech reappeared as late as 5000 years ago in northern Germany, Denmark, and southern Sweden. If one assumes a generation interval of 100 years (which is probably too short), this period adds up to 50 generations as the time required for the evolution of adaptive strategies and the adjustment to certain ecosystems. It has been shown that in North America some oaks and other deciduous trees appeared in the northern portions of their present ranges, only 3500 years ago, a maximum of 35 generations. Many population genetic studies, such as with *Drosophila*, have continued for much more than 100 generations.

Several methods are available for estimating climatic changes, in northern regions. One of the most interesting techniques is that used by WISEMAN (1954), who starts with the assumption that the CO_2 content of *Foraminifera* is higher in warm water than in cold water. He therefore investigates Foraminifer skeletons from different sediment cores taken at sea bottom to obtain the picture reproduced in Fig. 47.

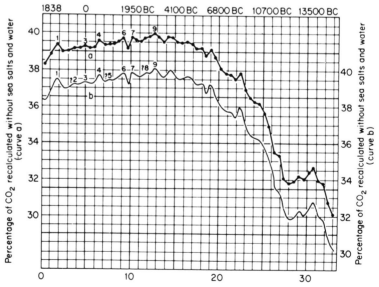

Fig. 47. Carbon dioxide measurements from sediment core taken in the equatorial Atlantic, 01°10′N, 19°50′W, depth ca. 4350 meters, by the Swedish "Albatros" expedition. (After DUNBAR, 1958, from WISEMAN, 1954)

According to this illustration, there was a temperature maximum about 6000 years ago, which was preceded by a period of 6000–7000 years marked by a steep rise in temperature. The prevailing trend since then is a slow decline in temperature.

We shall not concern ourselves here with theories of alternating glaciation of the northern hemisphere. Different explanations of the origin of glaciation and methods allowing the most accurate dating of temperature development can be disregarded, as all authors agree on one point: temperature development in the northern hemisphere permitted immigration of the existing tree species only in relatively recent times (as expressed by generation intervals), and the evolution of ecosystems in these areas may not have progressed very far when measured against the possibilities of evolution of the much older tropical ecosystems. Consequently, the number of species should be much smaller at this extreme point of the north–south ecocline, as it is.

In connection with this age theory of ecosystems and its bearing upon species diversity, a comparison of the forests in northern Europe and North America with those in northern Asia should be of interest. The Siberian taiga and forests at its southern boundary cover large areas and are, in fact, mixed forests of larch, pine, spruce, birch, and aspen, much to the dismay of the forester, whereas the distribution of the few species in northern Europe and northern North America is more parapatric, that is according to a mosaic pattern depending upon site.

This fact confirms one of the six theories mentioned by PIANKA (1966) concerning the basis of diversity along north–south gradients. There is no need to discuss another theory, that regarding productivity. It is obvious that the climatic conditions of the temperate and boreal zones (lower temperature, cessation of growth in winter) lead to a lower production of matter than in the humid tropics and subtropics. In addition, the turnover rate of organic materials should also increase as the ecosystem reaches becomes more complex. This theory thus has two aspects: on the one hand, higher productivity, leads to specialization, and in the other, it promotes the turnover of organic materials and probably also production.

It has often been stressed that species in extreme climates need to develop special adaptations to ensure their existence. For tree species of the far north this includes tolerance of low winter temperatures, including winter drought, etc., and adaptations to short periods of vegetative growth. This is certainly true but does not permit the deduction that only a few species are able to acquire these adaptations and to exist in the far north, for there is also need to adapt to long-term fluctuations of climate and other factors; furthermore, experimental proof is lacking.

Many authors who stress adaptation forget that, especially in the extreme northern regions, one finds the most pronounced oscillations of annual weather. Observations on the phenology of geographic sources of tree species reflecting adaptation to vegetative periods of different length constitute the standard repertoire of forestry provenance research. In addition to the observations discussed earlier, those of GATHY (1960) on clinal variation of growth initiation (and height growth) in Norway spruce may be mentioned. Up to the 53° N latitude, growth starts earlier with latitude, but then the relationship is reversed — giving an indication of the possibilities this species possesses for adaptation to different climates. WANG and PERRY (1958), working in Florida, observed provenances of *Betula papyrifera* and *B. alleghaniensis* from Canada, Alaska, and other parts of the United States. They found that the beginning and end of winter dormancy varied clinally, but again the northernmost sources were an exception. They required no winter chilling to flush after ceasing growth in response to shorter days. Differentiation in requirements for winter dormancy and its stages is also discussed by WORRALL and MERGEN (1967), again in a study of Norway spruce and related to adaptation to specific conditions at the source. The role of photoperiod in relation to trees has been thoroughly discussed by NITSCH (1959), who also considered light quality and other, usually neglected factors.

Less attention has been given to variation within sources than to variation between sources, though such studies could provide further insight. In addition to the experiments cited in the chapter on adaptation, the experiment on geographic sources in Sitka spruce by LINES and MITCHELL (1966) may be mentioned. In 12 sources, ranging in origin from 43 to 60° N latitude of the almost linear range of the species, the authors found a large variance for growth initiation within sources and a smaller variance among them. The northernmost sources exhibited the largest variance within populations. Such comparisons, of course, are always somewhat uncertain when undertaken in an unusual environment. Thus EVANS (1939 — to mention one of the oldest studies in this field) found in two clones of timothy with extreme differences in flowering time mean flowering dates of June 13 and August 12, respectively, when grown at 38° N latitude, but dates of July 12 and 21 at 55° N latitude. In any case, it becomes apparent that the assumption of a narrow within-population variance in northernmost areas resulting from extreme selection pressure is not necessarily valid. We can now explain the limitation of this thesis: the farther one goes north, the greater is the oscillation of annual weather cycles and the more "coarse-grained" the environment. This should lead to the adoption of mixed strategies, and even favor genetic polymorphisms.

How rapidly adaptation is possible to comparatively northern regions is indicated by a large number of experiments, and again one of the earlier results will be cited. SYLVEN (1937) compared three white clover varieties from Denmark and Germany with native varieties in southern Sweden. All imported varieties were at first inferior to the native ones. Yet after only one generation two varieties were equal to the native varieties. The third variety, still inferior to the native ones, characteristically was the only one selected for high yield under the conditions of a different climate. However, the implications of these results must be restricted to the extent that the influence of preconditioning in this experiment cannot be estimated. LEDIG and PERRY (1967) have indicated its significance in similar experiments with trees. They found in *Pinus taeda* high heritabilities and also genotype–environment interactions of photosynthetic rate, but also a large influence of preconditioning. The role of preconditioning or acclimatization outside the influence of selection is still not well known.

Theories concerning reasons for the small number of species at the northern end of the north–south ecocline need not be discussed. These reasons are found by a reversal of arguments for wealth of species in tropical and subtropical forests as given in the previous section. DUNBAR, too, views this similarly. To demonstrate this he uses a graphic presentation of CONNELL and ORIAS (1964), which is reproduced in Fig. 48.

The dearth of species in forests (and the flora in general) of subarctic regions can also be related to the lack of specialized animals as pollen vectors. It has already

Rate of production of new species

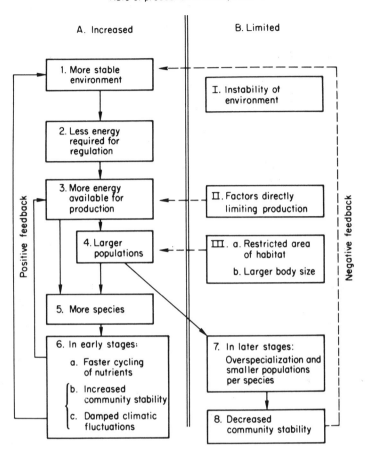

Fig. 48. Model for the production and regulation of species diversity in an ecological system, after CONNELL and ORIAS (1964). See text. Solid lines indicate an increase, dashed lines a decrease, in diversity

been shown that animals specialized for pollen and fruits of certain species find best conditions in the forests of the humid tropics and subtropics. As a reverse argument, it may be stated the specialization of animal vectors may in turn facilitate speciation. Of course this viewpoint is closely related to the theory of the "maturity" of ecosystems; i.e., both processes are strongly related to the time factor. However, this viewpoint acquires a certain importance of its own through the necessity of the participating animal and plant species to first adapt to the extreme conditions of the abiotic environment given in the subarctic climatic zone.

Selection toward stability of ecosystems will probably also suffer from the dearth of species. Selection for stability might mean, as a rule, simultaneous selection for diversity, as we have already seen, and this at the level of participating species as well as at the level of genotypes within species. To this comes the extreme oscillation of climatic niche variables. " . . . the general rule of evolution toward complexity . . . hold(s) well in ecological evolution, in which it may reasonably be postulated that the simple ecosystems, being subjected to violent oscillation in numbers of individual species, have this built-in flaw of the danger of self extinction, and that there will be a selection toward greater stability, in the sense already defined, and hence toward enhanced viability. It has long been a tenet of ecological theory that stability is given to ecosystems by the development of more complex food chains or food webs, and thus by increase in the numbers of species in the system . . . Evolution toward greater stability, therefore, in favouring increasing numbers of species must, in Arctic situations, work contrary to ecological adaptations to the highly oscillating environment, which tends to keep the number of species small. There are other elements in which the two processes come into conflict." (DUNBAR, 1968).

This citation, too, is given as an objection to the philosophy of the holistic forest organism, which is not interested in details. It is incomprehensible today how this philosophy hoped to achieve ecosystem stability in ecologically extreme regions by using methods, that were successful in the conditions of greater evolutionary stability of the humid tropics and subtropics, and also of temperate climates, simply because the methods were suited to the evolutionary status of these ecosystems. It would be interesting to find out from representatives of this school how they propose to meet the problem of cyclic oscillation of lemming population size (except by spraying poisons), which may be critical for regeneration in the far north (Fig. 49).

It is quite clear, and representatives of forest science and practice know it well, that there is a need to adjust every management activity in the extreme boreal and subarctic regions to the evolutionary status of these regions. Recommendation of "ideas" foreign to science and unsuitable in practice, which ignore the special evolutionary status of these regions, reveal the ideological character of silvicultural holism.

Forestry in these regions must accept this evolutionary condition. At best, evolution in the direction of greater stability could be supported. But how should this

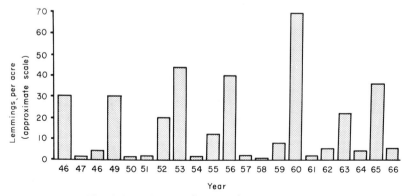

Fig. 49. Lemming population cycles. (From SCHULTZ, 1969)

actually be carried out? We shall not attempt an answer to this question here, but leave it to the future — to the growing number of scientifically trained foresters.

c) Forest Ecosystems of the Temperate Zones

Typical of forest ecosystems of the temperate zones are successional series that are initiated by catastrophes and lead to stable climax forest or to subclimax communities, which occupy a terminal position in succession but differ in character from stable climax communities. Such successions, of course, also take place in the humid tropics and subtropics (for additional examples here see RICHARDS, 1964) and forests of the far north, but are most typical of some forest communities of the temperate climatic zone. Climax associations are also found here that are remarkably stable and contain a wealth of species[3].

Probably the best survey of forest communities in a large area is given in the list of "Forest Cover Types of North America" (ANONYMUS, 1967), which was compiled by the Society of American Foresters. A cover type is defined as a forest type that at present occupies a certain area, regardless of whether permanently or temporarily. However, successional trends are in many cases described and partly explained. Of interest here is only the so-called secondary succession, i.e., vegetation development after catastrophes affecting large areas. Primary succession — evolution of ecosystems during and after retreat of the ice in the last glaciation — will be disregarded.

One example is probably representative of many others. MARTIN (1959) describes two successions in the Algonquin Park of Ontario, namely, a hydrosere in moister lower sites and a xerosere on upland sites. The hydrosere is interpreted as the representative of primary succession, beginning with the retreat of Pleistocene glaciers. The xerosere is equated with typical secondary succession that is set in motion by fire or cutting. Table 28 gives an indication of species composition in eight consecutive stages of succession.

In the first stage *Populus tremuloides*, *Betula papyrifera*, and *Acer rubrum* are clearly dominant. These three species are no longer represented in the last stages. *Abies balsamea* participates with large proportions, mainly during the middle stages; *Pinus strobus* enters even later; and *B. alleghaniensis* and *Tsuga canadensis* reoccupy their habitat toward the end of the succession.

In extreme conditions climax communities may consist of few species and feature hardly any transitional stages. An example is the *Juniperus occidentalis* forest in eastern and central Oregon and in California (ANONYMUS, 1967). This occupies areas between the sagebrush steppe and the forests of *Pinus ponderosa* – *P. jeffreyi* beginning at elevations of 1300 to 1700 m above sea level. The trees are widely spaced and there is an undergrowth of shrubs. Another type bordering on sagebrush steppe or prairie is dominated by *P. ponderosa*, which often occurs in pure stands over extensive areas. As precipitation increases species admixtures appear depending upon bordering communities, such as *Pseudotsuga menziesii*, *Abies concolor*, *Larix occidentalis*, and others. Species diversity increases as precipitation rises even more. These examples may be sufficient to indicate that the position of a

---

[3] The authors wish to thank Prof. DIETRICH MÜLDER, Göttingen, for discussions and suggestions.

Table 28. Basal area of principal trees over 1.5 in. d.b.h. for stages in the xerosere (percent). (From MARTIN, 1959)

Species	Betula Populus	Betula Populus a. Abies-Picea	Abies Picea	70-ft Pinus	200-ft Pinus	Abies-Picea a. Hardwoods	Hardwoods	Tsuga
Populus tremuloides	16	16		5				
Betula papyrifera	35	31	1			2		
Acer rubrum	17	7	2			7		
Abies balsamea	1	18	68			16		4
Picea glauca	1	9	13		1	3		1
Picea mariana		8	14					
Pinus strobus	1	8		65	78	3		
Pinus resinosa				24				
Pinus banksiana				4				
Acer saccharum	9					30	65	3
Betula lutea	3				12	37	18	21
Fagus grandifolia							5	
Tsuga canadensis					9		1	62
Thuja occidentalis						2		8
Others	17	3	2	2		3	8	1

"forest cover type" in this part of western North America relative to the ecocline (here determined by moisture) influences, species diversity and the number of stages in secondary succession. Of course, forest cover types also exist in dry areas with relatively many tree species, such as the *Pinus–Juniperus* type at the edge of the Rocky Mountains and the Sierra Nevada. This also borders directly on sagebrush and prairie. Possibly the decisive factor is also the position of the community in primary succession, which is another word for evolution of the ecosystem: the longer the chain of primary succession, the greater is the chance for the development of an ecosystem with a wealth of species (see above). Of course the ecological conditions for the development of a multitude of species also play a role, such as niche diversity in the area, properties of the gene pool of the immigrant species, etc.

Such successions in forest communities at the extreme edge of ecoclines are not of great interest here. In principle they resemble the simple successions at the northern end of the north–south ecocline, which are also determined by climatic extremes. They are mentioned here to distinguish them from the secondary successions in ecosystems with a wealth of species, such as the succession in Algonquin Park mentioned earlier. A second distinction must be made against possible really "stable" climax forests of the humid tropics and subtropics. In forestry literature, associated with the idea of the stable climax forest, there appears the assumption that nothing changes since a "stable" equilibrium cannot be disturbed. This is probably not so. Rather, under the conditions of natural forests, the frequency of catastrophes and the area affected should be the decisive criteria. Such criteria call for a quantitative rather than a qualitative expression of the differences.

The main difference between subclimax and climax types seems to consist in the ability of the climax community to regenerate itself without the detour of succession and to maintain itself over long periods. The reasons could be interior (perhaps evolution of the ecosystem has progressed sufficiently) or exterior (the environment is stable, catastrophes rare, etc.). It is often impossible to decide whether a certain succession terminates in a stable climax community or a less stable subclimax. Some authors are of the opinion that certain successions preceeding to a subclimax only need not be maintained by fire or other catastrophic events, such as the type in northern Idaho, Montana, and adjacent British Columbia that is dominated by *Pinus monticola*. It is claimed that the last stage of the succession in these types (predominantly hemlock) is not stable: regeneration is not possible for a sufficiently long period even if forest fires, which usually set succession in motion, fail to occur. The reason for the lack of regeneration of the terminal community could be a thick raw humus or similar obstacle.

Again, this is a question with which we are not very much concerned. It is a question that is more closely related to the evolution of ecosystems. One could ask, for example, what the stability of an ecosystem actually means. The answer could be that every ecosystem is stable if its species composition does not change in historical times (in longer time periods it would surely change if our conception of evolution is at least fairly correct).

Such species composition could be developed equally well by repeated succession, inasmuch as the difference between such succession and more continuous, (small-area) regeneration is merely a quantitative one. Even if one chooses the additional criterion of the equilibrium — i.e., not only species constancy but simultaneously also unchanged frequencies of the species — fluctuations in time and space must be admitted depending upon the size of sample plots or areas used as a reference basis. Over large areas and measured in the longer periods of generation intervals, ecosystems with successions ending in the subclimax stage under certain conditions could also be designated as stable. In any case, too little is known about the principles of evolution of whole ecosystems to exclude the possibility that the succession ending in a subclimax (a stage forced upon the ecosystem by the inner dynamics of the system) is an optimal strategy of whole ecosystems. In environments characterized by periodic catastrophes, this could well apply. LOUCKS (1970) has examined the partly controversial theories concerning principles of the evolution of ecosystems, the bases of which have already been discussed. He investigates particularly the problem of ecosystems with periodically recurring secondary cycles of succession. It is not possible here to discuss all details of the interesting study of this author and thus his synopsis will be cited:

"The response in species diversity associated with successional change in vegetation, or in a more general sense, species diversity as a function of time in any system of primary producers, has been the subject of much speculation but little direct study. All evidence available shows that pioneer communities are low in diversity, that in mesic environments the peak in diversity in forest communities can be expected 100–200 years after the initiation of a secondary successional sequence (when elements of both the pioneer and the stable communities are present), and that a downturn in both diversity and primary production takes place when the entire community is made up of the shade-tolerant climax species".

"The natural tendency in forest systems toward periodic perturbation (at intervals of 50–200 years) recycles the system and maintains a periodic wave of peak diversity. This wave is associated with a corresponding wave in peak primary production. Specialization for the habitats in the early, middle, and later phases of the cycle has figured prominently in species-isolating mechanisms, giving rise to the diversity in each stage of the forest succession. It is concluded that any modifications of the system that preclude periodic, random perturbation and recycling would be detrimental to the system in the long run."

The conclusions of the author, supported by his own investigations in forest ecosystems of Wisconsin, draw a picture for secondary successions that is related to the typical one in primary successions. The first stages are characterized by the dominance of a very few species qualified for this role by their genetic system and adaptation. As the ecosystem becomes more mature, species diversity increases and specialization to niches is more obvious. But, in contrast to ecosystems with a more permanent climax, there is a need for all participating species to adjust themselves to a repeated loss and reinvasion of parts of the range. In other word, both pioneer species and species occurring in subsequent stages of succession must be prepared for recolonization. This should find its expression particularly in the ability to distribute seeds rapidly over large areas and to establish seedlings.

Pioneer species are clearly defined in this respect. All of them possess special adaptations for this role, which are independent of the type of secondary succession. They are prolific seed producers and their dispersal devices allow distribution over large distances. This is different, however, for species occupying positions near or at the climax stage. Where large-scale disruption of the ecosystem is the exception rather than the rule, there is no need for a rapid recovery of lost territory since the largest part of the species range still exists under normal circumstance. But if larger parts of the range are lost at short intervals and if the duration of the climax or of the climax-line phase is relatively short, the need for recolonization has priority. For example, species of *Thuja* or *Tsuga*, which are considered to be climax species, have small seeds that are capable of flying. The seedlings are extremely tolerant of shade and therefore survive during the first stages of succession; i.e., they possess characteristics usually related by most authors to the size of seed and storage of nutrients as typically expressed in climax species of humid forests of the tropics and subtropics. Consequently, the compromise in optimization achieved between seed size and seed dispersal by near-climax species of the temperate zone might differ from that achieved by most species in tropical ecosystems.

In the absence of experimental results no further conclusions can be drawn and we must limit ourselves to the indication that position of a species in succession is not adequate to assess certain components of its genetic system. Comparisons of typical climax and pioneer species in previous sections assumed a single evolutionary type of ecosystem: the ecosystem with occasional catastrophes that lead to successions but where the climax persists for a considerable period of time. Conceptually this idea is probably applicable to many of the well-known ecosystems. Yet there are also other ecosystems where periodic recycling is typical and where charcteristic species of intermediate stages may exist. Perhaps the occurrence of such species is typical for ecosystems where there is a need for such recycling. In any case, this question will be left open. That a problem has arisen here at all is a consequence of the development where reasoning in terms of ecosystems begins to replace reasoning in terms of the dialectic forest organism. The latter philosophic trend has clouded the view, at least in central Europe, and inhibited analysis by considering the stable climax of forest ecosystems only. It was thought that this climax could possibly be disturbed but that it was always reached. For these reasons, it will take a considerable time before we shall be able to define even the theoretical basis for a comparison of ecosystems with secondary succession.

Aspects of primary succession are illustrated by the so-called colonizing species. Excluded from consideration will be the species regenerating under the conditions of modern forestry, which will be treated in the next chapter. The colonizing species discussed here are those that occupy previously inaccessible territory and migrate into an ecosystem that is in the first stages of primary succession, such as that on the Island of Krakatau following volcanic eruption; or those migrating into an ecosystem subjected to human influence offer the best opportunities to colonizing species since they are always relatively young and, with rapidly changing human influence, offer a variety of conditions for evolution. In view of the widespread occurrence of such ecosystems, it is surprising that the international symposium on the influence of man on vegetation (TÜXEN, 1966) dealt only with conventional ideas and neglected evolutionary aspects of such ecosystems entirely, although another symposium two years earlier had been specifically dedicated to the special genetic problems of colonizing species that are of the greatest significance to these ecosystems (BAKER and STEBBINS, 1964). Therefore, we must rely essentially on the contributions to the earlier symposium. Thus we are dealing here not with the typical pioneers of secondary succession (the special features of which were probably best defined, since this was done by means of a mathematical model, by MACARTHUR, 1960), but with species occurring in the beginning stages of primary successions or those that migrate into disturbed or established ecosystems. Both should probably be treated separately.

We begin with colonization in established ecosystems where primary succession is far advanced. Many investigations have shown that the species of such ecosystems are especially adjusted to certain ecological niches in all phases of the development of their individuals. To name but one example: GOODMAN and KREFTING (1960) have investigated the ecology of establishment of *Betula alleghaniensis* in the Upper Peninsula of Michigan. The species occurs in the later stages of cyclic succession. Prerequisites for establishment (or some of the known prerequisites) include soil conditions, protection against direct sunlight and late frost, and avoidance of weed competition by control of incoming light. All of these conditions are met in a certain stage of succession. To compete, a newly arrived species must fit into an existing niche or be capable of modifying the whole ecosystem. Both are possibilities.

We shall first examine the situation where one species "colonizes" in a neighboring ecosystem by means of its different competitive ability. The discussion of this situation is based upon the studies of SAKAI (1969) and COX (1968). The first author assumes that characters of a population correlated with competitive ability must lead to its division into different subpopulations, which probably is to be understood as adaptation to different ecological niches. This niche restriction in turn creates a new basis for the invasion of other niches. Because there are regional differences in niche frequencies, this then becomes a reason for migration of the species. The second author concludes that in certain situations selection favors genetic mechanisms leading individuals of species, which are normally stationary in one region, into temporarily suitable neighboring regions. A species may achieve an occasional net gain if the cost of migration is exceeded by a gain in reproduction and survival. Thus the evolution of migration is partly attributable to competition within as well as among species. The results of these two authors are essentially compatible and each author supports them by examples from both the plant and animal kingdoms.

In principle, therefore, we are dealing here with adaptive strategies of populations leading to colonization (migration). In his introduction to the above mentioned symposium, WADDINGTON (1964) emphasizes this particularly, as well as the possibility that different strategies may lead to the same result, such as, for instance, minimax strategy. This is in agreement with our thesis that the opportunities offered by the ecosystem being invaded must be taken into account. However, the genetic system of the invading species also may be of decisive importance.

The last point is particularly stressed by ALLARD (1964). His view is that there is a disproportionately large number of self-fertilizers among the colonizing species. This favors the development of locally specialized subpopulations as is evidenced by two colonizing species (*Bromus mollis* and *Avena fatua*) in California. In spite of cytological stability there is large variation in both species, including variation between neighboring subpopulations, and in adaptive characters. Hence such species are characterized by the presence of genetically differentiated subpopulations in the same area (see S. WRIGHT, 1931), and are also well suited for selection within populations of colonizing species in the same area characterized by much smaller proportions of cross-fertilization. Homozygozity within the subpopulations is therefore more advanced and the species is further away from the optimal ratio between variability and homogeneity that is typical for colonizing species. Hence these species are no longer capable of utilizing the opportunities offered by heterozygote advantage. This is important, especially in ecologically marginal areas; several experiments have indicated that the fitness values of the homozygotes are strongly correlated with annual weather but those of the heterozygotes less so (compare discussion of polymorphisms in the section on northern forest ecosystems). In predominantly self-fertilizing species, where the optimal self-fertilization rate depends on environment, the parallel necessities of local specialization and maintenance of genetic variation may be realized best. The differences between the genetic systems of obligate self-fertilizing species (better described as species with a low rate of cross-fertilization) and the ideal random mating, cross-fertilizing species are slight. Therefore, the genetic systems of colonizing, predominantly self-fertilizing species represent a compromise solution between the high potential for recombination of the cross-pollinators and the population stability of the self-fertilizers. This facilitates adaptation to a mosaic of microniches. Additional conclusions in ALLARD's paper, although of interest here, cannot be given.

In his contribution to the same symposium, LEWOTIN (1964) expects the following characteristics in optimal colonizing species: effective dispersal, large somatic plasticity, and high interspecific competitive ability. Questions regarding the optimal genetic system play a secondary role (note: if the first and second characteristics are not considered part of the genetic system). Third in importance is variability (this really is part of the genetic system). It is best to search for episodes of colonization by a species since the history of some colonizing species consists only of colonization episodes. Three questions should be asked. The first concerns the effects of selection in a population with unlimited growth. The second pertains to possible changes in selective influences during certain periods of colonization, e.g., during the period of expansion. Finally, the third question involves the possibilities of selection among subpopulations. In sum, this amounts to a search for the conditions of selection that provide for an optimal strategy. A comparison of characters such as age to sexual maturity, to period of decline in number of progenies, to last progeny, etc., indicated that the success of colonizing species depends upon rate of development. For this character there should be little or no genetic variance, but there should be large genetic variance for characters such as number of eggs in animal species.

At this point one may note that for the success of colonizing species, the two different situations distinguished here — immigration into an established ecosystem or occupation of a new, open ecosystem — differ greatly. The example of birch in North Germany (mentioned earlier) will illustrate this. In the frequently disturbed ecosystems of the North German Plains, *Betula pendula* consists of a large proportion of bushy, rapidly growing, early and prolifically flowering individuals. In forest ecosystems of the adjacent mountain region, the individual's survival in competition is important in order to take part in succession, and hence late-flowering, tall (and possibly long-lived) individuals predominate. However, these relationships have been considered by LEWONTIN, at least indirectly, when he emphasizes the significance of competitive ability and indicates the models of game theory that should be used to explore these relationship.

The special characteristics of weedy species are considered by BAKER (1964) and HARPER (1964). Weedy species are special types of colonizing plants. The first author classifies them as agrestal and ruderal plants and thereby emphasizes the connection between the concept of a weed with the habitat disturbed by man.

To our knowledge, there are no typical weeds among the tree species that fit this definition and so the highly interesting account of these two authors will not be discussed further. However, it is recommended for study to the readers since it contains significant ideas useful for the explanation and clarification of the concepts used here.

An especially interesting aspect of colonizing species is discussed in the same symposium by WILSON (1964). Rapidly colonizing species, which include weeds and other species such as trees migrating into disturbed ecosystems, and in artificial forests, come into contact with new enemies, as well as with their old enemies (which may have been exported) but in a new environment. According to investigations of the author on many important organisms in nine different countries of the world, the host–parasite equilibrium in such cases is determined by the imported genetic variability of the host in the original habitat, by selection of populations that support the trend toward the equilibrium in the new habitat, and by the possibility of selection in direction of the equilibrium where overspecialization of the host in the natural habitat has taken place. The many examples of the author support the ideas given in the section on host–parasite systems.

At this stage the reader will have realized that the detailed study of colonizing species overlaps to a large extent the study of ecosystems that have been subjected to human influences to a significant degree. It is therefore necessary to discuss these human influences separately, as has been undertaken by in the following chapter. The problems to be discussed there will probably be best understood if seen from the viewpoint of primary succession, which may have been set into motion by disturbance of the environment or by artificial introduction of tree species into areas that were previously inaccessible to them.

VI. How Man Affects Forest Ecosystems

The effects of man on forest ecosystems have been, and continue to be, very great. As DARLING (1956) has pointed out, "Man advances materially and ultimately in his civilization by breaking into the stored wealth of the world's natural ecological climaxes." To this advance forest ecosystems have contributed abundantly, and continue to do so. However, in the process, regions of the world have been denuded of forests as a result of man's activities, and history has recorded that, though the immediate exploiter has gained materially, mankind has suffered harshly as a consequence. Much human misery through subsequent centuries may be attributed to the elimination of the great forest resources of the Mediterranean basin at the end of antiquity, and of Central Europe during the 12th and 13th centuries (DARBY, 1956; SARTORIUS and HENLE, 1968). Today it is possible to see the ruins of great cities in the midst of deserts. The land has changed drastically and regions once highly fertile and capable of supporting large populations can now support only a few nomadic tribes at a subsistence level. It has been speculated that the major cause of the spread of deserts through once fertile regions has been climatic changes resulting in gradual desiccation of these regions. However, accumulating evidence now suggests that man and not climate must be held accountable for the encroachment of deserts into the fertile lands of ancient empires. The headwaters of the Tigris and the Euphrates lie in areas that in the past were densely populated, and that have been overgrazed and deforested. The erosion that has resulted has caused an ever-increasing silt load, which blocked irrigation canals and filled in the Persian Gulf to a distance of 180 miles (290 km) out from where the river emptied in Sumerian times. The silt-laden flood waters now carry soil without interruption from the highlands to the sea, and the lands of Mesopotamia in modern times support only a fraction of their former population (DASMANN, 1968). Similarly, the mountains of Lebanon, once covered with forests of cedar, are now barren and almost devoid of soil, valueless both for man and beast. These great forests helped to make the Phoenicians mighty; their elimination impoverished the country through succeeding ages.

Man's impact on forest ecosystems changes with his social development in time and space. Shedding his ancient fear of the wild forest, and now an ecological dominant (DARLING, 1956), modern man is less a wilful destroyer, and more of an exploiter and manager of forests. Furthermore, he is beginning to rehabilitate large areas of land denuded of forests by his forefathers. Still, however, in many parts of the world, and particularly in nonindustrialized countries, many forest ecosystems continue to be eliminated.

Though some restorative measures are under way in the developing countries of Latin America, Asia, and Africa, they are not sufficient to offset the destruction of forest ecosystems that is taking place. All too frequently, forest annihilation has ceased only when there is no more forest to be clear cut (HOLDRIDGE, 1956; SARTORIUS and HENLE, 1968). In the long run the social and economic consequences of this could be as great and as harsh as those inflicted on Central European and Mediterranean countries following the elimination of the great indigenous forests.

In discussing the results of forest destruction, SARTORIUS and HENLE (1968) conclude: "Thus, it is no empty menace when foresters and ecologists warn, time and again, against the nefarious and inescapable downward spiral which is set in motion when a society permits massive extermination of its forests. Disappearance of the tree cover engenders not only pedological and biological micro-reactions but, in the long run, macro-reactions beginning with economic disruption and leading to social decline and, finally, to the collapse of a whole civilization."

There can be little doubt that fire has been man's principal weapon in his fight with the forest. With the evolution of man and his dispersion across the earth went an increase in the incidence of fire. "Even to palaeolithic man, occupant of the Earth for all but the last 1 or 2 percent of human time, must be conceded gradual deformation of vegetation by fire." (SAUER, 1956) Fire, therefore, has been an environmental pressure on the forest ecosystems of the earth for so long that it is now impossible to understand fully their distribution, history, and, for certain species, even genetic structure, without taking fire into consideration. Fire can be caused by agencies other than man, but the evidence increasingly indicates that in its effect on many forest ecosystems, and indeed the landscape of the habitable regions of the earth, fire has been primarily an anthropogenous factor (STEWART, 1956).

Both wittingly and unwittingly, man has played a major role, amply documented, in the evolution of modern forms of plants and animals. The degree to which he has played a similar role in relation to tree species is less well known and inadequately documented. There is little doubt, however, that the prehistoric and historic activities of man have considerably influenced the distribution, composition, and genetics of modern forest ecosystems, and when potsherds, wells, and ditches are found beneath so-called virgin forests, the resulting incredulity is merely a measure of the extent to which the subject has been ignored both by anthropologists and, with much less excuse, by foresters.

Of the various activities of man through the ages that have profoundly affected forests and forestry, and in which fire has played a major role, the most important are hunting, grazing by domesticated animals, and agriculture. Massive industrialization, and urbanization have also been of great importance. Modern forestry, with its emphasis on the rehabilitation of degraded lands, sustained yield, and multiple use, and on conservation and tree breeding, represents man's positive impact on forest ecosystems, and, one may hope, in time will enhance not only the gross national product of nations, but also the physical beauty of man's habitat. These activites of man and their effects on forest ecosystems are the subject matter of the following sections.

1. Hunting, Grazing, Agriculture

DAY (1953) has discussed the effects of Indian peoples on the forest ecosystems of northeastern United States. Indians created large openings for their villages and fields, and over much of the region set fire to the forest to improve traveling and visibility, and to drive or enclose game. Their activities destroyed the forest in some areas, and modified it over much larger areas. Much of the park-like woods or savanna available for immediate cultivation by early settlers was the product of repeated Indian burning, which reduced dense forest of pine and spruce to open grown scrub oak. From the evidence presented, DAY (1953) concluded that an area that was wooded when first seen by white men was not necessarily primeval, and that an area for which there was no record of cutting is not necessarily virgin. He suggested that a knowledge of local archeology and history should be part of the ecologist's equipment. The effects of North American Indians on forest ecosystems have not been confined to the northeastern United States. The Plains Indians very considerably extended the prairies in historical times by deliberate burning of the forest's edge, and boreal forest ecosystems have been subjected to deliberate burning by natives to improve hunting (DARLING, 1956; STEWART, 1956).

There is evidence that such man-caused fires, even in northern coniferous forests, can have major effects over vast areas. For the ten-year period, 1943–52, the average annual number of fires in Canada was about 5100 and the forested area burned each year amounted to approximately 2400 square miles (6243 km^2). Statistics show that lightning is responsible for about 17% of these fires, that 75% can be attributed to human carelessness, and the remaining 8% to incendiarism or unknown causes (TUNSTELL, 1956). According to KIIL and CHROSCIEWICZ (1970), fire has been the single most important disturbance in the temperate forest of North America. These authors add that fire has been particularly extensive in the boreal forests of Canada and the main regulatory force in determining the composition and geographical distribution of many forest types.

It is clear that prior to the arrival of colonists in North America, forest ecosystems had already been affected by the hunting, pastoral, and agricultural activities of the native population. The view that the great prairie regions of North America are entirely of climatic origin is increasingly questioned, and a number of authors have suggested that the use of fire by Indian peoples has been a major factor in the modification of forest ecosystems and in the establishment of prairie and savanna-like regions over vast areas of the continent. The fact that the forest will colonize such lands and that normal forest succession is reestablished following reduction in man's interference is advanced as evidence that these treeless or nearly treeless regions once supported natural forest ecosystems. For example, following the destruction of the Plains Indians and their sustaining buffalo herds the forest invaded vast areas of prairie land. Parts of northwestern Illinois and southwestern Wisconsin, now heavily forested where not cultivated, were 80–90% prairie grassland when first visited by Europeans (GLEESON, 1932; DARLING, 1956; STEWART, 1956).

Early maps of the vegetation of Minnesota showed isolated forest stands scattered throughout grassland in the southwestern third of the state. Each stand tended to be of crescent form. They fitted the easternly margins of lakes, ponds, and marsh, and possessed a more or less complete association of forest undergrowth. DAUBENMIRE (1968) concluded from the distribution of these stands, and their plant associations, that the forest had once been continuous, then broken up and the remnants reduced in size by fire spread by westerly winds. In time these fires, originating in extensive prairies to the west, advanced further into the forest, allowing small stands to persist only on the lee sides of natural fire barriers. When the incidence of fire was reduced, the forest reinvaded grassland.

COTTAM (1949) investigated the changes that took place in the vegetation of portions of the prairie–forest border region of southwestern Wisconsin since settlement by Europeans during the mid-19th century. The typical vegetation at that time was oak opening and prairie. Evidence indicated that these oak openings, that is, populations of widely spaced oak trees with an understory of prairie plants and forest shrubs, were created by the advance of prairie into forest as a result of fire and that the maintenance of these openings was effected by recurrent fires set by Indians (CURTIS, 1956). Under the influence of fire protection following settlement, the remnant oak openings were transformed in the space of 100 years to dense closed-canopy forests. Some of the old, open-grown trees remained and were eventually surrounded by tall, relatively unbranched forest-grown trees. A survey carried out by the original settlers showed a density of 14 trees per acre 4 inches (10 cm) or more in diameter. A survey approximately 100 years later showed 143 trees per acre of this size class. There was an increase in the number of species present, and COTTAM (1949) concluded that the woods were advancing toward a climax condition indigenous to the area.

When agro-urban effects are persistent, though varying in intensity, over a long period of time, the result is almost total destruction or drastic modification of the original forest ecosystems. The remnants of the original forest come to be restricted to sites not of immediate interest to man, such as infertile sandy soils, swamps, rocky outcrops, and hedgerows where successional development is impossible. Such sites retain a high proportion of pioneer species that become ecologically dominant and less susceptible to environmental change (CURTIS, 1956). Frequently these species thrive in environments influenced by man, and at the expense of those species comprising the original climax forest.

The first settlers to move into what is now known as the southern pine region of the United States encountered a climax forest of oak, hickory, and other hardwoods interspaced with loblolly pine. Cotton farming spread inland to the Piedmont plateau, and by 1840 87% of the land was under cultivation. However, by the end of the 19th century, owing to agricultural failures, civil war, and subsequent economic depression, there was massive abandonment of farmland, much of which was subsequently invaded by loblolly pine. Loblolly pine, therefore, a vigorous pioneering species with maximum capacity for germination and growth in a wide variety of environments, actually increased in distribution under the impact of man (WAHLENBERG, 1960). Its continued dominance over extensive areas requires the intervention of man to disrupt natural successional trends that favor broad-leaved species such as oak and hickory at the expense of pine.

The effects of man on the forest ecosystems of Europe began in prehistoric times and has continued from many centers of high population density to the present time. Only fragments of virgin forest now exist that have not been disrupted by man. JONES (1945) has briefly reviewed the literature pertaining to these vestigial populations. As early as the 5th century B.C. the destruction of the Mediterranean forest of oak and pine was well advanced, the major damage being caused by grazing, particularly by goats. "Vegetational climaxes have been broken insidiously rather than by some traumatic act, and just as the cultivation of food plants involves setting back ecological succession to a primary stage pastoralism deflects succession to the xeric, a profound and dangerous change." (DARLING, 1956) The practice of transhumance was common, and the destruction of natural regeneration was extended over large areas. Massive exploitation of forests for shipbuilding, fuel, and many other uses, added to the effects of grazing and agriculture, resulted in a shift toward xeric conditions throughout mediterranean lands. Natural forest ecosystems were completely destroyed, and the modern forester in these regions must cope with a legacy of bare, eroded surfaces, and vast tracts of brushwood communities (DARBY, 1956; SARTORIUS and HENLE, 1968; DASMANN, 1968).

This process of denudation of plant cover has continued to the present day, and old sanctuaries, protected for one reason or another through the ages, and now under the pressure of increased population, are being invaded. For centuries a unique sample of the original flora of Greece was preserved on the Mount Athos Peninsula under the control of the monasteries. In 1947 an invasion of peasants with 30000 domestic animals took place. Two years later the vegetation, which included 12 endemic plant species, had been browsed off (MELVILLE, 1970).

DARBY (1956) has described the clearing of the woodland in Europe. Prior to the advent of neolithic agriculture, central and western Europe appears to have been completely covered with broad-leaved forests of oak, elm, beech, lime, and hazel. Following the development of neolithic agriculture, much forest land was cleared by fires and maintained in this condition by grazing domestic animals. Over most of central and western Europe land clearing had passed its peak by the beginning of the 14th century. The next two centuries were ones of stagnation and there is evidence that during this period many forest ecosystems reestablished themselves on abandoned agricultural land. This period of relative stability was followed in the late Middle Ages by a rapidly expanding demand for wood, which has continued to the present day. The existing forests of Europe, therefore, differ greatly, both in distribution and structure, from the natural forests that would exist had there been no effects of man. One of the most striking and obvious effects has been the increase during the 19th and 20th centuries in the percentage of conifers. Throughout eastern and northern Europe there is evidence of the effects of man on forest ecosystems from the earliest times to the present day, and a number of authorities suggest that the treeless plains of steppes are the product of man's activities rather than the natural interaction of soil and climate. Shifting agriculture was common in northern coniferous forest, particularly among Finnish people. Fire was the principal tool used to clear forest land, and four to six years of cropping then followed. The subsequently abandoned land regenerated to birch and alder rather than to conifers. Similar practices existed in Sweden from prehistoric times. The northern coniferous forests, therefore, though not so greatly disrupted as those of central and western Europe, have not escaped agro-urban effects. Much of these forests have been cleared at least once by man's actions, and fire has undoubtedly played a major role in determining their modern distribution and composition (DARBY, 1956).

DARLING'S (1947) studies of the ecology of the Scottish Highlands, which is maintained in its present condition by pastoralism, provide a classic example of the changes that occur following the destruction of a forest ecosystem under the impact of man and his domestic animals. The terrain of the region is composed of acid rocks and steep slopes. Precipitation is high, and according to Darling the natural product of the region, as it was before the impact of man, is cellulose in the form of

Fig. 50. Tree savanna *Combretum splendens* in foreground, and *Erythrina tomentosa*, *Dombeya umbraculifera*, and *Bauhinia reticulata*, Kenya. (Photo R. S. Troup)

timber. However, during the last 150 years the region has been relegated to protein production in the form of sheep. The natural ecological climax of the region — oak, birch, alder, and pine, and accompanying fauna — to a considerable extent made good the deficiencies of the habitat. Deforestation, on the other hand, and the continued elimination of natural regeneration by pastoralism, resulted in its progressive deterioration. Darling (1956) has expressed the view that such deterioration is inevitable where pastoralism is conducted for commercial ends, and that where pastoralism over a long period has not damaged the habitat, it is markedly nomadic in character.

It is a commonly observed ecological phenomenon that many of the vast treeless, or virtually treeless, areas in Africa possess all the climatic requirements favoring the development of natural forest ecosystems. Despite this, however, savanna regions (Fig. 50) are continually being extended at the expense of high forest (Figs. 51, 52). Authoritative students of this phenomenon in tropical and subtropical regions of the world are virtually unanimous in their conclusion that the primary agency causing the advance of savanna at the expense of high forest is man using fire rather wantonly as his tool, and it is believed that practically all African savanna vegetation, where not forming stages in secondary successions, is a biotic climax determined by fire (Richards, 1957). "Man has transformed much of the central African equatorial forest into grassland by primitive agriculture, fire and grazing. By overgrazing and repeated burning, he has moved the zone of thornbush to actual desert." (Bartlett, 1956) This advance toward xeric, unforested conditions has been so universal in Africa that one authority has suggested that we are witnessing slow stages in the drying up and degeneration of tropical Africa (Aubréville, 1947).

So much of the original plant cover of Africa has been changed as a result of man's activities that it is difficult, if not impossible, to map its vegetation on

Fig. 51. Pure Mutabo stand, *Isoberlinia tomentosa* in Zambia. Such stands are typical of dry forest ecosystems in tropical Africa. In the past they have been managed under selection systems but increasingly they are clear cut, and the logged area planted with exotics

Fig. 52. Tropical rain forest with large *Lophira procera* and *Nauclea diderrichii*. Okomu Forest Reserve, southwest of Benin, Nigeria. (Photo E. W. JONES)

climatic climax concepts, and many workers have preferred to map the natural vegetation as it exists (SHANTZ and TURNER, 1958). According to AUBRÉVILLE, the dry closed forests, which in the past covered the drier regions of Africa, have ceased to exist save for a number of vestigial ecosystems found in relatively uninhabited areas, and protected by accidents of terrain. These vestigial or relic ecosystems appear to be analogous to those found in certain north American prairie lands as described by DAUBENMIRE (1968). In the fragmented ecosystems those species of trees and shrubs commonly occurring in isolated stunted groups on wooded savanna are found in closed communities. Various tree associations are frequent in these ecosystems, e.g., *Isoberlinia-Uapaca-Monites, Burkea-Erythrophleum-Tetrapleura-Prosopis, Parkia-Pterocarpus, Brachystegia-Isoberlinia, Cryptosephalum, Detarium-Parinari-Afzelia, Combretum-Terminalia; Anogeissus-Boswellia* (AUBRÉVILLE, 1947). STEBBING (1937) has made similar observations in regard to tree species found on degraded savanna lands, and concluded that these species are of true high-forest origin.

The effects of man are not confined to dry deciduous forests in tropical countries. Succession in tropical rain forest and moist deciduous forest in general follows a similar pattern throughout the world unless disrupted by the activities of man. Herbaceous weeds, shrubs, and small, rapidly-growing, short-lived trees are followed by larger trees typical of second growth forest, usually light-demanding and rapid-growing, and finally by the establishment of a climax association. In both moist deciduous forests and in tropical rain forests exploitation, grazing, cultivation, and fire have all played a role in disrupting forest succession. "Undoubtedly the greatest damage and destruction of rain forests, and the factor that has had the strongest effect in creating the conditions that exist today, has been man, both directly felling the forest to clear land for cultivation, through commercial cuttings, and through carelessness in the use of fire." (HAIG et al., 1958) RICHARDS (1957) has stated that unless determined efforts are made to halt the destruction, the entire tropical rain forest may disappear within our lifetime, except for a few inaccessible areas, and such forest resources artificially maintained as sources of timber. In Asia aggressive fire-resistant species such as teak and sal have colonized large areas, and huge expanses of grass and bamboo forests have been established as a result of shifting agriculture and repeated fire (HAIG et al., 1958).

It has been estimated (SHANTZ, 1948) that the present forests of tropical Africa occupy about 531 million hectares. The adjacent savanna and grasslands derived primarily from the destruction of forests is about 991 million hectares, most, if not all, of which is capable of producing forest under proper management. SHANTZ (1948) concluded that the present forests of tropical Africa have already been reduced to a third of what they probably were originally, and are shrinking rapidly at the present time. A more recent estimate has been given by McCOMB and JACKSON (1969), who estimated the area of African savanna to be five million square miles (1300 million hectares).

Much of Latin America lies within the tropics and the anthropogenous forces that have disrupted forest ecosystems throughout Africa have been equally disruptive on this continent. Many of the forest ecosystems of Mexico, Guatemala, El Salvador, Costa Rica, Venezuela, and Chile have been devastated, and the resulting problems of soil erosion, silting of rivers, and lowering of water cables are evident everywhere (VOGT, 1948). HOLDRIDGE (1956) has stated that the main cause of forest reduction until recent times in Central America has been rapid agricultural expansion and that the major part of the timber resources of the cleared forests in Central America has either rotted on the ground or been burned.

Shifting agriculture has played a major role in the disintegration of many of the forest ecosystems in tropical and subtropical countries. Land under shifting agriculture is capable of supporting a population of less than 20 persons per square mile (260 ha), and at this level the system is fairly stable. When population density increases to critical levels, as it has done almost everywhere, the fallow period is drastically shortened, soil deterioration occurs, and forest succession is permanently arrested. It is estimated that in Africa south of the Sahara the area of closed tropical high forest has shrunk by at least 100 million hectares as a result of shifting agriculture. The situation is similar in tropical America (BARTLETTS, 1956; NYE and GREENLAND, 1960; KING, 1968).

Numerous authors have discussed the effects of fire by primitive man on the forest ecosystems of tropical and subtropical countries (STEBBING, 1937; KUHNHOLTZ-LORDAT, 1939; AUBRÉVILLE 1947, 1949, 1956; BARTLETT, 1956; SAUER, 1956; STEWART, 1956; SHANTZ and TURNER, 1958), and there seems little doubt that man-made fire has for thousands of years affected the distribution and composition of these ecosystems. Clearly, naturally caused fire has always been a major ecological factor influencing the distribution and composition of vegetation in many forest ecosystems (VOGL, 1969; MUTCH, 1970). But since man discovered fire and its uses, fire as an ecological factor has enormously increased in intensity and scope. A corresponding intensification of an already present adaptive ability to survive the hazards of fire may have occurred in a number of forest ecosystems. Many tree species occupying man-made savanna are pyrophytes with inherent, fire-resistant characteristics that allow them to survive and regenerate in regions where fires are commonplace. As AUBREVILLE (1956) has pointed out, if these trees could not resist short-lived fires that take place annually during the dry season, they would have disappeared long ago.

Fire-resistant trees tend to have thick, corky bark and often possess a greater mass below ground than above, and on burning can regenerate vegetatively. On the other hand, species in non-fire-dependent tropical rain forest ecosystems usually have remarkably thin bark. Furthermore, buds are also much less protected, and, except for size, often look very similar in the resting and actively expanding condition (RICHARDS, 1957). Fire-resistant species such as teak and sal frequently exhibit a characteristic known as die-back. A whippy shoot is produced and dies back later in the season. The rootstock continues to increase in size and vigor until eventually a more vigorous leading shoot is sent up that is capable of surviving fire and other hazards (HAIG et al., 1958). TROUP (1921) has described the essential role of fire in the establishment of natural regeneration in the moister types of teak forest in Burma. The characteristics of fire-dependent pine species have already been referred to. As might be expected, fire-resistant tree species frequently characterize tropical and subtropical regions between high forest and savanna or between high forest and thorn shrub (BARTLETT, 1956).

One of the most informative and penetrating analysis of fire in relation to the evolution of forest ecosystems is that given by MUTCH (1970) in his study of wildland fires and ecosystems. The following is a summary of this analysis. Species have developed reproductive mechanisms, such as underground rhizomes, root sprouting, and serotinous cones, and anatomical mechanisms, such as thick bark and epicormic sprouting, in order to enhance their survival advantage under the selection pressure of fire. This being so, fire-dependent plants could also possess characteristics fixed through natural selection that increase the flammability of ecosystems in which they are prevalent. Although plant communities may be ignited accidentally or randomly, the character of burning is not random, but is associated with particular ecosystems. For example, the predominant eucalyptus forest in southeastern Australia is a pioneer association resulting from, and perpetuated by, frequent forest fires, which prevents succession to temperate rain forest type. Many pine forests, including ponderosa pine, are similarly established and perpetuated. Both eucalyptus and ponderosa pine produce an energy potential on the forest floor that helps to insure a high flammability in these stands. Given an ignition source and favorable environmental conditions, a widespread, fairly intense fire is inevitable. Therefore, these ecosystems can be perpetuated through periodic burning because plant characteristics enhance fire propagation. Tropical-rain-forest litter, on the other hand, provides an example of a non-fire-dependent ecosystem. Litter was collected from all three ecosystems and subjected to physical analysis and combustion tests in the laboratory: (1) *Eucalyptus obliqua* leaves from the warm temperate zone (36° 26′S latitude, 143° 03′E longitude), (2) Ponderosa pine needles from the cool temperate zone (46° 55′N latitude, 114° 05′W longitude), and (3) tropical hardwood leaves from the equatorial zone (11° 30′N latitude, 106° 01′E longitude). The tropical hardwood litter sample included leaves of *Anacardiaceae*. The laboratory tests conducted on this material suggested that the chemical composition of plants establishes the energy base upon which fire intensity is determined. A hypothesis was formulated that goes beyond the commonly accepted fire climate–fuel moisture basis of wildland fire occurrence. This hypothesis states that fire-dependent plant communities burn more readily than non-fire-dependent communities because natural selection has favored development of characteristics that make them more flammable. Thus: "Wildland fires must be studied and managed as integrated events associated with the ecosystem. The vegetation brings certain properties to the ecosystems that condition the fire history, and the fire history determines, in part, the maintenance, regression, or succession of plant communities. The proposed hypothesis attempts to bring order and design into the study and management of fires in context with the ecosystem." (MUTCH, 1970). See Fig. 53.

Because of man's effects on natural forest ecosystems, frequently only vestigial, fragment populations of these systems remain. Thus a knowledge of the man-induced forces that have influenced the present distribution and structure of these populations is important in relation to the selection and breeding of the species of which they are composed. CURTIS (1956) gave a case history covering the first century of use of a township of land in Green County in Wisconsin (Fig. 54). The vegetation in 1831 before agriculture settlement began was mostly upland deciduous forest dominated by basswood, slippery elm, and sugar maple. Forest cover was progressively destroyed and by 1954 was reduced to 3.6% of the original. In the same area, as a result of the drying up of springs in the original headwaters, permanently flowing streams had decreased by 36% by 1935.

Fig. 53. Wind and fire damage in virgin stand of *eucalyptus regnans*, Tanjil Valley, Victoria, Australia

The pattern of progressive destruction of natural forest ecosystems described by CURTIS has occurred throughout Europe many centuries ago in and close to densely populated areas, and is presently taking place on the periphery of great urban areas in Canada and other parts of the world where natural forest ecosystems still exist.

Change in Wooded Area, Cadiz Township, Green Country, Wisconsin (89°54′W, 43°30′N), from 1831 to 1950 (CURTIS, 1956)

	1831	1882	1902	1935	1950
Total acres of forest	21,548	6380	2077	1034	786
Number of wood lots	1	70	61	57	55
Average size of wood lot in acres	21,548	91.3	34.0	18.2	14.3
Total wooded area as a percentage of 1831 condition	100	29.6	9.6	4.8	3.6
Total periphery of wood lots in miles	—	99.0	61.2	47.2	39.8
Average periphery per wooded acre in feet	—	82	155	241	280

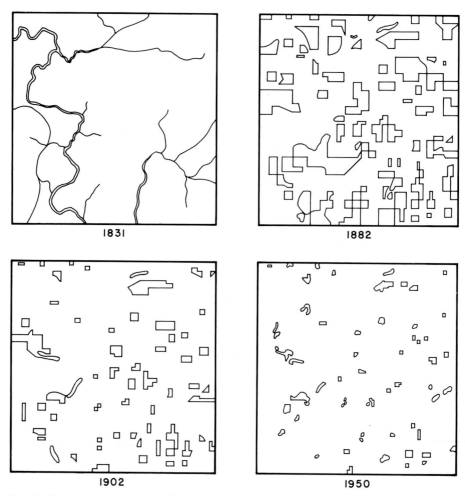

Fig. 54. Changes in wooded area of a township in Green County, Wisconsin, during the period of European settlement. The township is six miles on a side. Outlined areas represent the land remaining in or reverting to forest in 1882, 1902, and 1950. (After CURTIS, 1956)

The small size and fragmented nature of vestigial populations of the original forest ecosystem reduces the probability of seed and pollen exchange between populations. Due to increased xeric conditions and general disturbance, there may be a loss of one or more species with a resulting increase in the number of stems of the species that remain. Invariably the more conservative species of the population are eliminated; i.e. species that are more demanding in their requirements for germination and growth and with low tolerance of fluctuations in the moisture regime. "They make up the most advanced communities of a given region from the standpoints of degree of integration, stability, complexity and efficiency of energy utilization. They are climax plants in the basic sense of the word." (CURTIS, 1956) For example, the replacement of yellow birch in mixed-wood stands in eastern

Canada and the United States following disturbance is a commonly observed phenomenon. Inventory has shown a decline of 16% for yellow birch in eastern United States in the years 1963–1968. The change has been attributed to the inability of this intolerant species to compete with other species in unmanaged second-growth stands (HAIR and SPADA, 1970). Similar changes in species composition occur in tropical and subtropical forest ecosystems, and the view has been expressed that all secondary tropical forest species originally existed in small numbers in primary rain forest, and large-scale clearing and burning have resulted in the selection of those species able to survive and thrive in the changed environmental conditions (RICHARDS, 1957 citing VAN STEENIS, 1937). The survival advantage is shifted to those species with maximum tolerance of a wide variety of environmental conditions. These latter species are generally more susceptible to monoculture than climax species, and hence the emphasis on such species in the establishment of man-made forests.

Periodic disturbance of forest ecosystems and the reduction of population size, for example, by the increase and contraction of agricultural activity, may result in a survival advantage being conferred not only on individual, vigorous pioneering species, but also on populations of a given species not considered a pioneer under natural conditions. As already indicated, the small size and increased isolation of remnant stands reduce the exchange of seed and pollen, and a reduction in species composition can result from increased xeric conditions. Cross-pollination may be confined to relatively small numbers of trees in isolated stands with resulting increased probability of the evolution of deviant types by random gene fixation (CURTIS, 1956). Under these circumstances, and with the gradual elimination of the more conservative climax genotypes, the selection advantage may shift in heavily disturbed stands to those genotypes that have pioneer tendencies. For example, white spruce plantations established from seed collected in natural populations in a region of the Ottawa River valley where agro-urban effects have been operative for a relatively long period exhibit strikingly vigorous growth when planted over a wide range of environments (NIENSTAEDT, 1968; TEICH, 1970; CORRIVEAU and BOUDOUX, 1971). This growth vigor cannot be fully explained in terms of the climatic environment at place of origin, and it is possible that the apparent success of this particular strain as a plantation tree is more easily explained in terms of agro-urban effects on the original forest ecosystem of the area. These effects may have resulted in the evolution of a deviant pioneering type more suited to monoculture than populations from relatively undisturbed ecosystems established through normal forest succession.

ANDERSON (1956), writing of man as a maker of new plants and new plant communities, concluded that man has been a major force in the evolution of both plants and animals. To what extent this conclusion is valid for long-lived tree species is difficult to say, as research in this field in relation to forest trees has been limited. However, the synthetic theory of evolution as restated by HARDY (1965) suggests that the colonizing potential of an animal species in new environments is a profound source of evolutionary change. As has already been suggested, a shift in survival advantage to less specialized genotypes of a given tree species may take place as a result of man's intensive and continual intervention in forest ecosystems and the creation of new environments. CHAPMAN (1950) has sug-

gested that lightning fires were responsible for the undoubted genetic adjustment of long-leaf pine to fire. STEWART (1956), however, has pointed out that the
incidence of lightning-caused fire is too low in the southeastern pine forest for that
natural agent to result in evolutionary changes, and suggests that this change is due
to the increased incidence of fire caused by early man. "If ten to twenty-five
thousand years would suffice for the genetic adjustment of the southern pines, the
adjustment could be due to man-made fires" (STEWART, 1956). In this context it
should be noted that there is increasing evidence that highly variable, heterozygous
tree species can adapt to new conditions with much greater rapidity than was
previously thought.

The effects of man on the forest ecosystems of the world not only have resulted
in changes in distribution, species composition, and genetic structure, but also have
caused the extinction or near extinction of a number of tree species. When an
animal or bird is threatened with extinction it is very quickly a matter of public
concern. The threatened permanent loss of a plant species, however, has not had a
similar effect on the public consciousness. Yet, as MELVILLE (1970) has pointed out,
the rate of species loss is increasing, and for the majority of disappearing or threatened plants the hand of man can be seen in their decline. It is only in the last few
years that the extent of the problem has begun to emerge, and according to MEL-
VILLE (1970) the numbers of endangered species and other Taxa of flowering plants
must be many thousands.

GORDON (1968) has suggested that *Picea chihuahuana* may not long escape
extinction, and KRUG (1967) pointed out that because of the way the original
forests are being treated *Araucaria angustifolia* as a timber-producing tree will
probably vanish within a few years. The report of the Food and Agricultural
Organization (FAO) Panel of Experts on Forest *Gene Resources* (FAO 1969) listed
the following species threatened with extinction or severe depletion of the gene
pool: Central and South America, *Araucaria angustifolia, Pinus caribaea* var. *baha-
mensis;* Mexico, *Pinus maximartinezii;* Mediterranean region, southern Europe,
and Near East, *Abies nebrodensis, Cupressus dupreziana, Pinus brutia* var. *elder-
ica;* southeast Asia *Pinus merkusii;* Africa, *Aucoumea klaineana.* As recognized in
the report, there is little doubt that this list is incomplete and that the second report
of this panel will include other species.

There are also tree species not presently of major commercial importance that
are in danger of extinction or severe depletion of the gene pool. MELVILLE (1970) has
listed a number of these species, of which the following are representative. There
were four endemic tree species of the genus *Hibiscadelphus* in the Hawaiian Archi-
pelego. Two species are now extinct, the last surviving tree of the third has been
propagated, and the fourth survives as a small grove in the Volcanoes National
Park on the island of Hawaii. The species *Vateria seychellarum* dyer was at one
time a common timber tree, and produced a good-quality timber comparable to
that of other Dipterocarps from Malaya. It is endemic to the Seychelles, and it is
recorded that now only three trees survive on Mahé. Nothing is recorded of its
biology, and no protective measures have been taken to ensure the survival of this
species. Altogether, MELVILLE (1971) lists 25 tree species at present in danger of
extinction or severe depletion of the gene pool and it should be noted that this list is
continually being added to.

2. Modern Forestry

Since the end of World War II there has been widespread recognition of the degree to which man has devastated the face of the earth. This recognition has led to considerably improved forestry practice in many countries and a serious reexamination of conventional attitudes concerning the use of forests by native people. Many of the problems of modern forestry, particularly in tropical and subtropical countries, are related to past effects of primitive man on natural forest ecosystems of the world. Nevertheless, studies such as that of KING (1968) on agri-silviculture have indicated the inherent wisdom of certain uses of the forest by native people, and their validity in relation to economic conditions in developing countries. An era of rehabilitation of degraded forest lands has been initiated, and forest conservation and silviculture have achieved in many countries a degree of importance in the public consciousness that they have not had previously. Furthermore, forests are increasingly regarded as a renewable resource with multiple uses rather than as an expendable, single-use resource, and there is a greater acceptance of the view that well-managed forests on the periphery of great urban areas have a very high social value. Finally, man-made forests are rapidly increasing in size and importance, and genetic principles, used so effectively in agriculture and horticulture, now form part of modern silvicultural practice in many countries. Though these developments have considerably extended the scope and complexity of the literature pertaining to forestry, there are, nevertheless, certain trends in modern forestry, and their effects on forest ecosystems, that may be identified and described. This is the object of the following section, and since forestry is worldwide in scope, the discussion will be confined to major trends, and to representative forest ecosystems.

a) Exploitation and Natural Regeneration

The classical silvicultural systems (Fig. 55) that evolved in Europe in the 18th and 19th centuries were built about the central problem of natural regeneration of a relatively small number of species. Frequently the success of such systems depended on the fact that the method of exploitation was dictated by the silvicultural system used. In more recent times, however, and under the economic pressures of demand, methods of exploitation have developed quite independently of silvicultural systems and increasingly these systems are being dictated by the method of exploitation. Still, variations of the classical European shelterwood and selection silvicultural systems linger on in the tropics and parts of North America, even when the conditions for their success are no longer present. Exploitation has, of course, proceeded apace in many parts of the world without any attempt to ensure the regeneration of the exploited species. It was, and still is in a number of countries, taken for granted that they would regenerate naturally and in sufficient abundance to restock the exploited area — an assumption that had some validity prior to the era of clear-cutting, and when extraction was by horse, buffalo, or elephant, but which under the conditions of modern exploitation has little foundation.

For many years foresters in North America have relied to a great extent on natural regeneration to restock exploited forest ecosystems. "Economic conditions have discouraged elaborate and painstaking care in seeing that reproduction be-

Fig. 55. Natural regeneration by group silvicultural system in spruce and silver fir. Note young group of natural regeneration joined to older group behind. Schifferschafts forest, Forbach, Baden Black Forest. (Photo R. S. TROUP)

comes established over the vast extensively managed spruce-fir stands in Canada and in the United States with the result that reproduction is left largely to chance" (BAKUZIS and HANSEN, 1965). There is little doubt that natural regeneration has occurred over most cut-over areas, but it is equally certain that it is frequently inadequate and that there has been a massive shift in species composition, and a concomitant reduction in the range of occurrence in commercial quantities of a number of important species (Fig. 56).

Forests of white and red pine once covered the northeastern part of the United States and adjacent regions in Canada, and provided the raw material for the great softwood lumber industry since colonial times. However, owing to overexploitation, only remnants of these great forests remain today. Throughout the range of these two species most pine sites have been restocked with aspen, birch, or other northern hardwoods of some commercial importance (JOSEPHSON and HEIR, 1956). These latter species have in turn been seriously depleted to the point of exhaustion in certain regions (KIRK, 1970). According to JONES (1945), the original virgin forest of eastern and central America has been destroyed so completely that only fragments now remain.

Fig. 56. Logged and burned Engelmann spruce and alpine fir stand regenerating to lodgepole pine. Southwest of Cranbook, British Columbia

HOSIE (1953) reviewed the literature pertaining to natural regeneration in Ontario. This review, which was based on regeneration surveys conducted in the province during the period 1918–1951, concluded that there was a considerable area of cut-over land, particularly on the better sites, that was not reproducing adequately either as to species or as to quality. Furthermore, for many valuable commercial species little progress has been made in developing methods of logging that would insure adequate regeneration, and seeding and planting would be necessary. One of the most comprehensive surveys of natural regeneration following exploitation in Canadian forest ecosystems is that completed by CANDY (1951). Surveys were carried out over a three-year period in major forest ecosystems from the Atlantic to the Rocky Mountains. The method employed was the stocked quadrat system, using quadrats one milacre in size. In all, 583000 quadrats were examined. There was a broad clinal pattern in degree of stocking from the Maritimes in the east, where stocking was adequate, to the prairie provinces in the west, where stocking was generally inadequate. A reduction in the incidence of white and red pine, and yellow birch, was apparent in the eastern regions following conventional logging. Black spruce swamp type regenerated well, but in drier sites stocking of black spruce was inadequate, and generally replaced by balsam fir. Regeneration following fire in logged areas was found to be disastrously low. Most conifer reproduction, particular in the eastern part of the surveyed area, was to balsam fir. Later studies of natural regeneration following logging in forest ecosystems east of the Rocky Mountains have in general confirmed CANDY's findings, and most concluded that increased reliance must be placed on the artificial restocking of cut-over areas (WEBBER et al., 1969; HUGHES, 1970; DE VOS and BAILEY, 1970. (See Fig. 57)

Fig.57. Logged and burned black spruce stand near Chibougamau, Quebec. No regeneration
nine years after logging

In Europe natural regeneration by the uniform shelterwood system is still
generally preferred for *Fagus sylvatica,* though some planting is done, and some
forests of *Picea abies* are still regenerated naturally by shelterwood methods in
southern Germany, Switzerland, and Czechoslovakia (LEWIS, 1967). There is a
trend, however, toward the abandonment of elaborate stand manipulation to
obtain natural regeneration, and productivity rather than successful natural
regeneration of the stand is increasingly considered the criterion of good management. Artificial regeneration is now the rule rather than the exception in Europe.

Only a few of the large man-made pine forests in the southern hemisphere have
now reached the regeneration stage, though it is anticipated that by the end of the
century this will be a major problem. In South Africa *Pinus patula* is regenerated
naturally following felling in 30–40-year-old stands, and in Chile *P. radiata* is
commonly regenerated naturally. In New Zealand this species is regenerated both
naturally and artificially. In South Africa natural regeneration of *Acacia mearnsii*
usually follows clear-cutting. In nearly all instances of natural regeneration of
man-made forests, site treatment is requisite to successful stocking, and frequently
supplementary planting is necessary (LEWIS, 1967).

Natural regeneration in evergreen tropical rain forests results in the establishment of a climax condition characterized by a multiplicity of species (seldom less
than 40 and sometimes well over 100), great density, and a lofty multilayered
structure with an abundance of lianas and creepers (RICHARDS, 1957; HAIG et al.,
1958). The process of natural regeneration in such ecosystems is not well understood, and much of the literature dealing with so-called natural regeneration of
tropical rain forests "refers to the reproduction of a few economic species under conditions rendered more or less unnatural by the exploitation of timber"

(RICHARDS, 1957). Yet, as RICHARDS (1957) has stressed, before regeneration under these artificial conditions can be understood, or controlled scientifically, much more must be known about what happens under undisturbed conditions, and information about this is extremely scanty. However, it should be pointed out here that there are many foresters, both in the tropics and in the north temperate zone, who would not fully agree with this point of view. Many would argue that studies of regeneration following disturbance are at least of equal validity and have provided a great deal of indispensable information.

Tropical rain forests, because of their heterogeneity and low productivity, are in general of low commercial value, and natural regeneration, when it is sought by foresters, is regulated by a variety of shelterwood and selection systems. The presence of advance growth of the desirable species is often essential to the success of the system used, though many successfully regenerated species are obligate light-demanders that cannot exist, except in casualty gaps, prior to felling. In many cases, however, and despite the success obtained with certain species, the silvicultural system employed has not always yielded the desired results, and it has been stated (STEENBERG, 1969) that although methods of tending advance growth are known, it is doubtful whether it is really known how to induce natural regeneration of a desirable species in a truly mixed tropical high forest. Furthermore, the tending of advance growth of tropical forests is in many cases a very costly operation. Tropical shelterwood systems have not been applied in French-speaking tropical countries on the same scale that they have been applied elsewhere, and enrichment planting practices have been much more common (HAIG et al., 1958). To the extent that the tropical shelterwood system requires an intensive manipulation of the natural forest, it is a regeneration method hardly less artificial than that employed by enrichment planting practices.

Both teak and sal, and particularly the former, are the most valuable species of moist deciduous forest ecosystems, and problems related to their silviculture and management have dominated forestry practice for many years in Southeast Asia (HAIG et al., 1958). Teak is associated with a diversity of other tree species, e.g., *Dipterocarpus* spp., *Tuberculatus* spp., *Pentacme suavis*, and *Shorea robusta*, in the drier regions of its distribution, and *Cedrela* spp., *Shorea assamica*, and *Michelia champeca* in the wetter. Frequently bamboos form the main constituent of the undergrowth. Teak may form only 12% of the growing stock, but on occasion it may be as high as 50% with a variety of species of different economic value making up the balance. The perpetuation of teak in such forests in the past has been insured by various combinations of selection and shelterwood systems, and on occasion by arificial regeneration. In Indonesia most teak forests are regenerated by systems of agri-silviculture (Fig. 58). The death of the bamboo undergrowth following flowering and its subsequent elimination by fire can result in the abundant natural regeneration of teak. Sal is characteristic of many moist deciduous forests in India, and may form 60–90% of the growing stock. Its most common associates in the upper canopy are *Pterocarpus marsupium*, *Terminalia tomentosa*, and *Eugenia jambolana*. Bamboo is usually absent in such forests. Various combinations of shelterwood and selection systems are used to obtain natural regeneration of sal, though in suitable areas artificial regeneration by systems, of agri-silviculture is common. In the dry deciduous tropical ecosystems in Southeast Asia teak and sal are usually

Fig. 58. Teak plantation established by method of agri-silviculture. Note nurse crop of *Leucaena glauca*. Tjepu, Java. (Photo H. G. CHAMPION)

managed by a system of coppice with standards. The artificial regeneration of both these species is likely to increase, for despite the variety of the silvicultural systems used to obtain natural regeneration the results frequently do not justify the cost and efforts made (HAIG et al., 1958).

There are no large areas of moist deciduous forests in Africa comparable to those that exist in Asia, which are typified by the presence of teak and sal. In those regions of Africa where such forests might be expected because of climatic patterns, wooded savanna prevails (HAIG et al., 1958). In tropical America the degree of disruption by man has been such that it is difficult to delineate the distribution of moist deciduous forest ecosystems, though where these forests are extant they contain a number of valuable commercial species, such as *Swietenia* and *Cedrela* spp. Both of these species are regular and prolific seed producers, light-demanding, and capable of vigorous early growth. Hence they both regenerate well on cultivated, cleared, or burnt land. An example of the pioneering abilities of both these species has been given by LAMB (1967) for British Honduras where, following a forest fire, natural regeneration resulted in the establishment of 70–80 mahogany and cedar saplings per acre over large areas. *Brachystegia–Isoberlinia* woodlands, typical of dry forest ecosystems in Africa, have been managed in the past under selection systems. Increasingly, however, such woodlands are being cleared and planted, often with exotics. A similar trend is apparent in the dry deciduous forests in Latin America (HAIG et al., 1958).

DAWKINS (1961), in discussing present and future trends in tropical forestry, has indicated that when demands on the forest are intense — that is, when optimum utilization is needed — intensive replacement takes precedence over extensive improvement. This is the position in the Asiatic tropics, where artificial regeneration is the rule. In many other tropical countries, however, demand is light and concentrated on a small numer of species only. In such circumstances extensive improvement techniques, incorporating new methods rather than modifications of traditional practice, will continue to have validity for a long time to come in the tropical regions of Africa and Latin America. Methods of improving stand composition embracing both natural and artificial regeneration have been discussed by AUBRÉVILLE (1957), HOLDRIDGE (1957), DAWKINS (1961), LAMB (1967b, 1969a), and CATINOT (1969).

In summary, it may be said that modifications of classical European silvicultural systems have been applied with varying degrees of success throughout exploited forest ecosystems in north temperate and tropical countries. However, the trend is increasingly toward artificial regeneration, using modern methods of improving stand composition. The process is hastened by a more general acceptance of the view that: "The concepts of stand regeneration, as they are accidentally realized in the European selection forests are not suitable for far reaching generalizations. They are based upon specific phytogeographical, ecological, and even socio-political conditions which may apply to part but not to all the forests of the world, and even not to most European forests" (VAN MIEGROET, 1967).

Successful natural regeneration of a desirable species following exploitation is the result of the interaction of a large number of variables, e.g., the reproduction cycle of the desired species, the size and time of the cut, the degree of disturbance of the soil and advance growth, the effect of fire and weather patterns following logging. Seed years can be extremely erratic, and seed yields low in quality and quantity resulting in inadequate stocking. Also, the modern practice of clear-cutting large areas with massive machinery, though in the short run economically sound, is biologically inimical to natural regeneration of most species, a number of pine species being the exception. Remaining seed sources in most instances are too sparse and the inadequacies of natural seed sources thus greatly augmented. Furthermore, environmental conditions unfavorable to the survival and growth of established seedlings are the rule rather than the exception in large cut-over areas.

Throughout the world there is a trend toward mechanized logging (Fig. 59) and where mechanized logging has already taken place there is continual innovation, and a persistent drive to find ever-more-sophisticated logging equipment. (See ANONYMOUS, 1970 for a review of modern trends in mechanized logging systems.) No stage in this development is seriously influenced by such considerations as the effect in a given forest ecosystem of a particular logging machine or logging system on natural regeneration following logging. Economic pressures are forcing the increasing mechanization of all forest operations even before the results of biologic studies, whether favorable or unfavorable, are known (SILVERSIDES, 1964). The most recent development of the whole tree logging system is a case in point. With this system the felled tree is skidded to a central roadside point, where it is limbered, barked, and slashed. Thus the entire nutrient content of the tree is taken from the site. The long-term effects of this logging system on forest ecosystems is virtually unknown. As usual, attempts to determine the effects come only after the system has been developed. If it is demonstrated by research that the effects are bad, the system will in no way be affected. Action will be taken to offset the effects by adding fertilizier rather than eliminating the cause by changing the logging system. Thus

Fig. 59. Timberjack grapple skidder operating in a white spruce stand near Prince George, British Columbia. (Photo R. J. HATCHER)

virtually all the major factors affecting natural regeneration, including a number controlled by man such as logging equipment and systems, vary both in time and space, and in many instances it is impossible to determine the probability of successful natural regeneration on the basis of past experience. When adequate natural regeneration of a desired species does occur following logging by modern methods, it is more often due to the chance combination of favorable circumstances rather than to the efforts of silviculturists. It is not surprising, therefore, as LEWIS (1967) has pointed out, that "a notable feature of modern regeneration practice is how little natural regeneration is used." Though this observation was made in relation to the regeneration of man-made forests, it is increasingly applicable to natural forest ecosystems.

In the tropics, as in other forest ecosystems, natural regeneration of many noncommercial species is not, of course, a major problem. Frequently, however, desirable commercial species are shade-tolerant, slow-growing species of climax forest ecosystems, and thus the ones most difficult to regenerate following disruption of the ecosystem. The shift in species composition following exploitation in many tropical ecosystems, such as that described by LAMB (1967) for tropical lowland forests in Trinidad, is analogous to that which occurs in the temperate forest zones following exploitation. Invariably the survival advantage is transferred to pioneer, rapid-growing, intolerant species, e.g., aspen and willow or pine at the expense of spruce in Canada, and *Byrsonima spicata* and *Didymopanex morototoni* at the expense of *Manilkara* in Trinidad.

Many of the difficulties encountered by silviculturists, whether in the tropics or the north temperate zone, are frequently derived from attempts at securing natural regeneration of climax forest species in conditions that favor shade-intolerant, rapid-growing, pioneer species, or at arresting succession in order to maintain a forest in a seral stage. In either case, adequate information, both as to natural forest succession within the ecosystem and the biology of the species being subjected to treatment, is indispensable. More often than not, this information is lacking. These difficulties are likely to continue, particularly in the tropics, until the biology and ecology of species are better understood, and markets are available for species presently of no commercial value. In the meantime, the trend toward permanent elimination or massive disruption of natural forest ecosystems by modern methods of exploitation is unmistakable. Inevitably, many logged areas must be restocked by artificial methods if they are to remain productive. Modern exploitation, there-

Fig. 60. Aerial photo showing strip cutting in white spruce on North Western Pulp and Power Co. limits at Hinton, Alberta. The residual strips that appear dark an the photograph are left as a source of seed until the intervening strips have been regenerated with established seedlings. Note the fringe timber left along the major watercourse that serves as a buffer between the river and the cutover area. (Photo Western Pulp and Power Co.)

fore, is an important factor in the continual reduction of the world's natural forest ecosystems, and the establishment of man-made forests. A significant change, compatible with economic realities, would require a wider understanding and acceptance of the ecosystem concept (BAKUZIS, 1969) in forest resource management, and also the intrusion of biological considerations and conservationist philosophy into modern methods of exploitation (Fig. 60).

The one factor that is not likely to result in significant change, despite its importance in many silvicultural research programs, pertains to small, experimental cuts, the results of which rarely influence the major trend of modern exploitation.

b) Man-made Forests

Many factors influence the establishment of man-made forests, which may be defined as a forest crop raised artificially either by sowing or planting, but to date the most important are those associated with economic value. Ideally, therefore, man-made forests possess the most important characteristics for economic wood production. These characteristics may be summarized as follows: (1) suitability of the wood produced for the end purposes proposed; (2) homogeneity and size of growing stock; (3) large volumes per unit area, allowing intensive working and short logging hauls; and (4) accessibility in relation to markets. Some natural forests possess these characteristics in varying degrees (for example, northern coniferous forests) and others (such as the forests of tropical countries) generally lack these characteristics. Other characteristics of plantations, although not necessarily important for economic wood production, very often influence decisions to clear and plant rather than manage existing natural forest. Plantations are easily seen, and, when successful, politically impressive. They may be clear-cut with greater impunity, and are easier to administer than manipulated natural forest. Finally, under favorable conditions plantation growth rates in the tropics generally exceed those of the natural forest, often by a factor of ten. On the other hand, in northern coniferous forests — for example in the boreal forest zone of Canada — northern plantations are not likely to have such an advantage over the natural forest. These facts considerably influence the relative importance of man-made forests in the north temperate and tropical zones (FAO, 1967a). It should be noted, however, that reported large differences between plantation growth rates and those of the same species in natural stands are often more apparent than real. Frequently the growth rate of a successful plantation of a particular species on a rich site is compared with the growth rate of the same species in an untended, polyspecific natural stand of which the species is only a minor constituent, and which occupies a different site.

It has been estimated (FAO, 1967a) that man-made forests planned between 1965 and 1985 will cover approximately 80 million ha a rate of establishment of 4 million ha per year, and that these forests will yield a world average of $5/m^3/ha/yr.$ with an average rotation of 40 years.

Many pulpwood plantations, of course, such as those of eucalyptus and tropical pine, are being managed on an 8–15-year rotation. But, assuming an average rotation of 40 years, the four million hectares to be planted in 1975, the mid-point of the

period, would, make its full contribution to the world's increment of wood in the
year 2015, when it would produce the equivalent of an additional annual increment
of 20 million m³. On the other hand, the annual growth in the world's demand for
industrial wood alone in the year 2015 has been estimated at more than 60 million
m³ (FAO, 1967a). Furthermore, fuelwood continues to make up as much as half of
all the wood used in the world today. In Africa and Latin America nearly nine
tenths of all wood used is fuelwood; in Asia, excluding Japan, it is two thirds; in
Europe and the U.S.S.R., more than a quarter; and in north America, a tenth. With
increasing urbanization and industrialization there will be a decrease in the use of
fuelwood by some countries. Nevertheless, because of the huge expansion of rural
populations in developing countries, it has been estimated that there will be an
overall increase in the demands for fuelwood for some time to come, and that
demand is likely to reach a world total of 1199 million m³ in 1975 (FAO, 1967b).
Thus it is clear that the new man-made forests to be planted under present plans are
likely to contribute only a relatively small part of the additional total wood supplies
that the world will need, and that the remainder must come from natural forests.

Man-made forests in North America are concentrated almost entirely in the
United States, which possesses approximately ten million hectares of man-made
forests as opposed to Canada's 295 000 hectares (FAO, 1967b). There are a number
of facts that help to explain this discrepancy between the two countries. First,
Canada's forests are not fully developed. For example, the annual cut in 1964 was
less than half the allowable cut (CAYFORD and BICKERSTAFF, 1968), whereas the
United States is expected to face a wood deficit by the end of the century. Second,
the rapid growth rates and short rotations that can be achieved in man-made
forests in the southern pine region of the United States have been a considerable
stimulus to the extension of man-made forests throughout this region. It has been
estimated that the region will be producing 368 million m³ of softwood by the year
2000 (CLIFF, 1969) (Fig.61). Third, Canada has continued to depend heavily on
natural regeneration for restocking cut-over forest lands. For example, in British
Columbia, where 400 000 acres (161 877 ha) are cut over annually, large areas are
left unplanted in the hope that they will regenerate naturally (GILMOUR, 1970).
However, as already indicated, there is evidence that natural regeneration is not
occurring on many cut-over and burned-over areas in Canada, and that where
natural regeneration does occur stocking is unsatisfactory, and there is a considerable shift in species composition. It has been estimated that an annual total of
500 000 acres (202 347 ha) of cut-over and burned-over forest land in Canada is not
regenerated (CAYFORD and BICKERSTAFF, 1968). It is clear that if Canada is to maintain its accessible lands under productive forests the establishment of man-made
forests in these regions must be considerably increased. This increase is likely to
take place, therefore, in the southern, more accessible, parts of the country. It has
been estimated that by 1985 there will be about 5½ times as much land under man-
made forests in Canada as there was at the end of 1965. The largest areas will be in
British Columbia, Quebec, and Ontario (CAYFORD and BICKERSTAFF, 1968). In both
Canada and the United States, the trend, examplified particularly in the southern
pine region and in Ontario, is toward the establishment of man-made forests with a
resultant reduction in the number of species used, an increase in site treatment, and
the use of fertilizers, pesticides, and genetically improved and less variable planting

Fig.61. Eleven-year-old loblolly pine progeny test just prior to second thinning. This progeny test was first thinned at 7.5 years of age. Growth between thinnings averaged nearly four cords/acre/year (approximately 22 m³/ha/yr.). Test conducted by Westvaco Corporation in the southern pine region of the U.S.

stock, together with the mechanization of cone harvesting, sowing, planting, and tending operations.

Throughout Europe the establishment of man-made forests has resulted in a large increase in the area under coniferous species, both native and exotic. For example, the amount of coniferous species in the growing stock of the Federal Republic of Germany has been increased from 30% in the beginning of the century to 70% in 1960 (FAO, 1967b). The major trend appears to be to replant with the same species following logging, provided that yields continue to be satisfactory, though *Larix* and *Quercus* species, and on occasion *Pinus sylvestris*, are being replaced by other conifers such as *Picea abies*, *P. sitchensis*, *Pseudotsuga menziesii*, and *Pinus nigra* var. *calabrica* where sites are suitable for them (LEWIS, 1967) (Fig.62). When possible, higher productivity has been sought by concentrating planting, frequently with poplar species and hybrids, in the warmer, southern regions of the continent. In Italy, for example, though occupying only about 6% of the forest area of the country, poplar provides 40% of the output of industrial wood (FAO, 1967b). In general it can be said that the principal objective of Euro-

Fig. 62. Thriving, 40-year-old Sitka spruce plantation in Glenbrauter Forest, Argyll, Scotland. This species is an important exotic and a major commercial species in Britain and Ireland. It is also an important exotic in a number of other western European countries and in Scandinavia. (Photo L. W. THOMAS, British Forestry Commission)

pean forestry is to maintain, and possibly increase, productivity of existing forests, both man-made and natural, and to enhance the recreational and overall social value of these forests. In a number of other European countries, however, such as Ireland, Britain, Spain, Portugal, Greece, and Turkey, the area under forest is likely to increase as unproductive and degraded land is brought into productivity by the establishment of man-made forests. It has been argued that the recreational value of forests in certain European countries is of greater significance in economic terms than cellulose production (RICHARDSON, 1970).

Fig. 63. *Pinus patula* plantation, Ruungi Peak, Southern Rhodesia. Note patch of indigenous forest in foreground. (Photo H. G. CHAMPION)

According to FAO predictions (FAO, 1967a, b), the trend toward man-made forests in general could coincide with a tendency to site them nearer the equator wherever rainfall is sufficiently adequate to allow trees to take advantage of an almost continuous growing season. Some 15 m³/ha/yr. can be obtained from fast-growing conifers when planted in tropical and subtropical zones, and 20–30 is not uncommon on the best sites with *Pinus radiata, P. patula,* and *P. caribaea* (Fig. 63). In east and central Africa, eucalypts have yielded volumes as high as 35–55 m³/ha/yr. In warm temperate climates, such as Italy and Turkey, and with fall irrigation, 20–30 m³/ha/yr. is common for poplars, and 30–40 m³ on especially good sites. Eucalypts are capable of similar yields if planted in areas of good rainfall, or with irrigation in the hotter conditions of the tropics or subtropics. Indeed, this genus is superior to almost any other on good sites in tropical countries. Other tropical hardwoods are also capable of rapid growth — *Acrocarpus, Gmelina, Terminalia* spp., *Cedrela, Maesopsis,* and *Tectona* (Fig. 64). This last species, however, is by far the most important tropical species planted to date. There are now about one million hectares of teak plantations, mainly in Indonesia and Burma, as well as smaller areas in a number of other tropical countries. Eucalypts have formed exotic plantations for over 100 years, and are now important species in the forest economy of many countries (PENFOLD and WILLIS, 1961). There are now about two million hectares of Eucalyptus plantations, principally in Latin America and Africa. The species most commonly planted, or likely to be satisfactory for future use, are *E. globulus, E. camaldulensis, E. viminalis, E. tereticornis, E. grandis, E. robusta, E. citriodora, E. gomphocephala, E. microtheca, E. deglupta, E. dalrympleana, E. bicostata,* and *E. occidentalis* (PRYOR, 1967). *E. globulus* is remarkable in that although it has a relatively limited range of natural distribution, it has been grown

Fig. 64. Two-and-a-half-year-old plantation of *Cedrela odorata* L. Okhessa Forest, Ubiaja, Nigeria. (Photo A. F. A. LAMB)

successfully throughout the world. It is the most frequently occurring species in PENFOLD and WILLIS' list (1961) of eucalypts as exotic.

There is a general predominance of conifers over broadleaved species in the planting programs of many countries, and of the coniferous species used to establish man-made forests the genus *Pinus* is by far the most important over a wide range of latitudes. The success of *Pinus radiata* in New Zealand has been outstanding. However, it has also proved an important exotic in Chile and Australia. Other pine species planted widely include *P. patula* (Africa), *P. pinaster* (Spain, Italy, South Africa, Australia) *P. elliottii* (southern United States, Africa, South America, Australia), *P. sylvestris* (Europe and Asia), *P. merkusii* (Indonesia), and *P. halepensis* (Mediterranean). A number of pine species are also widely planted in mainland China (FAO, 1967 a, b).

For the foregoing reasons it is likely that pines and eucalypts will continue to feature prominently in the establishment of man-made forests in tropical and subtropical countries. As LAMB (1967) has pointed out, "The rising demand for the better land for agriculture in the lowland tropics and the low increment and difficulties in natural regeneration of the hardwood forests except in isolated instances have caused foresters in many countries to think of conifers and eucalypts as a

means of meeting the increasing timber and pulpwood needs in their countries."
According to LAMB (1967), the lands available for the establishment of man-made
forests in tropical countries lie below 1219 m at the equator and down to sea level at
the tropics of Cancer and Capricorn. They comprise large areas in east and west
Africa, the Deccan of India, Ceylon, Northern Territories, North Queensland,
Australia, and smaller areas of Malaysia and Indonesia. They include much of the
Amazon basin, the Guianas of South America, lowland Venezuela, Colombia,
Equador, and Panama. A number of tropical pines from Central America, including
the Caribbean and the Bahamas, and Southeast Asia, which are considered to have
potential throughout many of these areas, are presently being tested in international species and provenance trials. The species *Pinus caribaea, P. cubensis, P.
occidentalis, P. oocarpa, P. strobus* var. *chiapensis, P. tenuifolia, P. tropicalis, P.
kesiya,* and *P. merkusii,* and their important provenances have been appraised by a
number of authors in relation to this problem (HUGHES, 1967; LAMB, 1967; BURLEY,
1969a, 1969b; IYAMABO, 1969; KEMP, 1969; McCOMB and JACKSON, 1969; LAMB and
COOLING, 1970).

The success and rate of rehabilitation of tropical savanna lands and degraded
high forest will depend to a significant degree on the extent to which information
on suitable tree species and provenances is collected, codified, and published. In
this regard the series of monographs published in recent years by the Commonwealth Forestry Institute, Oxford, on fast-growing timber trees of the lowland
tropics are of considerable value and usefulness. The species treated to date are
Gmelina arborea (LAMB, 1968a), *Cedrela odorata* (LAMB, 1968b), the Araucarias
(NTIMA, 1968), *Pinus merkusii* (COOLING, 1968), and *Terminalia ivorensis* (LAMB and
NTIMA, 1971).

Zambia offers a representative example of what may be achieved in the establishment of man-made forests in a tropical country using exotic pine and eucalypt
species. Zambia is a land-locked country in central Africa with an area of 290000
square miles, and severely limited natural forest resources. The need for exotic
plantations has been apparent for years, and in 1964 the country begun a plantation
program designed to supply its industrial needs. The obstacles to the successful
establishment of man-made forests in Zambia are formidable. There is a dry season
of six to eight months without any precipitation at all. The soils are deficient in
boron, resulting in die-back of eucalypts, and high termite populations have been a
major problem in forestry. By 1964, as a result of what appears to have been a well-
designed and vigorously conducted program of applied research, solutions had
been suggested to all major problems, and it was decided to apply research results
on an operational scale even before they had been fully proven. Market studies and
yield forecasts indicated that Zambia's industrial needs could best be supplied by
establishing 30000 acres (12148 ha) of eucalypts, mainly *E. grandis,* on a 12-year
rotation, and 75000 acres (30352 ha) of pines, mainly *P. kesiya,* on a 30-year rotation. The selection of both of these exotic species was based on the growth performance of introduced populations. Before areas are allocated for plantations, soil
surveys are carried out. These surveys are initially extensive, one soil pit per
90 acres (36 ha) is examined, and on areas found generally suitable a more intensive
survey, one soil pit per 23 acres (9 ha), is completed. Data from these latter soils pits
are used to prepare soil maps at a scale of 1:20000, which form the basis of site
selection for plantation purposes. It has been found that soil moisture is the main

factor limiting the growth of tree crops, and that soil depth and texture are the main factors controlling its availability. When a site is finally selected for planting, the indigenous Miombo woodlands, mostly *Brachystegia* species, are cleared and the ground ploughed. Tubed planting stock is used, and insecticides and fertilizers, including trace elements, are applied. Weeding is essential, and thinning regimes must be rigidly adhered to. Almost the entire process of plantation establishment and tending is mechanized (COOLING and ENDEAN, 1966; ALLAN, 1967; ENDEAN, 1967; MOSTYN, 1967; SANDERS, 1967). Plantations in Zambia in 1965 had an area of 4700 ha, and it is planned to extend this to 44500 ha by 1985. It is believed that this rapid expansion in the establishment of man-made forests in Zambia is perfectly realistic given the continuation of a sound research program and a reasonable continuity of staffing (FAO, 1967a).

The destructive effects of undirected, shifting agriculture on natural forest eco-systems when population density reaches a certain level have been noted. However, when the process forms part of a management plan, a system of agri-silviculture can be developed, which, under certain conditions, may allow the integration of local customs and needs of native peoples with the requirements for successful timber production. Agri-silviculture has been defined as a method of raising forest crops in combination with agricultural crops (KING, 1968), and the number and variety of the names applied to this system are a measure of its universality throughout tropical and subtropical countries. KING (1968) listed 27 names, the best known of which in the English-speaking tropics is Taungya, and added that the list was far from exhaustive. Historically, of course, the system was not confined to the tropics, and as late as the latter part of the 19th century it was prevalent both in Finland and Germany (KING, 1968). An aspect of agri-silviculture that is common to most countries is that the system begins with clear-felling and burning of the forest cover. This is followed in most countries by the establishment of agricultural crops on the cleared land. On occasion the agricultural crop is planted after the tree crop, and more rarely both are planted together. Though many different kinds of light-demanding tree species are planted on the cleared areas, such as *Nauclea diderri-chii, Lovoea trichilioides, Khaya ivorensis,* and *Entandrophragma* spp. (KING, 1968; LAMB, 1969), teak is by far the most common. The value of the system of agri-silvi-culture in the establishment of man-made forests in the tropics has been cogently argued by KING (1968), who concluded that "a great deal of social upset has already been caused by the abandonment of shifting cultivation, and until tropical econom-ies become self-sustaining, until they are able to afford the fertilizers, machinery, and implements that are necessary for tropical sedentary agriculture, until storage facilities are improved, and until new genetic strains suited to tropical conditions are evolved, until then, agri-silviculture may offer some help in the solution of tropical land-use problems, and the development of a stronger rural and national economy." The validity of this conclusion is evident in Kenya where, in 1965, 8124 cultivators made available 8018 acres (3245 ha) of land for afforestation out of a total planting program of 10571 acres (4278 ha), while making a useful contribu-tion to Kenya's food requirements. The following species were used: *Pinus patula, P. radiata, Cupressus lusitanica, Eucalyptus,* and *Araucaria* spp. (FAO, 1967c).

Enrichment planting is the method of restocking tropical forest ecosystems that have been progressively depleted of commercially desirable species, although often still densely stocked with species of no immediate commercial value. Where agri-silviculture is not possible, it avoids the heavy cost of clear-felling unproductive natural forest and the establishment of man-made forests. The method has also been applied in savanna woodlands in English-speaking African countries but not in French (CATINOT, 1969; LAMB, 1969). Initially the objective was to introduce into natural stands a predetermined percentage of commercial species without establishing uniform, close plantations of these species, and without eliminating existing valuable species of the natural forest ecosystem. Gaps left after logging were planted with desirable species. Eventually, however, a more systematic approach was developed which necessitated the opening up and planting of lines, or strips, which in time were spaced at intervals equal to or slightly greater than estimated crown diameter of the final crop (LAMB, 1969). Thus defined, the criteria for success of the method of enrichment by line planting have been enuciated by DAWKINS (1961).

In French-speaking tropical countries the method of enrichment by line planting has evolved to what is now termed close planting (CATINOT, 1969). Essentially this means the conversion of a natural stand into a man-made forest, which, at the end of the rotation, will be uniformly and densely stocked.

The virtual abandonment of enrichment planting in its original form in French-speaking countries is the result of a number of factors. It was found necessary to provide more light for the introduced species than had been initially thought, and this necessitated the destruction of a greater number of trees of the natural stand that formed the upper canopy between lines. With the introduction of fast-growing, light-demanding commercial species, such as *Aucoumea, Terminalia,* and *Triplochiton,* less and less attention was paid to the remnants of the original forest ecosystems that were sacrificed to the ecological requirements of the introduced species (CATINOT, 1969). Dominant trees were removed and planting lines widened until finally a man-made forest replaced the natural forest ecosystem. According to CATINOT (1969), enrichment planting in its original form is no longer being carried out in French-speaking tropical countries, and all silvicultural operations now under way pertain to the establishment of close plantations. A similar situation prevails in English-speaking tropical countries, although in Ceylon, Fiji, the Solomon Islands, and Uganda the method of enrichment continues in use on a commercial scale (LAMB, 1969).

Exotic pines have played a major role in successful establishment of man-made forests in a number of nontropical countries. Monterey pine is one of the most valuable of these pines, and its importance in the southern hemisphere is indicated by the following data (SCOTT, 1960):

Country	Share of world total of major stands of Monterey pine, percent	Percentage of all exotic conifers in the country listed
New Zealand	37	60
Chile	32	99
Australia	20	76
Spain	8	100
Union of South Africa	3	8

In its native habitat in southern California south of San Francisco, where it occupies only about 10000 acres (4047 ha) along a narrow coastal strip, the species is not of economic importance. Yet the average mean annual increment in much of its exotic range is 350 ft³/ac. (24.5 m³/ha) on a rotation of 25 years, and it has been estimated that the cost of Monterey pine pulpwood on stump in Chile is between one fifth and one sixth of the corresponding cost of Scandinavian pulpwood (SCOTT, 1960). GOLFARI (1967) has concluded that where the climatic pattern, as determined from average climatic water balance data, is similar to that at the natural habitat in California, Monterey pine has been successful as an exotic (Fig. 65), and where it differs significantly it has been a failure (Fig. 66).

Despite the undoubted success achieved to date for a number of well-known species, the accelerated introduction in recent years of pines and other exotics for the establishment of man-made forests in tropical countries is not without serious hazards. Extensive failures are likely until much more knowledge is available concerning the biology of the species introduced. On occasion poor performance may result, not because the species has been planted outside its climatic range of tolerance, but because of edaphic conditions in the new habitat, e.g., a deficiency of a trace element or absence of appropriate mycorrhiza. For example, in Zambia the addition of boron has been found a prerequisite to the establishment of productive *Eucalyptus grandis* plantations, and as MIKOLA (1969) has pointed out the introduction of exotic pines to many countries begins with a long succession of failures until mycorrhizal infection is brought in, wittingly or unwittingly. Furthermore, the conversion of species-rich, natural forest ecosystems of relatively low productivity to a single rapid-growing species, frequently exotic, in the long run can produce results opposite to what had been expected. The risk of disease is high (BAKSHI, 1967) and losses due to insect damage, disease, and windthrow can be so great in such man-made forests as to considerably offset expected gain in productivity as a result of rapid growth. A number of such examples have been referred to by VAN MIEGROET (1967), who suggested that too narrow a conception of profitability can be one of the greatest obstacles to improved production in the long run. Nevertheless, it is to be expected that the extension of single-species, man-made forests of low genetic variability (Figs. 61, 62, 63) throughout the world will continue to take place in many instances at the expense of species-rich, natural forest ecosystems of high genetic variability (Figs. 67, 52). Furthermore, without due regard to the natural ecological requirements of the species used, this trend will lead inevitably to the increasing use of herbicides, insecticides, and fertilizers, and the development of a highly artificial forest crop as dependent on man for its maintenance as corn or wheat. The hazards of such a development have been underlined by concerned ecologists, and amply demonstrated in relation to agricultural crops, and foresters would do well to be aware of them. The point of view expressed in BARCLAY-ESTRUP's (1972) critical review of the use of herbicides in forest management is equally applicable to insecticides and fertilizers.

Man-made forests have been established for many purposes besides that of cellulose production, and although such forests do yield wood their principal function may be the rehabilitation of unproductive land, either desert, swamp, or man-made as a result of mining, the provision of shelter belts and windbreaks, water conservation, prevention of soil erosion, and finally the general enhancement of

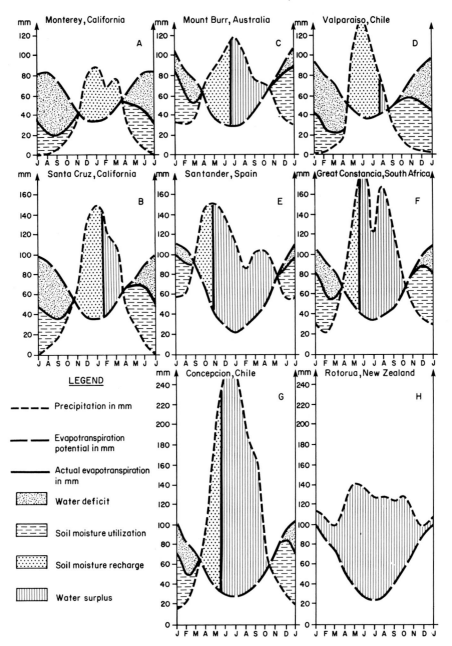

Fig. 65. Water-balance diagrams for six regions where *Pinus radiata* has been successfully introduced. Graphs a and b represent the climatic patterns in the natural habitat of the species.
(After GOLFARI, 1967)

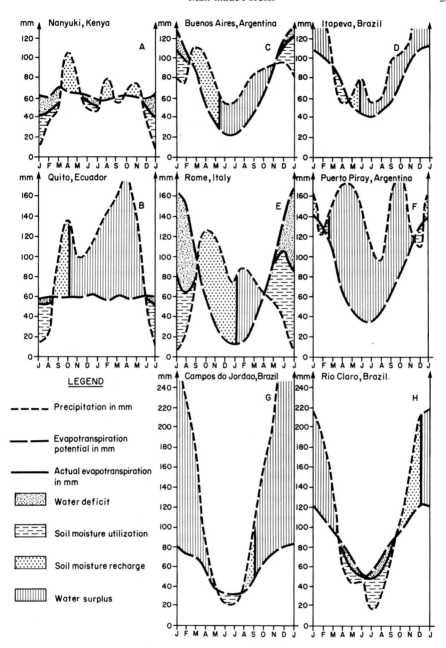

Fig. 66. Water-balance diagrams for eight regions where *Pinus radiata* has been unsuccessfully introduced. Note that the climatic pattern of these regions is significantly different from that of the natural habitat of the species. (After GOLFARI, 1967)

Fig. 67. Natural stand of Sitka spruce with western hemlock and western red cedar. Fourteen miles east of Prince Rupert, British Columbia

environmental quality, both rural and urban. The Sahara is one of the largest areas of marginal, unused land in the world and in many African countries its borders are encroaching on productive land. The stabilization of the desert's edge and coastal sand dunes can be attained by the establishment of forest plantations, frequently following fixation of moving dunes by physical or biological means (STONE and GOOR, 1967). In Tunisia the species *Acacia cyanophylla, A. cyclopis, Pinus pinaster, P. pinea, P. halepensis, Eucalyptus camaldulensis,* and *E. gomphocephala* are used for this purpose (BEN AISSA, 1967). It is estimated that the area of swamps and peatlands in the world exceeds 200 million hectares, of which one third to one fourth is completely treeless. Techniques for draining and planting such areas have been developed, and are being applied on an increasing scale, particularly in the north temperate and boreal forest zones (MIKOLA, 1967; STANEK, 1968).

The rehabilitation of degraded lands can be a formidable undertaking. It is estimated that in the United States nearly one million hectares of restoration planting are needed on areas that have been denuded by surface mining for coal, sand, gravel, and other minerals. Surface mining is currently increasing this area at a rate of 30000 ha annually (BACON, 1967). A number of tree species have proved to be particularly good in the rehabilitation of such lands. *Pinus contorta* has been successfully planted on slag heaps in Great Britain, *Betula pendula* on very arid waste in Europe, and poplar and willow hybrids on loose spoil material not strongly acid. *Pinus strobus* has been used on strip-mined land of the Appalachian coal field. In the tropics *Eucalyptus camaldulensis* has grown well on spoil lands from tin mines in Nigeria, and *Acacia auriculaeformis, Eucalyptus deglupta, Fagrea fragrans,* and *Pinus merkusii* have been used for similar purposes in Malaya (KNABE, 1967).

Clearly the first crop of pioneer tree species cannot always offset the cost of reclamation of such sites, and frequently the principal objective will be the creation of a fertile soil from raw waste material, and the clothing of unsightly mounds of slag and spoil.

In recent years there has been a trend in industrialized nations toward what it increasingly called environmental forestry. Foresters, occupied for years solely with the production of cellulose, are now called upon to be concerned with landscape architecture, trees for urban environments, and recreational facilities close to urban centers. Though some progress has been made in this regard (see GOMBRICH, 1971 for literature review), to date, this trend is often more apparent in words than in practical accomplishment, and although concern is publicly expressed and research projects in environmental forestry are proposed, virgin forests, where they still exist on the periphery of large urban centers, continue to be eliminated. Much time is likely to pass, therefore, before environmental forestry will significantly enhance the beauty of man's habitat, or even significantly reduce its progressive deterioration, and for a long time to come man-made forests will be established solely for commercial purposes. If both natural and man-made forests are to serve several purposes other than the purely commercial, a much greater emphasis must be given at an operational level to democentric forestry (JEFFRY et al., 1970) than is currently apparent in many of the great forested countries of the world.

c) Genetics in Forestry Practice

A number of stages are discernible in the application of genetic principles in modern forestry practice. At first, emphasis is given to the assessment of genetic variation within the most important commercial species. The methods employed in such genecological studies, and their fundamental importance in relation to the silviculture and breeding of tree species, have been discussed by a number of authors (LANGLET, 1963; STERN, 1964; LINES, 1967; ROCHE, 1968; CALLAHAM, 1970). In countries not well endowed with a diversity of productive species, provenances of diverse exotics may be introduced and tested under a variety of conditions. Following these investigations, and frequently depending on their stage of development and the amount of information they have yielded, single-tree selection for desirable characteristics is initiated. The selected material is propagated, either vegetatively or from seed, and subsequently subjected to progeny trials. Superior tested material is then mass-produced. Under certain circumstances inter- and intraspecific hybridization may form an important, and occasionally the major part, of the overall improvement program (Figs. 68, 69, 70).

Following investigations of variation, either within or between species and under different test conditions, criteria are drawn up that guide the movement of seed and seedlings from their place of origin. These criteria can be continually modified and refined on the basis of information occurring from long-term field trials, frequently referred to as provenance trials. The establishment of criteria governing the movement of seed and planting stock of a particular species is of major importance in programs of reforestation or afforestation, as it is the best way of insuring that the genetic potential of the species is correctly used by the silviculturist. Furthermore, and assuming that the studies have been well designed, information is pro-

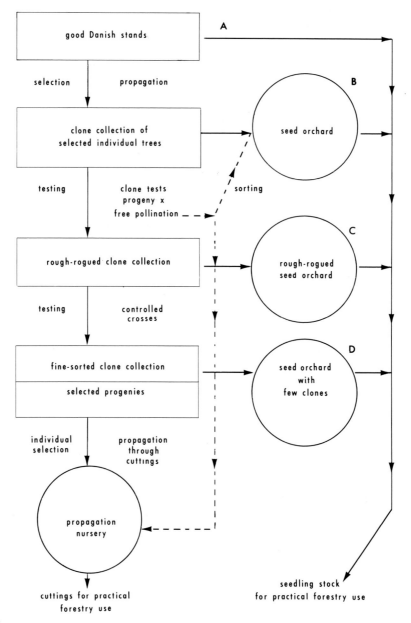

Fig. 68. Diagrammatic representation of an applied breeding program for the improvement of Sitka spruce in Denmark. (After BRANDT, 1970)

vided that will allow the initiation of a program of selection and breeding. A knowledge of the vegetative and reproductive cycle of the species in different natural ecosystems based on phenological observation (LIETH, 1970) greatly strengthens the data obtained by conventional provenance trials. Both the method and the value of

Fig. 69. The assessment of genetic differences between progenies of Scots pine obtained by controlled crossing. The method of testing progenies and rearing planting stock in plastic greenhouses has been developed in Finland with considerable success, and could be equally successful in countries with similar climates, such as Canada. The test shown here has been implemented by the Foundation of Forest Tree Breeding in Finland (see 1970 report of this Foundation)

such observations in natural stands have been demonstrated by SARVAS (1962, 1965) and KOSKI (1970). Such data have other long-term practical implications in that they enable the rationalization of seed orchard establishment by providing information on the probability and degree of contamination from local pollen sources. However, despite its importance in modern forestry practice, surprisingly little work of this nature has been carried out in natural forest ecosystems. Thus in some countries seed orchards continue to be established despite the lack of quantitative data on the reproductive and vegetative cycles of the provenances used, and on the potentially contaminating sources of the species in the vicinity of the seed orchard.

The worldwide movement of forest tree seed from its natural source, and the development of cultivars by selection and breeding, in recent years have emphasized the need for international standards of seed certification. Seed registration and seed certification schemes now exist in a number of countries (BARNER, 1963; BARBER, 1969; KENNEDY, 1969; WANG and SZIKLAI, 1969), and minimum standards for provenance and progeny testing for certification purposes have been proposed (STERN, 1969). Internationally acceptable tree seed certification standards are gradually being established. The most important of these standards are those presented

Fig. 70. The assessment of genetic differences in grafts of hybrid Larch (*Larix sibirica* X *decidua*) in Finland. The test shown here has been implemented by the Foundation of Forest Tree Breeding in Finland (see 1970 report of this Foundation)

in the Organization for Economic Cooperation and Development (OECD) scheme for the certification of reproductive materials moving in international trade (BARBER, 1969).

As already noted, numerous forest tree species have proved to be extremely successful when grown well outside their natural range. In Britain and Ireland, and in many parts of western Europe and Scandinavia, a number of western North American tree species are of major commercial importance, such as Sitka spruce, Lodgepole pine, and Douglas fir. Slash pine and loblolly pine from southeastern United States have been successfully planted in Africa and Latin America and in subtropical Queensland and New South Wales, and Monterey pine from California is a major commercial tree species in New Zealand and Australia, and is of importance in East Africa and Chile. Scots pine, Norway spruce, and hybrid poplars of European origin form successful plantations in northeastern United States and Canada. The introduction of exotics for reforestation of lowland regions of tropical countries that contain large areas of savanna woodland is a major development of recent years.

In time, and depending on how representative a sample of the gene resources of the species was originally introduced and its subsequent history as a seed source, the genetic structure of an introduced species can deviate significantly from that of the species in its natural habitat. Many New Zealand populations of Monterey pine may have become more heterozygous than those of the natural stands in California (BANNISTER, 1965), and seed selected from plantations of Sitka spruce in Denmark invariably produce planting stock more vigorous in growth than seed of the same provenances imported from western North America (BRANDT, 1970). A similar situation appears to prevail in France for Sitka spruce (LACAZE, 1970). Eastern white pine was introduced into Italy about 150 years ago. In one region in particular where the species regenerates naturally, and where natural selection has been in operation for only a little over a 100 years, geographically distinct populations have already evolved from the original introduction (VECCHI, 1969). Land races are not confined to introduced species. They can occur in regions where the natural forest ecosystems have been heavily and repeatedly disrupted by man's intervention over a long period of time, and where regeneration occurs from isolated vestigial populations of the original ecosystems, for example, in the southern pine region of the United States. In this region many pine stands have arisen from seeding into abandoned farmlands, often from one or very few trees growing along fence rows, near farm buildings, or along ditches (ZOBEL, 1972). There are numerous examples of the development of land races of forest tree species, such as those referred to above (VECCHI, 1969). They are essentially the product of man's actions, either deliberate or inadvertent, resulting in the evolution of new ecosystems of the species.

Plantations of Monterey pine in New Zealand, derived from bulked seed of diverse provenance, are likely to be more heterozygous than natural populations at the place of origin (BANNISTER, 1965). In most instances, however, plantations established from a small number of selected individuals will inevitably, for most species, represent only a part of the spectrum of variation of the species throughout its natural range. Breeding programs based on clonal, single-tree selection in such plantations will reduce variation even further. In the long run, if relatively few clones are the major source of seed for reforestation programs, highly heterozygous natural forest ecosystems will be replaced by man-made forests with dangerously narrow genetic bases. Such forests, in common with many agricultural crops, could require massive protective measures to maintain their productivity and ensure their perpetuation. For these reasons, the maximum number of clones compatible with random mating are usually included in seed orchards. KELLISON (1969) has suggested 25 as a minimum for southern pine, and two to four times that number as being desirable. Programs of inter- and intraspecific hybridization are also relied upon to provide a spectrum of variation that may be drawn upon to broaden the genetic base of the breeding stock.

It has been pointed out that a balanced tree-improvement program has both production and research functions, and that short-term objectives may be achieved by production seed orchards and long-term objectives research orchards. The operational or short-term objectives "consist of intensive selection and breeding to produce large quantities of desired stock for immediate use, often with a resultant reduction in the genetic base for certain important characteristics" (ZOBEL, 1972). On the other hand, "research orchards are not operational; they are long-term, having the objective of keeping the genetic base broad and purposely preserving genotypes that may not have immediate use in the applied program" (ZOBEL, 1972). This is an important distinction, which, when made in practice, will ensure that forest gene resource conservation forms an integral part of the tree improvement program.

Monterey pine in New Zealand may be taken as a case history of a species successfully introduced and subjected to conventional techniques of selection and breeding. Natural ecosystems of this species are found in California where it has no commercial value as a producer of cellulose. In New Zealand the species is fast growing, showing on average sites a mean annual increment of 21 m³/ha. A tree-breeding program was begun in 1953, with single-tree selections being made in 30-year-old plantations. Some 450 acres of clonal seed orchards were established by 1969, and it is estimated that 10000 pounds of orchard seed will be harvested annually by 1976. An assessment of a ten-year-old progeny test of some of the selected clones indicated that an improvement 63% in stem straightness, 45% in branching characteristics, and 14% in volume production had been achieved (THULIN, 1969).

There is ample evidence that genetic impoverishment of many tree species has occurred and is occurring as a result of the destruction of the forest ecosystems of which these species are a part. Furthermore, it is certain that in many countries forest genetics and tree-improvement programs will not be operative in time and on a sufficient scale to ensure the preservation in clone banks and seed orchards of an adequate sample of the spectrum of variation within the important tree species. Numerous foresters are aware of this situation, and of the need for the conservation of forest gene resources by insuring the survival of the natural ecosystems in which the species are found, and by the establishment of seed banks, clone banks, and seed orchards. Furthermore, united action is occurring at an international level that has a direct bearing on this problem (BARNER, 1963; FAO, 1969, 1972; FRANKEL and BENNETT, 1970; HEDEGART, 1971; BURLEY and KEMP, 1972). Under the auspices of the International Union for the Conservation of Nature and Natural Resources (IUCN), a data book is being compiled that lists plant forms, including tree species, in danger of extinction or severe genetic impoverishment (MELVILLE, 1970). (A typical entry for a tree species in this data book is reproduced on page 276.) The work of the FAO Panel of Experts on Forest Gene Resources has already been referred to (FAO, 1969). This agency also plans to issue periodically a newsletter dealing with the exploration, conservation, and utilization of forest gene resources (FAO, 1972). A new working party of the International Union of Forest Research Organization (IUFRO) dealing with forest gene resource conservation has been formed, and the activities of this working party will complement the work of the FAO in this regard.

The setting aside of forest land in national parks and the establishment of national systems of natural forest areas serve the purpose of forest gene resource conservation in certain countries. Plant gene resource conservation *in situ* is also one of the main objectives of the work of the International Biological Program (IBP/CT) (NICHOLSON, 1968). This method of forest gene resource conservation has also been recommended in the second report of the FAO Panel of Experts on Forest Gene Resources, which stated that conservation of forest gene resources within their natural range should be carried out *in situ* wherever possible (FAO, 1972). However, the conservation of forest ecosystems in which the species to be preserved is found at any given time does not mean that the species will continue to survive and perpetuate itself in those ecosystems. Successional trends may in time result in its elimination and substitution by other species. Moreover, standard fire protection procedures prevailing in many forested countries, which reduce the incidence and effects of fire, in the long run could affect the occurrence of particular species in the research areas (HEINSELMAN, 1971). For example, it has been shown that indiscriminate fire protection is detrimental to natural regeneration of teak in Burma, and could result in the complete elimination of the species in large areas of the most important class of teak-bearing forest (TROUP, 1921). Conservation *in situ*, therefore, in relation to programs of selection and breeding, must be regarded for some forest ecosystems as an interim measure that will insure the availability of desirable gene pools of a particular species until such time as their gene pools can be conserved by other means. In north temperate zones this point has less force. Because of the relative slowness of succession, the elimination of a particular species in a given north temperate forest ecosystem as a result of natural succession may be discounted for all practical purposes of gene resource conservation.

It will be many years before tree-breeding programs for all important tropical and north temperate tree species will result in the conservation of gene resources in clone banks and seed orchards, and in the production of commercial quantities of seed of the correct provenance. In the meantime, the elimination of the world's remaining natural forest ecosystems continues, and evolutionary centers, sources of great genetic variability and new forms of plant life, are being massively disrupted or destroyed (RICHARDS, 1957; ROCHE, 1971). Therefore, in countries where natural forest ecosystems are extant and are being heavily exploited, and where other means of conservation are not well developed, the conservation of forest gene resources *in situ* is essential. In the long run, of course, the conservation of natural forest ecosystems is desirable for many reasons besides the conservation of large heterogeneous gene pools: (1) to allow the comparison of those ecosystems with managed, utilized and artificial forest ecosystems; (2) for outdoor museums and areas of study, especially in ecology; and (3) for education in the understanding and enjoyment of the natural environment and for the intellectual and esthetic satisfaction of mankind (RICHARDS, 1957; NICHOLSON, 1968, WEETMAN, 1970).

Considerable progress has been made in Finland to ensure the conservation of forest gene resources for the major species of the country. Stands of natural origin, representative of each part of the country, have been selected and protected from felling. No stand is selected that may be contaminated by pollen from adjacent plantations. In most cases, standard stands consist of a sample plot 100 by 100 m with a surround 100 m wide. Following measurement and mapping, and the preparation of descriptive data, each stand is permanently registered. All seed for experimental purposes, including genecological research, is collected from these standard stands, and kept separate by trees if sufficient seed is available. It is always possible, therefore, to return to these standard stands to obtain seed that proves to be of particular interest to the researcher or the silviculturist. As often as possible, seed from these standard stands is used to meet requests for seed samples from abroad. When a standard stand becomes overmature, it is regenerated either naturally or artificially using seed from the same stand, and only seed from the standard stand is used to establish local plantations (HAGMAN, 1972).

The Finnish standard stands clearly not only conserve forest gene resources, but form the basis for a rational development of an applied tree-improvement program. The value of these stands in this regard is greatly enhanced where the vegetative and reproductive cycles of these stands in different ecosystems, and the environmental factors influencing them, are known. As already indicated, researchers in Finland (SARVAS, 1962, 1967, 1969; KOSKI, 1970) have demonstrated both the method and practical value of these phenological studies in natural stands. It has been pointed out that there is a universal need for the establishment of similar systems of standard stands in forested countries where large areas of indigenous forests are still extent, but where exploitation and the establishment of plantations are rapidly increasing (HAGMAN, 1972).

Present Status of a Tree Species in Danger of Extinction
(Sample Sheet From Red Data Book (MELVILLE, 1970)

Vateria seychellarum Dyer

Bois de Fer

DIPTEROCARPACEAE

Description: A tree 25—30 m high with leathery elliptic leaves 12—20 cm long, with a piculate apex and 12—20 pairs of arching lateral veins. Flowers few together in short axillary racemes, the sepals ovate obtuse, the petals obovate, concave, 7 mm long, the stamens numerous with tapering anthers and short filaments. Fruits globose, 3—4 cm diameter, containing a single seed with large fleshy cotyledons.

Type: Seychelles, Mahé, near Port Gland, J. Horne. (BAKER, 1877).

Illustrations: See DYER (1903).

Habitat and ecology: Formerly one of the dominants in the rainforest at the foot of the hills but not extending very far up the slopes, growing on soils derived from weathering of granite.

Status: Critically endangered.

Present distribution: Three trees survived on Mahé in the Seychelles. Under British administration in 1962.

Former distribution: Endemic to the Seychelles, formerly a common timber tree on Mahé.

Reasons for decline: Exhaustive exploitation for timber and firewood. It occupied the zone cleared for cultivation of Cinnamon and other crops.

Biology and potential value: Nothing is recorded of its biology. It produced a good quality timber, comparable with that of other Dipterocarps from Malaya, which was used for carpentry and building as well as firewood. The bark, when incised, gives out a very fragrant resin which was used as incense and has potential value in the perfume industry. As an outlying member of the Dipterocarpaceae it has considerable scientific and phytogeographical interest.

Cultivation: It could be regenerated from seed. Cut stools coppice readily.

Protective measures already taken: None.

Measures recommended: Strict protection should be given to the surviving trees. Propagation and cultivation in botanic gardens is needed urgently to ensure survival. It should be grown by tropical forestry research institutes as part of the Dipterocarp gene pool and its economic potential should be investigated.

References: BAKER, J.G., Flora Mauritius and Seychelles 526 (1977).
      DYER, W. T. T., in Hooker Icones Plantarium 28 t. 2759—60 (1903).
      VESEY FITZGERALD, L. D. F. F., J. Ecol. 28, 465—483 (1940).

References

AALDERS, L. E., CRAIG, D. L.: General and specific combining ability in seven inbred Strawberry lines. Can. J. Gen. Cyt. **10**, 1–6 (1968).

ADAMS, W. T.: Competitive relationships among loblolly pine (Pinus taeda L.) seedlings. M.S. Thesis, N.C. State Univ., Raleigh (1970).

ADAMS, W. T.: Competitive relationships among loblolly pine (Pinus Taeda L.) seedlings. Thesis, Grad. Fac. N.C. State Univ. Dept., of Forestry (1971).

ALLAN, R. E., PRITCHETT, J. A., PATTERSON, A.: Juvenile and adult plant growth relationships in wheat. Crop Sci. **8**, 176–178 (1968).

ALLAN, T. G.: Industrial plantation establishment methods in Zambia. In: FAO World Symposium on Man-made Forests and their Industrial Importance. Vol. II, pp. 1043–1056. Rome: FAO 1967.

ALLARD, R. W.: The relationship between genetic diversity and consistency of performance in different environments. Crop Sci. **1**, 127–133 (1961).

ALLARD, R. W.: Genetic systems associated with colonizing ability in predominantly self-pollinated species. In: Genetics of Col. Spec., pp. 49–76. New York: Academic Press 1964.

ALLARD, R. W., JAIN, S. K.: Population studies on predominantly selfpollinating species. II. Analysis of quantitative genetic changes in a bulk hybrid population. Evolution **16**, 90–101 (1962).

ALLARD, R. W., WORKMANN, P. L.: Population studies in predominantly selfpollinating species. IV. Seasonal fluctuations in estimated values of genetic parameters in Lima Bean populations. Evolution **17**, 470–480 (1963).

ALLARD, R. W., WEHRHAHN, C.: A theory which predicts stable equilibrium for inversion polymorphism in the grasshopper Moraba scurra. Evolution **18**, 129–130 (1964).

ALLARD, R. W., HARDING, J., WEHRHAHN, C.: The estimation and use of selective values in predicting population change. Heredity **21**, 547–563 (1966).

ALTENKIRCH, W.: Zum Vorkommen von Tortrix viridana L. in Portugal. Ztschr. Angew. Zoologie **53**, 403–415 (1966).

ANDERSON, E.: Introgressive Hybridization. New York: John Wiley 1949.

ANDERSON, E.: Man as a maker of new plants and new plant communities. In: Man's Role in Changing the Face of the Earth (THOMAS, JR., W. L., Ed.), pp. 763–777. Chicago: Univ. of Chicago Press 1956.

ANDERSSON, E.: The sources of effective germ plasm in hybrid maize. Ann. Mo. Bot. Gard. **31**, 355–366 (1944).

ANDERSSON, E.: Pollen and seed setting studies of an asyndetic spruce and some normal spruces; and a progeny test of spruce. Sv. Papperstidning 4–17 (1947).

ANDERSSON, E.: Nagra data om pollenvariationen och pollenfertiliteten hos gran och tall. Sv. Papperstidning 1–37 (1954).

ANDERSSON, E.: Pollenspridning och avstandisolering av skogsfröptantager. Sv. Papperstidning 35–100 (1955).

ANDERSSON, E.: Seed stands and seed orchards in the breeding of conifers. World Cons. For. Gen. Tree Impr. Stockholm **2**, 1–18 (1963).

ANDERSSON, E.: Cone and seed studies in Norway spruce (Picea abies L. Karst.). Studia For. Suecica no. 23 (1965).

ANDERSSON, E.: Barrträdens blomning och frösättning. Symp. Skogshögsk. Stockholm 1966.

ANDERSSON, E.: Meiosstörningar i pollenmodercell hos gran. Manuskript. Zit., after ANDERSSON et al. 1969 (1969).

ANDERSSON,E., EKBERG,I., ERIKSSON,G.: A summary of meiotic investigations in conifers. Stud. For. Suecica no. 70 (1969).

ANDRESEN,J.W.: A multivariate analysis of the Pinus chiapensis-monticola-strobus phylad. Rhodora 68, no. 773 (1966).

ANDREWARTHA,H.G., BIRCH,L.C.: Distribution and Abundance of Animals. Chicago: Univ. of Chicago Press 1954.

ANONYMUS: Versuche zur Kontrolle von Fomes annosus. Wiss. Ber. Pflanzenschutz der Balt. Region, Riga 149–150 (1961).

ANONYMUS: Forest cover types of North America. Publ. Soc. Amer. For., Washington, D.C. (1967).

ANONYMUS: Where it's at with machines for fully mechanized logging systems. Canadian Industries (1970).

ANONYMUS: XVth annual report N.C. State University Cooperative Tree Improvement and Hardwood Research Programs. Raleigh, N.C.: N.C. State Univ. 1971.

ANTONOVICS,J.: Evolution in closely adjacent plant populations V. Evolution of self-fertility. Heredity 23, 219–238 (1968a).

ANTONOVICS,J.: Evolution in closely adjacent plant populations VI. Manifold effects of gene flow. Heredity 23, 507–524 (1968b).

ARASU,N.P.: Self-incompatibility in Angiosperms: a review. Genetica 39, 1–24 (1968).

ASHTON,P.S.: Ecological studies in the mixed Dipterocarp. forests of Brunei state. Oxford For. Memoirs no. 25 (1964).

ASHTON,P.S.: Speciation among tropical forest trees: some deductions in the light of recent evidence. Biol. J. Linn. Soc. 1, 155–196 (1969).

ATSATT,R.R.: Angiosperm parasite and host: coordinated dispersal. Science 149, 1389–1390 (1965).

ATSATT,P.R., STRONG,D.R.: The population biology of annual grassland hemiparasites I. The host environment. Evolution 24, 278–291 (1970).

AUBREVILLE,A.M.A.: The disappearance of the tropical forests of Africa. Unasylva 1, 5–11 (1947).

AUBREVILLE,A.M.A.: Climats, Forêts et Désertification de l'Afrique Tropicale. Paris: Société d'Editions Géographiques, Maritimes et Coloniales 1949.

AUBREVILLE,A.M.A.: Tropical Africa. In: A World Geography of Forest Resources (HADEN-GUEST,S., WRIGHT,J.K., TECLAFF,E.M., eds.). New York: Ronald Press 353–384 (1956).

AUBREVILLE,A.M.A.: Sylviculture dans les forêts tropicales hétérogènes. In: Tropical Silviculture. Vol. II FAO Forestry and Forest Products Studies no. 13. Rome: FAO 46–56 (1957).

AYALA,F.J.: Reversal of dominance in competing species of Drosophila. Amer. Nat. 100, 81–83 (1966).

AYALA,F.J.: An evolutionary dilemma: fitness of genotypes versus fitness of populations. Con. J. Gen. Cyt. 11, 439–456 (1969).

AYALA,F.J.: Competition, co-existence and evolution. In: Ess. in Ev. and Gen. in Honor of Th. Dobzhansky, pp. 121–158. New York: Meredith 1970.

BACON,E.M.: Relation of man-made forests to soil, water, recreation, community development and multiple use of natural resources. In: FAO World Symposium on Man-Made Forests and Their Industrial Importance, Rome: FAO 141–164 (1967).

BAKER,H.G.: Hybridization and natural gene flow between higher plants. Biol. Rev. 26, 302–337 (1951).

BAKER,H.G.: Race formation and reproductive method in flowering plants. In: Symp. Soc. Exp. Biol. 7, 114–145 (1953).

BAKER,H.G.: Reproductive methods as factors in speciation in flowering plants. Cold Spring Symp. Quant. Biol. 24, 177–191 (1959).

BAKER, H. G.: Characteristics and modes of origin of weeds. In: Gen. of Col. Spec., pp. 147–168. New York: Academic Press 1964.

BAKER, H. G., STEBBINS, G. L. (eds.): The Genetics of Colonizing Species. New York, London: Acad. Press 1964.

BAKSHI, B. K.: Diseases of man-made forests. In: FAO World Symposium on Man-Made Forests and Their Industrial Importance, pp. 639–661. Rome: FAO 1967.

BAKUZIS, E. V., HANSEN, H. L.: Balsam Fir (Abies balsamea (Linnaeus) Miller): A monographic Review. Minneapolis: Univ. of Minnesota Press 1965.

BAKUZIS, E. V.: Forestry viewed in an ecosystem perspective. In: The Ecosystem Concept in Natural Resource Management, pp. 189–258, (VAN DYNE, G. M., Ed.). New York, London: Academic Press 1969.

BANCROFT, H.: The arborescent habit in Angiosperms. New Phytol. 29, 153–275 (1930).

BANNISTER, M. H.: Variation in the breeding system of Pinus radiata. In: Genetics of Colonizing Species, pp. 353–374 (BARKER, H. G., STEBBING, G. L., Eds.). New New York, London: Academic Press 1965.

BARBER, H. N.: Adaptive gene substitution in Tasmanian eucalypts. Evolution 9, 1–14 (1955).

BARBER, H. N.: The processes of natural selection. Proc. X. Int. Congr. Gen. Montreal 13–14, 1958.

BARBER, H. N.: Selection in natural populations. Proc. X. Int. Congr. Bot. Edenburgh 226, 1964.

BARBER, H. N.: Selection in natural populations. Heredity 20, 551–572 (1965).

BARBER, J. C.: Control of genetic identity of forest reproductive materials. In: Second World Consultation on Forest Tree Breeding, pp. 1289–1300. Rome: FAO 1969.

BARCLAY-ESTRUP, P.: Herbicides: a critical look at their use in forest management. Can. Field-Nat. 86, 91–95 (1972).

BARENS, B. V., BINGHAM, R. T., SQUILLACE, A. E.: Selective fertilization in Pinus monticola Dougl. II. Results of additional tests. Silv. Gen. 11, 103–111 (1961).

BARKER, J. S. F.: Population density and interspecific competition between Drosophila melanogaster and Drosophila simulans. Proc. XII. Int. Congr. Gen. 1, 228 (1968).

BARNER, H.: Basic principles of origin certification. World Consultation on Forest Genetics and Tree Improvement. Invited paper FAO/FORGEN 63-8/6. Stockholm 1963.

BARNES, B. V.: The clonal growth habit of American aspens. Ecology 47, 439–447 (1966).

BARTELS, H.: Genetic control of multiple esterases from needles and macrogametophytes of Picea abies. Planta 99, 283–289 (1971).

BARTON, L. V.: Seed Preservation and Longevity. London, New York: Hill 1961.

BATEMAN, A. J.: Number of S-alleles in a population. Nature 160, 337 (1947 a).

BATEMAN, A. J.: Contamination in seed crops. II. Wind pollination. Heredity 1, 235–246 (1947 b).

BAWA, K. S., STETTLER, R. F.: Needed: information on breeding systems in tropical tree species. Proc. Sec. World Cons. For. Tree Breeding 1969, 997–1003.

BEARDMORE, J. A., DOBZHANSKY, Th., PAVLOVSKY, O.: An attempt to compare the fitness of polymorphic and monomorphic experimental populations of Drosophila pseudoobscura. Heredity 14, 19–33 (1960).

BEDDOWS, A. R.: Flowering behaviour, compatibility and major gene differences in Holcus lanatus. New Phytol. 60, 312–324 (1961).

BELL, C. R.: Incidence of polyploidy correlated with ecological gradients. Evolution 18, 510–511 (1964).

BEN AISSA, J.: Fixation et reboisement des dunes littorales en Tunisie. In: FAP World Symposium on Man-Made Forests and Their Industrial Importance, pp. 1087–1097. Rome: FAO 1967.

BENNETT, E.: Historical perspectives in genecology. Scott. Plant. Breed. Stat. Rec. 1964, 49–115.

BENSON, L., PHILLIPS, E. A., WILDER, P. A.: Evolutionary sorting of characters in a hybrid swarm. I. Direction of slope. Amer. J. Bot. **54**, 1017–1026 (1967).

BERG, R. L.: The ecological significance of correlation pleiades. Evolution **14**, 171–180 (1960).

BERGMANN, A.: Variation in flowering and its effect on seed cost. Tech. Rep. 38, N.C. State Univ. School of For., Raleigh (1968).

BIBELRIETHER, H.: Unterschiedliche Wurzelbildung bei Kiefern verschiedener Herkunft. Forstw. Centralbl. **83**, 129–140 (1964).

BINGHAM, R. T. (Ed.): NATO Symposium on White Pine Blister Rust Resistance. Moscow, Idaho, 1969 (in press) 1972.

BIRCH, C. L.: The genetic factor in population ecology. Amer. Nat. **94**, 5–24 (1960).

BISHOP, J. A., KORN, M. E.: Natural selection and cyanogenesis in white clover, Trifolium repens. Heredity **24**, 423–430 (1969).

BODMER, W. F., PARSONS, P. A.: Linkage and recombination in evolution. Adv. Gen. **11**, 1–100 (1962).

BODMER, W. F., CAVALLI-SFORZA, L. L.: A migration matrix model for the study of random genetic drift. Genetics **59**, 565–592 (1968).

BONNIER, G.: Nouvelles observations sur les cultures expérimentales à diverses altitudes. Rev. Gen. Bot. **32**, 305–360 (1920).

BORMANN, F. H.: Changes in the growth pattern of white pine trees undergoing suppression. Ecology **46**, 269–277 (1965).

BORMANN, F. H., GRAHAM, B. F.: The occurrence of natural root grafting in eastern white pine, Pinus strobus L., and its ecological implications. Ecology **40**, 502–509 (1959).

BOSSERT, W. H.: Simulation of character displacement in animals. Ph.D. Thesis, Harvard Univ., Div. Engl. Math. (1963).

BOVEY, P., MAKSYMOV, J. K.: Le problème des races biologiques chez la tordeuse grise du Mélèze, Zeiraphera griseana. Vierteljahresschr. Naturf. Ges. Zürich **104**, 263–274 (1959).

BRADSHAW, A. D.: Population differentiation in Agrostis tenuis Sib. III. Populations in varied environments. New Phytol. **59**, 92–100 (1960).

BRADSHAW, A. D.: Evolutionary significance of phenotypic plasticity in plants. Adv. Gen. **13**, 115–155 (1965).

BRANDT, K.: Statusopgørelse for Sitka spruce. Dansk Skorforen Tidsskr. **55**, 300–329 (1970) (English translation by L. GOMBRICH).

BRANSCHEIDT, P., PHILLIPPI, X.: Befruchtungs-biologische Untersuchungen an Zwetschgen und Pflaumen IV. Gartenbauwissenschaften **14**, 38–47 (1940).

BRAUN, H. J.: Die Organisation des Stammes von Bäumen und Sträuchern. Wiss. Verl. Ges. Stuttgart (1963).

BRAYSHAW, T. C.: What are the blue birches? Can. Field Nat. **80**, 187–194 (1966).

BREESE, E. L.: The genetical structure of populations of Lolium perenne. Proc. X. Int. Congr. Bot. Ed. **1964**, 228–229.

BRIAN, M. V.: Segregation of species in the ant genus Myrmica. J. An. Ecol. **25**, 319–337 (1956a).

BRIAN, M. V.: Exploitation and Interference in interspecies competition. J. An. Ecol. **25**, 339–347 (1956b).

BRIGGS, D., WALTERS, S. M.: Plant variation and evolution. World Univ. Library (1970).

BRINAR, M.: Einige morphologische Charakteristiken der Buche und ihre Abhängigkeit vom Relief und der genetischen Divergenz. Ljubljana (1968).

BROWER, L. P., BROWER, L. V. Z.: The relative abundance of model and mimic butterflies in natural populations of the Battus philenor mimicry complex. Ecology **43**, 154–158 (1962).

BRÜCKNER, F.: Eine Komplementärwirkung der verschiedenen Allele für Mehltauresistenz bei Gerste. Ztschr. Pflanzenzüchtung **58**, 122–127 (1967).

BÜNNING, E.: Entwicklungs- und Bewegungsphysiologie der Pflanze. Berlin, Göttingen, Heidelberg: Springer Verlag, 3. Aufl., 1963.

BURDON, R. D.: Clonal repeatabilities and clone-site interactions in Pinus radiata. Silv. Gen. **20**, 33–38 (1971).

BURLEY, J.: Review of variation in Slash Pine (P. taeda L.) in relation to provenance research. Comm. For. Rev. **45**, 322–338 (1966).

BURLEY, J.: Methodology for provenance trials in the tropics. Unasylva **23**, 24–28 (1969a).

BURLEY, J.: Breeding tropical pines. In: Second World Consultation on Forest Tree Breeding, pp. 1061–1075. Rome: FAO 1969b.

BURLEY, J., KEMP, R. H.: Centralized planning and international co-operation in the introduction and improvement of tropical tree species. Invited paper, Second General Congress, SABRO, New Delhi (1972).

BUSH, G. L.: Sympatric host race formation and speciation in the genus Rhagoletis (Diptera, Tephritidae). Evolution **23**, 237–251 (1969).

BUSSE, W. J.: Vom Umsetzen unserer Waldbäume. Thar. Forstl. Jahrb. **81**, 118–130, (1930).

CAIN, A. J., SHEPPARD, P. M.: The theory of balanced polymorphism. Amer. Nat. **88**, 321–326 (1954).

CAIN, A. J., CURREY, J. D.: Area effects in Cepea. Phil. Trans. **246**, 1–81 (1963).

CALLAHAM, R. Z.: Geographic variation in forest trees. In: Genetic Resources in Plants — Their Exploration and Conservation, pp. 43–48 (FRANKEL, O. H., BENNETT, E., eds.). Oxford, Edinburgh: Blackwell Scientific Publications 1970.

CANDY, R. H.: Reproduction on cut-over and burned-over land in Canada. Research Note no. 92. Forestry Branch, Canada (1951).

CANNON, G. B.: Intraspecies competition, viability, and longevity in experimental populations. Evolution **20**, 117–131 (1966).

CARLISLE, A., TEICH, A. H.: The Hardy-Weinberg law used to study inheritance of male inflorescence color in a natural Scots pine population. Can. J. Bot. **48**, 997–998 (1970).

CARLQUIST, S.: The biota of long-distance dispersal. IV. Genetic systems in the flora of oceanic islands. Evolution **20**, 433–455 (1968).

CARPENTER, I. W., GUARD, A. T.: Some effects of cross-pollination on seed production and hybrid vigor of tuliptrees. J. For. **48**, 852–855 (1950).

CARSON, H. L.: The genetic characteristics of marginal populations of Drosophila melanogaster. Cold Spring Harb. Symp. Quant. Biol. **20**, 276–287 (1959).

CARSON, H. L.: The population flush and its genetic consequences. In: Pop. Biol. and Evol. pp. 123–137. Syracuse: Univ. Syr. Press 1967.

CARSON, H. L., HEED, W. B.: Structural homozygosity in marginal populations of nearctic and neotropical species of Drosophila. Proc. Nat. Ac. Sci. Wash. **52**, 427–430 (1964).

CATINOT, R.: Results of enrichment planting in the tropics. In: Report of the Second Session, FAO Committee on Forest Development in the Tropics, pp. 38–43. Rome: FAO 1969.

CAYFORD, J. H., BICKERSTAFF, A.: Man-made forests in Canada. Forestry Branch Publication no. 1240. Dept. of Fisheries and Forestry, Canada (1968).

CEREPNIN, V. L.: Die Bedeutung von Herkunft, Gewicht und Farbe der Samen der Kiefer für die Züchtung. In: Züchtung von Baumarten im Östl. Sibirien. Nauka, Moskau (1964).

CHAPMAN, H. H.: Lightning in the longleaf. Amer. For. **56**, 10–12 (1950).

CHEESMAN, E. E.: Fertilization and embryonogy in Theobroma cacao. Ann. Bot. **41**, 170–215 (1927).

CHING, K. K., AFT, H., HIGHLEY, T.: Color variation in strobili of Douglas-fir. Proc. Wast. For. Gen. Ass. Olympia Wash. 37–43 (1965).

CHIRA, E., BERTA, F.: Eine der Ursachen 1 für Kreuzungsunverträglichkeit zwischen Arten der Gattung Kiefer. Biologia, Bratislava **20**, 600–609 (1965).

CHYLARECKI, H., GIERTICH, M.: Cone variation in Picea abies (L.) Karst from Poland. Zaklad Dend. I Arbor, Korn 19–21 (1967).

CLARKE, B.: The evolution of morph-ratio clines. Amer. Nat. **100**, 389–402 (1966).

CLARKE, B., O'DONALD, P.: Frequency-dependent selection. Heredity **19**, 201–206 (1964).

CLARKE, C. A., SHEPPARD, P. M.: The evolution of dominance under disruptive selection. Heredity **14**, 73–87 (1960).

CLARKE, C. A., SHEPPARD, P. M.: Disruptive selection on a metrical character in the butterfly Papilio dardanus. Evolution **16**, 214–226 (1962).

CLATWORTHY, J. N., HARPER, J. L.: The comparative biology of closely related species living in the same area. J. Exp. Bot. **13**, 307–324 (1962).

CLAUSEN, J.: Stages in the Evolution of Plant Species. Ithaca, N.Y.: Cornell Univ. Press 1951.

CLAUSEN, J., KECK, D. D., HIESEY, W. M.: Experimental studies on the nature of plant species I: The Effect of Varied Environments in Western North American Plants. Washington: Carnegie Inst. Publ. no. 520, 1940.

CLAUSEN, J., HIESEY, W. M.: Balance between coherence and variation in evolution. Science **130**, 1413–1414 (1959).

CLAUSEN, K. E.: New data on distribution of the paper birch x bog birch hybrid in Minnesota. Min. For. Not. St. Paul, no. 81 (1959).

CLAUSEN, K. E.: Interspecific crossability test in Betula. Proc. Work. Group IUFRO Sect. 22 on Sex. Repr. of For. Trees, Finland (1970).

CLIFF, E. P.: Forestry in the United States. Unasylva **23**, 29–33 (1969).

COCKERHAM, C. C.: Variance of gene frequencies. Evolution **23**, 72–84 (1969).

COHEN, J. E.: A Model of Simple Competition. Cambridge, Mass.: Harvard Univ. Press 1966.

COLE, L. C.: The population consequences of life history phenomena. Quart. Rev. Biol. **29**, 103–137 (1954).

CONNELL, J. H., ORIAS, E.: The ecological regulation of species diversity. Amer. Nat. **98**, 399–414 (1964).

CONSTANTINO, R. F.: The genetical structure of populations and developmental time. Genetics **60**, 409–418 (1968).

COOK, L. M.: The edge effect in population genetics. Amer. Nat. **95**, 295–307 (1961).

COOK, L. M.: Coefficients of Natural Selection. London: Hutchinson 1971.

COOLING, E. N. G.: Pinus merkusii. Fast growing timber trees of the lowland tropics no. 4. Commonwealth Forestry Inst., Dept. of Forestry, Univ. of Oxford (1968).

COOLING, E. N. G., ENDEAN, F.: Preliminary results from trials of exotic species for Zambian plantations. Forest Research Bulletin no. 10, Ministry of Lands and Natural Resources, Zambia (1966).

COOPER, J. P.: Selection and population structure in Lolium III. Selection for date of ear emergence. Heredity **13**, 461–480 (1959).

COOPER, J. P.: Selection and population structure in Lolium. Heredity **14**, 229–240 (1960).

COOPERRIDER, M.: Introgressive hybridization between Quercus marilandica and Quercus volutina in Iowa. Amer. J. Bot. **44**, 804–810 (1957).

COPE, F. W.: Incompatibility in Theobroma cacao. Nature **181**, 279–281 (1958).

COPE, F. W.: The mechanism of pollen incompatibility in Theobroma cacoa L. Heredity **17**, 157–182 (1962a).

COPE, F. W.: The effects of incompatibility and compatibility on genotype proportions of Theobroma cacao L. Heredity 183–195 (1962b).

CORMACK, R. M.: A boundary problem in population genetics. Biometrics **20**, 785–793 (1964).

CORNER, E. J. H.: The evolution of tropical forests. In: Evolution as a Process (HUXLEY, J. S., HARDY, A. C., FORD, E. B., eds.). London: Allen and Unwin 1954.

CORRIVEAU, A., BOUDOUX, M.: Le développement des provenances d'épinette blanche de la région forestière des Grands-Lacs et du St-Laurent au Quebec. Rapport Information Q-F-X-15. Laboratoire de Recherches Forestières, Ste-Foy, Quebec (1971).

Cox,G.W.: The role of competition in the evolution of migration. Evolution **22**, 180–192 (1968).

Cram,W.H.: Parent-Seedling characteristics in Caragana arborescens. Sci. Agr. **32**, 380–402 (1952).

Critchfield,W.B.: Geographic variation in Pinus contorta. Harv. Univ., Maria Moores Found. Publ. no. 3, 118 pp. (1957).

Crocker,W.: Life-span of seeds. Bot. Rev. **4**, 235–274 (1938).

Crosby,J.L.: The evolution of genetic discontinuity: computer models of the selection of barriers to interbreeding between subspecies. Heredity **25**, 253–297 (1970).

Crossley,S.: An experimental study of sexual isolation within a species of Drosophila. D.Phil. Thesis, Oxford (cit. after Robertson 1966) 1963.

Crow,J.E.: Some possibilities for measuring selection intensities in man. Hum. Biol. **30**, 1–13 (1963).

Crow,J.E., Morton,N.: Measurement of gene frequency drift in small populations. Evolution **9**, 202–214 (1955).

Crow,J.F., Kimura,M.: An Introduction to Population Genetics Theory. New York: Harper and Row 1970.

Curnow,R.N.: The effect of continued selection of phenotypic intermediates on gene frequency. Gen. Res. **5**, 341–353 (1964).

Daday,H.: Gene frequencies in Trifolium repens, I, II. Heredity **8**, 61–75, 377–390 (1954).

Daday,H.: Genetic relationship between cold hardiness and growth at low temperature in Medicago sativa. Heredity **19**, 173–180 (1964).

Dansereau,P., Lemms,K.: The grading of dispersal types. Contr. Inst. Bot. Montreal, no.71 (1957).

Darby,H.C.: The clearing of the woodlands in Europe. In: Man's Role in Changing the Face of the Earth, pp.183–216 (Thomas,Jr.,W.L., ed.). Chicago: Univ. of Chicago Press 1956.

Darling,F.F.: Natural History in the Highlands. London: Collins 1947.

Darling,F.F.: Man's ecological dominance through domesticated animals on wild lands. In: Man's Role in Changing the Face of the Earth, pp.778–787 (Thomas,Jr.,W.L., ed.). Chicago: Univ. of Chicago Press 1956.

Darlington,C.D., Mather,K.: The Elements of Genetics. London: Allen and Unwin 1949.

Darlington,C.D.: Evolution of Genetic Systems, 2nd ed. Edinburgh: Oliver and Boyd 1958.

Darlington,H.T.: The seventy year period for Dr. Beal's seed viability experiment. Amer. J. Bot. **38**, 379–381 (1951).

Darnell,R.M.: Evolution and the ecosystem. Amer. Zool. **10**, 9–15 (1970).

Dasmann,R.F.: Environmental Conservation. New York: John Wiley 1968.

Daubenmire,R.: A seven-year study of cone production as related to xylem layers and temperature in Pinus ponderosa. Amer. Middl. Nat. **64**, 187–193 (1960).

Daubenmire,R.: Plant Communities. New York, Evanston, London: Harper and Row 1968.

Davidson,R.A., Dunn,R.A.: A correlation approach to certain problems of population-environment relations. Amer. J. Bot. **54**, 529–538 (1967).

Dawkins,H.C.: New methods of improving stand composition in tropical forests. Caribbean For. **22**, 12–20 (1961).

Day,G.M.: The Indian as an ecological factor in the northeastern forest. Ecology **34**, 329–346 (1953).

Deakin,M.A.B.: Continuous clines and population structure. Amer. Nat. **102**, 295–296 (1968a).

Deakin,M.A.B.: Sufficient conditions for genetic polymorphism. Amer. Nat. **100**, 690–692 (1968b).

De Byle,N.V.: Detection of functional intraclonal Aspen root connections by tracers and excavation. For. Sci. 386–396 (1964).

DEL SOLAR, E.: Selection for and against gregariousness in the choice of oviposition sites by Drosophila pseudoobscura. Genetics **58**, 275–282 (1968).

DENGLER, A.: Über die Befruchtungsfähigkeit der weiblichen Kiefernblüte. Ztschr. Forst- u. Jagdw. **72**, 48–54 (1940).

DENGLER, A.: Bericht über Kreuzungsversuche mit Trauben- und Stieleiche (Quercus sessiliflora Smith und Quercus pedunculata Ehrh. bzw. robur L.) und zwischen europäischer und japanischer Lärche (Larix europaea D.C. bzw. decidua Miller und Larix leptolepis Murray bzw. Kaempferi Sargent). Mitt. Herm. Gör. Ak. Forstw. **1**, 87–109 (1941).

DENGLER, A.: Über den Pollenflug und seine Ausfilterung innerhalb von Waldbeständen. Ztschr. Forstgen. **4**, 107–110 (1955a).

DENGLER, A.: Pollenflugbeobachtungen in der Umgebung von Waldbeständen. Ztschr. Forstgen. **4**, 110–113 (1955b).

DENGLER, A., SCAMONI, A.: Über den Pollenflug der Waldbäume. Ztschr. f. d. Gesamte Forstwesen **76**, 136–155 (1944).

DESSUREAUX, L.: Heritability of tolerance to manganese toxicity in Lucerne. Euphytica **8**, 260–265 (1959).

DE VOS, A., BAILEY, R. H.: The effect of logging and intensive camping on vegetation in Riding Mountain. For. Chron. **46**, 49–55 (1970).

DIETRICHSSON, J.: Some results from an anatomic investigation of Norway spruce provenances in four international spruce tests in Sweden and Norway. World Cons. For. Gen. Stockholm, Contr. 3/6 (1963).

DIETRICHSSON, J.: Proveniensproblemet belyst ved studier av vekstrytme og klima. Medd. Norske Skogforsoksvesen **19**, 505–656 (1964).

DINGLER, H.: Die Bewegung der pflanzlichen Flugorgane. München (1889).

DIVER, C.: The problem of closely related species living in the same area. In: The New Systematics, pp. 303–328 (HUXLEY, J., ed.). Oxford (1940).

DOBZHANSKY, Th.: Evolution in the tropics. Amer. Sci. **38**, 209–221 (1950).

DOBZHANSKY, Th.: Genetic diversity and fitness. In: Genetic Today. Proc. XI Int. Congr. Gen., Vol. 3, pp. 541–552. New York: Pergamon Press 1963.

DOBZHANSKY, Th.: Adaptedness and fitness. In: Pop. Biol. and Evol., pp. 109–121. Syracuse, N.Y.: Syracuse Univ. Press 1967.

DOBZAHNSKY, Th., EPLING, C.: Contributions to the genetics, taxonomy and ecology of Drosophila pseudoobscura and its relatives. Carn. Inst. Wash. Publ. **554**, (1944).

DOBZHANSKY, Th., SPASSKY, B.: Drosophila paulistorum, a cluster of species in statu nascendi. Proc. Nat. Ac. Sci. Wash. **45**, 419–428 (1959).

DOBZAHNSKY, Th., PAVLOVSKY, P.: How stable is balanced polymorphism? Proc. Nat. Ac. Sci. Wash. **46**, 41–47 (1960).

DOBZHANSKY, Th., HUNTER, A. S., PAVLOVSKY, O., SPASSKY, B., WALLACE, B.: Genetics of natural populations XXXI. Genetics of an isolated marginal population of Drosophila pseudoobscura. Genetics **48**, 91–104 (1963).

DOCTERS VAN LEEUWEN, W. M.: Krakatau. Leiden: Brill 1936.

DOGGETT, H., MAJISU, B. N.: Disruptive selection in crop development. Heredity **23**, 1–22 (1968).

DOGRA, F. D.: Pollination mechanisms in Gymnosperms. Adv. in Palynology, Licknow 142–175 (1964).

DORMAN, K. W., BARBER, J.: Time of flowering and seed ripening in southern pines. SEFES Sta. Paper 72 (1956).

DOYLE, J., O'LEARY, M:: Pollination in Pinus. Sci. Proc. Roy. Dublin Soc. **20**, 181–190 (1935).

DUFFIELD, J. W.: An evolutionary view of wood. J. For. **66**, 354–357 (1968).

DUGLE, J. R.: A taxonomic study of western Canadian species in the genus Betula. Can. J. Bot. **44**, 929, 1007 (1966).

DUMOUCHEL, W. H., ANDERSSON, W. W.: The analysis of selection-in experimental populations. Genetics 58, 435–449 (1968).

DUNBAR, M. J.: Ecological development in Polar Regions. Englewood Cliffs, N. J.: Prentice-Hall Inc. 1968.

DYAKOWSKA, J., ZURZYCKI, J.: Gravimetrio studies on pollen. Bull. Ac. Pol. Sci. 7, 11–16 (1959).

EAST, E. M.: The distribution of self-sterility in flowering plants. Proc. Amer. Phil. Soc. 82, 449–260 (1940).

EHRENDORFER, F.: Dispersal mechanisms, genetic systems, and colonizing abilities in some flowering plant families. In: Gen. of Col. Spec., pp. 331–352. New York: Academic Press 1964.

EHRLICH, P. R., RAVEN, P. H.: Butterflies and plants: a study in coevolution. Evolution 18, 586–608 (1965).

EHRLICH, P. R.: Differentiation of populations. Science 165, 1228–1231 (1969).

EHRMAN, L.: Hybrid sterility as an isolating mechanism in the genus Drosophila. Quart. Rev. Biol. 37, 279–302 (1962).

EHRMAN, L.: Direct observation of sexual isolation between allopatric Drosophila paulistorum races. Evolution 19, 459–464 (1965).

EICHE, V.: Spontaneous chlorophyll mutations in Scotch Pine. Medd. St. Skogsf. Inst. 13, 1–64 (1955).

EICHE, V.: Cold damage and plant mortality in experimental provenance plantations with Scotch Pine in Northern Sweden. Stud. For. Suec. no. 36 (1966).

EINSPAHR, D. W.: Sex ratio in quaking aspen and possible sexrelated characteristics. V. World For. Congr. Seattle, Wash., SP/6/II — USA (1960).

EISENHUT, G.: Blühen, Fruchten und Keimung in der Gattung Tilia. Flora 1947, 43–75 (1959).

EISENHUT, G.: Untersuchungen über die Morphologie und Oekologie der Pollenkörner heimischer und fremdländischer Waldbäume. Forstw. Forschungen 15. Hamburg: Paul Parey 1961.

EKBERG, I., ERIKSSON, G.: Development and fertility of pollen in three species of Larix. Hereditas 57, 303–311 (1967).

ELDRIDGE, K. G.: Breeding system of Eucalyptus regnans. Proc. Work. Group IUFRO Sect. 22 on Sex. Repr. of For. Trees, Finland (1970).

ELKINGTON, T.: Introgressive hybridization between Betula nana L. and B. pubescens Ehrh. in north-west Iceland. New Phytol. 67, 109–118 (1968).

ELLENBERG, H.: In: Aufgaben und Methoden der Vegetationskunde. Einführung in die Phytologie (WALTER, H., ed.). Stuttgart 1956.

ELLENBERG, H.: Bodenreaktion (einschließlich Kalkfrage). In: Handbuch der Pflanzenphysiologie, Vol. 4 (RUHLANDT, W., ed.) 1958.

ENDEAN, F.: Research into plantation silviculture in Zambia. In: FAO World Symposium on Man-Made Forests and Their Industrial Importance, pp. 1645–1663. Rome: FAO 1967.

ENEROTH, O.: Om granfröets spridningsvidd. Skogen (1929).

EPLING, C.: Actual and potential geneflow in natural populations. Amer. Nat. 81, 104–113 (1947).

EPLING, C., LEWIS, H., BALI, F. M.: The breeding group and seed storage. Evolution 14, 238–255 (1960).

ERIKSSON, G.: Temperature response of pollen mother cells in Larix and its importance for pollen formation. Stud. For. Suecica no. 63 (1968a).

ERIKSSON, G.: Meiosis and pollen formation in Larix. Thesis, matematisk-naturvetensk. fak. vid. Stockholms universitet (1968b).

ERIKSSON, G., SULIKOVA, Z., EKBERG, I.: Varför är frösättningen hos lärk sa lag? Sv. Skogsv. Tidskr. 691–697 (1967).

ERTELD, W.: Der Verlauf des Umsetzens bei der Fichte. Forstw. Holzw. 4, 301–309 (1950).

EVANS, M. W.: Relations of latitude to certain phases of growth of timothy. Amer. J. Bot. **26**, 212–231 (1939).

EWENS, W. J.: On the problem of self-sterility alleles. Genetics 50, 1433–1438 (1964).

FAEGRI, K., PIJL, L. VAN DER: The Principles of Pollination Ecology. Oxford: Pergamon 1966.

FALCONER, D. S.: Selection for phenotypic intermediates in Drosophila. J. Gen. **55**, 551–561 (1957).

FALCONER, D. S., ROBERTSON, A.: Selection for environmental variability of body size in mice. Z. Ind. Abst. Vererb. Lehre 87, 385–391 (1956).

FAO: Actual and potential role of man-made forests in the changing world pattern of wood consumption. In: FAO World Symposium on Man-Made Forests and Their Industrial Importance, pp. 1–50. Rome: FAO 1967 a.

FAO: Wood: world trends and prospects. FFHC Basic Study no. 16. Rome (1967 b).

FAO: Taungy in Kenya: the shamba system. In: FAO World Symposium on Man-Made Forests and Their Industrial Importance, pp. 1057–1068. Rome: FAO 1967 c.

FAO: Report of the First Session of the FAP Panel of Experts on Forest Gene Resources. Rome: FAO 1969.

FAO: Report of the Second Session of the FAO Panel of Experts on Forest Gene Resources Rome: FAO (in press) 1972.

FEDOROV, A.: The structure of the tropical rain forest and speciation in the humid tropics. J. Ecol. **54**, 1–11 (1966).

FEJER, S. O.: Growth and phenotypic stability under controlled environments in inbred and outorossed Lolium populations and individual plants. Ztschr. f. Pflanzenz. **58**, 151–196 (1967).

FENDRICK, I.: Entwicklung einer Indikator-Aktivierungsmethode zum Studium des Pollenflugs von Waldbäumen. Diss. T. U. Hannover (1967).

FENNER, F.: Myxoma virus and Oryctolagus cuniculus: two colonizing species. In: Gen. of Col. Spec., pp. 485–501. New York: Academic Press 1964.

FIRBAS, F.: Über die Wirksamkeit der natürlichen Verbreitungsmittel der Waldbäume. Natur und Heimat 6, 1–11 (1935).

FIRBAS, F., REMPE, H.: Über die Bedeutung der Sinkgeschwindigkeit für die Verbreitung des Blütenstaubs durch den Wind. Biokl. Beibl. 3, 15–26 (1936).

FIRBAS, F., SAGROMSKY, K.: Untersuchungen über die Größe des jährlichen Pollenniederschlags vom Gesichtspunkt der Stoffproduktion. Biol. Zentralbl. 66, 131–147 (1947).

FISHER, R. A.: Genetical Theory of Natural Selection. Oxford: Clarendon Press 1930.

FISHER, R. A.: The evolution of dominance in certain polymorphic species. Amer. Nat. **64**, 385–406 (1930).

FISHER, R. A.: The evolution of dominance. Biol. Rev. 6, 345–368 (1931).

FISHER, R. A.: The wave of advantageous genes. Ann. Eugen. 7, 355–369 (1937).

FISHER, R. A.: Number of self-sterility alleles. Nature 160, 797 (1947).

FISHER, R. A.: Gene frequencies in a cline as determined by selection and diffusion. Biometrics **6**, 353–361 (1950).

FISHER, R. A.: Polymorphism and natural selection. Ecology **46**, 289–293 (1958).

FLICHE, M. P.: Un reboisement. Etude botanique et forestière. Ann. Sci. Agronom. 1, 22–31 (1888).

FLOR, H. H.: Identification of races of flax rust by lines with single rust conditioning genes. U.S. Dept. Agr. Tech. Bull. 1077 (1954).

FORD, E. B.: Darwinism and the study of evolution in natural populations. J. Linn. Soc. London, Bot. 56, 41–48 (1958).

FORD, E. B.: Ecological Genetics. London: Methuen 1964.

FORD, H. D., FORD, E. B.: Fluctuation in numbers and its influence on variation in Melitea aurinia. Trans. Roy. Ent. Soc. London 78, 345–351 (1930).

FORDE, M. B.: Variation in natural populations of Pinus radiata in California, Part 4, Discussion. New Zeal. J. Bot. **2**, 486–501 (1964).

FOWLER, D. P.: Effects of inbreeding in Red Pine, Pinus resinosa Ait. III. Factors affecting natural selfing. Silvae Gen. **14**, 37–46 (1965).

FOWLER, P. D.: Effects of inbreeding in Red Pine, Pinus resinosa Ait. II. Pollination studies. Silvae Gen. **14**, 12–23 (1965).

FRANKEL, O. H.: Variation under domestication. Austr. J. Sci. **22**, 127–132 (1959).

FRANKEL, O. H., BENNETT, E. ed.: Genetic Resources in Plants — Their Exploration and Conservation. Oxford, Edinburgh: Blackwell Scientific Publications 1970.

FRANKLIN, E. C.: Artificial self-pollination and natural inbreeding in Pinus taeda L. Ph. D. Thesis, N.C. State Univ., Dep. of For., Raleigh (1968).

FRANZ, J. M.: Qualität und intraspezifische Konkurrenz im Regulationsprozeß von Insektenpopulationen. Ztschr. Angew. Entom. **55**, 319–325 (1964).

FRASER, A.: Colonization and genetic drift. In: Gen. of Col. Spec., pp. 117–122. New York: Academic Press 1964.

FROILAND, S. G.: The biological status of Betula andrewsii. A. Nels. Evolution **6**, 268–282 (1952).

FROST, S. W.: Insects and Pollinia. Ecology **46**, 556–558 (1965).

FRYXELL, P. G.: Modes of reproduction in higher plants. Bot. Rev. **23**, 135–233 (1957).

FUCHS, A., ANDEL, O. M. VAN: Physiological and biochemical aspects of hostpathogen interactions. Neth. J. Plant Path. **74**, Suppl. 1 (1968).

GALOUX, A.: Forêt, ecosystème et cybernétique Bull. Soc. Roy. For. Belgique 4–23 (1963).

GALOUX, A.: La variabilité génécologique de hêtre commun (Fagus silvatica) en Belgique. Stat. Rech. Eaux et For. Groenendaal-Hoeilhardt, Ser. A, no. 11 (1966).

GALOUX, A.: Diversification génécologique régional chez les espèces ligneuses feuillues. Proc. 14. Int. Congr. IUFRO, Munich, Sect. **22**, 14 (1967).

GALOUX, A.: Diversification génécologique régional chez les espèces ligneuses feuillues. Ecol. Plant. **4**, 1–14 (1969).

GALOUX, A., FALKENHAGEN, E.: Recherches sur la variabilité génécologique de l'Erable sycomore (Acer pseudoplatanus) en Belgique. St. Rech. Eaux et For. Groenendaal-Hoeilhaardt, Ser. A, no. 10 (1965).

GATHY, P.: L'expérience internationale sur l'origine des graines d'epicea (Picea abies Karst.), resultats en Belgique. St. Rech. des eaux et des For. Groenendaal-Hoeilaart, Ser. B, no. 24, 29 (1960).

GAUSE, G. F.: The Struggle for Existence. Baltimore: Williams and Wilkins 1934.

GERHOLD, H. D., SCHREINER, E. J., McDERMOTT, R. E., WINIESKI, J. A., eds.: Breeding pest resistant trees. Proc. NATO and NSF Study Inst. Penn. State Univ., Univ. Park, Pa. New York: Pergamon Press 1966.

GERSHENSON, S.: Evolutionary studies on the distribution and dynamics of melanism in the Hamster (Cricetus cricetus L.). Genetics **30**, 207–251 (1945).

GIBBS, J. N.: Resin and the resistance of conifers to Fomes annosus. Ann. Bot. Oxf. **32**, 649–665 (1968).

GIBSON, J. B., THODAY, J. M.: Effects of disruptive selection. VIII. Imposed quasi-random mating. Heredity **18**, 513–524 (1963).

GIBSON, J. B., THODAY, J. M.: Effects of disruptive selection. IX. Low selection intensities. Heredity **19**, 125–130 (1964).

GILBERT, J. M.: Forest succession in the Florentine Valley, Tasmania. Proc. Roy. Soc. Tasmania **93**, 129–151 (1959).

GILMOUR, J. R.: Methods and treatments used to secure natural regeneration in British Columbia. For. Chron. **46**, 451–452 (1970).

GILMOUR,J.S.L., GREGOR,J.W.: Demes: A suggested new terminology. Nature **144**, 333–334 (1939).

GLEESON,H.A.: The vegetational history of the Middle West. Ann. Assoc. Amer. Geographers **12**, 39–85 (1932).

GODMAN,R.M., KREFTING,L.W.: Factors important to yellow birch establishment in upper Michigan. Ecology **41**, 18–28 (1960).

GOHEEN,A.C.: The cultivated highbush-blueberry. Yearb. Agr., USDA Wash., D.C. 784–785 (1953).

GÖHRE,K.: Die Robinie und ihr Holz. Berlin: Akad. Verlag 1952.

GOLDSCHMIDT,R.: Lymantria. Bibl. Gen. 11 (1934).

GOLFARI,L.: Water balance according to Thornthwaite as a guide for establishing analogous climates: some examples in Pinus radiata D. Dom. In: FAO Symposium on Man-Made Forests and Their Industrial Importance, pp. 988–992. Rome: FAO 1967.

GOMBRICH,L.: Forestry and the environment. For. Abst. **32**, 8–13 (1971).

GONTARSKE,H.: Leistungsphysiologische Untersuchungen an Sammelbienen. Arch. Bienenkunde **16**, 110–125 (1935).

GOOR,A.Y.: Tree planting practices for arid zones. FAO Forestry Development Paper no. 16 (1963).

GORDON,A.G.: Ecology of Picea chihuahuana Martinez. Ecology **49**, 880–896 (1968).

GRANT,K.A.: Hummingbirds and Their Flowers. New York: Columbia Univ. Press 1968.

GRANT,K.A., GRANT,V.: Mechanical isolation of Salvia apiana and S. mellifera. Evolution **18**, 196–212 (1964).

GRANT,V.: Pollination systems as isolating mechanisms in angiosperms. Evolution **3**, 82–97 (1949).

GRANT,V.: Pollination. Encycl. Amer. (1953).

GRANT,V.: The regulation of recombination in plants. Cold Spring Harbor Symp. Quant. Biol. **23**, 337–363 (1958).

GRANT,V.: The Origin of Adaptations. New York: Columbia Univ. Press 1963.

GRANT,V.: The selective origin in incompatibility barriers in the plant genus Gilia. Amer. Nat. **100**, 99–118 (1966).

GRANT,V.: Linkage between morphology and viability in plant species. Amer. Nat. **101**, 125–139 (1967).

GRANT,V., GRANT,K.A.: Records of hummingbird pollination in the western American flora. I. Some California plant species. Aliso **6**, 51–66 (1966).

GRANT,V., GRANT,K.A.: Records of hummingbird pollination in the western American flora. II. Additional California records. Aliso **6**, 103–105 (1967 a).

GRANT,V., GRANT,K.A.: Records of hummingbird pollination in the western American flora. III. Arizona records. Aliso **6**, 107–110 (1967 b).

GREENWOOD,J.J.D.: Apostatic selection and population density. Heredity **24**, 157–161 (1969).

GREGORY,P.H.: The Microbiology of the Atmosphere. London: Leonhard Hill 1961.

GUPPY,H.B.: Observations of a Naturalist in the Pacific. II. Plant-Dispersal. London: McMillan 1906.

GUTIERREZ,M.G., SPRAGUE,G.: Randomness of mating in isolated polycross plantings of maize. Genetics **44**, 1075–1082 (1959).

HAACK,F.: Der Kienzopf, Pŧeridermium pini. Ztschr. Forst- u. Jagdw. **46**, 3–45 (1914).

HAGERUP,O.: Pollination in the Faroes in spite of rain and poverty in insects. Dan. Biol. Medd. **18**, 1–48 (1951).

HAGMAN,M.: Genetic mechanisms affecting inbreeding and outbreeding in forest trees; their significance for microevolution of forest tree species. In: Proc. XIV IUFRO Congr. München, vol. 3, 346–365 (1967).

HAGMAN, M.: Observations on the incompatibility in Alnus. Proc. Work. Group IUFRO Sect. 22 on Sex. Repr. of For Trees, Finland (1971).

HAGMAN, M.: On self- and cross-incompatibility shown by Betula verrucosa Ehrh. and Betula pubescens Ehrh. Comm. Inst. For. Fenn. 73, 1–125 (1971).

HAGMAN, M.: The Finnish standard stands for forestry. Proc. 13th Meeting Comm. For. Tree Breeding in Can. 1972 (in press) 1972.

HAIG, I. T., HUBERMANN, M. A., AUUNG DIN, U.: Tropical Silviculture FAO Forestry and Forests Products Studies no. 13, vol. 1. Rome (1958).

HAIR, D., SPADA, B.: Hardwood timber resources of the United States. Unasylva 24, 29–32 (1970).

HALDANE, J. B. S.: The Causes of Evolution London (1932).

HALDANE, J. B. S.: The theory of a cline. J. Gen. 48, 277–284 (1948).

HALDANE, J. B. S.: Suggestions as to quantitative measurements of rates of evolution. Evolution 3, 51–56 (1949).

HALDANE, J. B. S.: The measurement of natural selection. Proc. IX. Congr. Gen. a1, 480–487 (1954a).

HALDANE, J. B. S.: The measurement of natural selection. Caryologia 6 (suppl.), 480–487 (1954b).

HALDANE, J. B. S.: The statics of evolution. In: Evolution as a Process, pp. 109–121 (HUXLEY, J., ed.) (1954c).

HALDANE, J. B. S., JAYAKAR, S. D.: Polymorphism due to selection depending on the composition of a population. J. Gen. 58, 318–323 (1963).

HALKKA, O.: Geographical, spacial and temporal varaibility in the balanced polymorphism of Philaenus spumarius. Heredity 19, 383–401 (1964).

HALKKA, O., MIKKOLA, E.: Characterization of clines and isolates in a case of balanced polymorphism. Hereditas 54, 140–148 (1965).

HALL, M. T.: A hybrid swarm in Juniperus. Evolution 6, 347–366 (1952).

HALLAUER, A. R.: Inheritance of flowering in maize. Genetics 52, 129–137 (1965).

HALLER, J. R.: The role of hybridization in the origin and evolution of Pinus washoensis. Proc. IX Int. Congr. Bet. Montreal 2, 149 (1959a).

HALLER, J. R.: Factors affecting the distribution of ponderosa pine and jeffreyi pine in California. Madrono 15, 65–71 (1959b).

HANOVER, J. W.: Genetics of terpenes. I. Gene control of monoterpene levels in Pinus monticola. Heredity 21, 73–84 (1966).

HANSON, W. D.: Effects of partial isolation (distance), migration, and different fitness requirements among environmental pockets upon steady state gene frequencies. Biometrics 22, 453–468 (1966).

HARDING, J., ALLARD, R. W., SMELTZER, D. G.: Population studies in predominantly self-pollinated species IX. Frequency dependent selection in Phaseolus lunatus. Proc. Nat. Ac. Sci. Wash. 56, 99–104 (1966).

HARDING, J., ALLARD, R. W.: Population studies in predominantly self-pollinated species XII. Interactions between loci affecting fitness in a population of Phaseolus lunatus. Genetics 61, 721–736 (1969).

HARDY, A.: The Living Stream. London, Glasgow: Collins 1965.

HARLAND, G.: An alternation in gene frequency in Ricinus communis due to climate conditions. Heredity 1, 121–125 (1947).

HARLEY, J. L.: Mycorrhiza and nutrient uptake in forest trees. In: Physiology of Tree Crops, pp. 163–179 (LUCKWILL, L. C., CUTTING, C. J., eds.). London, New York: Academic Press 1970.

HARPER, J. L., CLATWORTHY, J. N., MCNAUGHTON, U. H., SAGAR, C. R.: The evolution and ecology of closely related species living in the same area. Evolution 15, 209–227 (1961).

HARPER, J. L.: The nature and consequence of interference among plants. In: Genetics Today, Proc. XI. Int. Gen. Congr. 2, 465–482 (1963).

HARPER, J. L.: Establishment, aggression, and cohabitation in weedy species. In: Gen. of Col. Spec., pp. 243–268. New York: Academic Press 1964.

HARPER, J. L.: The regulation of numbers and mass in plant populations. In: Pop. Biol. and Evol., pp. 139–158. Syracuse, N.Y.: Syracuse Univ. Press 1967.

HASKELL, G.: Selection, correlated response and speciation in subsexual Rubus. Genetica 30, 240–260 (1959).

HASKELL, G.: Seedling morphology in applied genetics and plant breeding. Bot. Rev. 27, 382–421 (1961).

HATHEWAY, W. H., BAKER, H. G.: Reproductive strategies in Pithecellobium and Entrelobium — further information. Evolution 24, 253–254 (1970).

HATTEMER, H. H.: Estimates of heritabilities published in forest tree breeding research. Proc. I. World Cons. For. Gen. For. Tree Improv. Stockholm (1963).

HATTEMER, H. H.: Genetic mechanisms allowing and equilibrium between a parasitic fungus and a forest tree species. XIV. IUFRO Congress, München (1967).

HATTEMER, H. H., HANSON, W. R., MERGEN, F.: Some factors in the distribution of European pine sawfly egg clusters in an experimental plantation of hard pines. Theor. Appl. Gen. 39, 280–289 (1969).

HAYWARD, M. D., BREESE, E. L.: The genetic organization of natural populations of Lolium perenne L. Heredity 23, 357–368 (1968).

HEDEGART, T.: The Thai-Danish teak improvement centre five years after initiation. Unasylva 25, 31–37 (1971).

HEDEMANN-GADE, E.: Om tall- och granfrös spridningsvidd. Skogen (1929).

HEIKINHEIMO, O.: Über die Besamungsfähigkeit der Waldbäume. Comm. Inst. For. Fenniae 17 (1932).

HEIMBURGER, C.: Forest tree breeding and genetics in Canada. Proc. Gen. Soc. Can. III, 41–49 (1958).

HEINSELMAN, M. L.: Preserving nature in forested wilderness areas and national parks. In: Recreation Symposium Proceedings. Northeastern Forest Experiment Station, Forest Service USDA, pp. 57–67 (1971).

HEINTZE, A.: Handbuch der Verbreitungsmittel der Pflanzen. Stockholm. Selbstverlag (1932/35).

HENDERSON, C. R., KEMPTHORNE, O., SEARLE, S. R., KROSIGK, C. M., VON: The estimation of environmental and genetic trends from records subject to culling. Biometrics 15, 192–218 (1959).

HENGST, E.: Das Umsetzen von Einzelbäumen in gleichaltrigen Buchenbeständen. Arch. f. Forstw. 11, 791–798 (1962).

HESLOP-HARRISON, J.: Variability and environment. Evolution 13, 145–147 (1959).

HESLOP-HARRISON, J.: Forty years of genecology. Adv. Ecol. Res. 2, 159–248 (1964).

HESLOP-HARRISON, J.: Reflections on the role of environmentally-governed reproductive versatility in the adaptation of plant populations. Trans. Bot. Soc. Edinb. 40, 159–168 (1966).

HESSELMANN, H.: Beobachtungen über die Verbreitungsfähigkeit des Waldbaumpollens. Medd. Stat. Skogsförsöksanst. 16 (1919).

HESSELMANN, H.: Nagra studier över fröspridninge hos gran och tall och kalhyggets besåning. Medd. St. Skogsförsöksanst. H. 27, 145–182 (1934).

HILDEBRAND, F.: Die Verbreitungsmittel der Pflanzen. Leipzig: Engelmann 1873.

HOENIGSBERG, H. F., CHEJNE, A. J., HORTOBAGJI-GERMAN, E.: Preliminary report on artificial selection towards sexual isolation. Z. Tierpsychologie 23, 129–135 (1966).

HOFFMANN, J.: Über die bisherigen Ergebnisse der Fichtentypenforschung. Archiv f. Forstw. 17, 207–216 (1968).

HOFFMANN,K., BERGMANN,J.H.: Klonspezifische Wirkung des Dalaponpräparates Omnidel-Spezial auf Kiefernpfropflinge. Soz. Forstw. **6**, 277–278 (1966).

HOFFMANN,K., SCHEUMANN,W.: Anthocyanverfärbung und Wuchsleistung von Fichtenkeimlingen. Arch. Forstw. **16**, 683–688 (1967).

HOLDRIDGE,L.R.: Middle America. In: World Geography of Forest Resources, pp.183–200 (HADEN-GUEST,S., WRIGHT,J.K., TECLAFF,E. M., eds.). New York: The Ronald Press 1956.

HOLDRIDGE,L.R.: The silviculture of natural mixed tropical hardwood stands in Costa Rica. In: Tropical Silviculture, vol.11, pp.57–66. Rome: FAO 1957.

HOLGATE,P.: Genotype frequencies in a section of a cline. Heredity **19**, 507–509 (1964).

HOLMBOE,J.: Nogle iaktagelser over fröspridning pa ferskvandis. Bot. Notis (1898).

HOLMES,F. W.: Virulence in Ceratocystis ulmi. Meth. J. Plant. Path. **71**, 97–112 (1965).

HOLMGREN,P.: Leaf factors affecting light-saturated photosynthesis in ecotypes of Solidago virgaurea from exposed and shaded habitats. Physiologia Plantarum **21**, 676–698 (1968).

HOSIE,R.C.: Forest regeneration in Ontario. Forestry Bull. Univ. of Toronto **2**, 1–134 (1953).

HOVANITZ,W.: Polymorphism and evolution. Symp. Soc. Exp. Biol. Cambridge **7**, 238–253 (1953).

HUGHES,E.L.: Regeneration after logging in the maritime provinces. Pulp and Paper Mag. Can. **71**, 72–78 (1970).

HUGHES,J.F.: Utilization of the wood of low altitude tropical pines. In: FAO World Symposium on Man-Made Forests and Their Industrial Importance pp.845–864. Rome: FAO 1967.

HURD,L.E., TELLINGER,M.V., WOLF,I.L., MCNAUGHTON,S.J.: Stability and diversity at three tropic levels in terrestrial successional ecosystems. Science **173**, 1134–1136 (1971).

HUTCHINSON,G.E.: Copepodology for the ornithologist. Ecology **32**, 571–577 (1951).

HUTCHINSON,G.E.: Patterns in ecology. Proc. Nat. Ac. Sci. Philadelphia **105**, 1–12 (1953).

HUTCHINSON,G.E.: Concluding Remarques. In: Cold Spring Harbor Symp. Quant. Biol. **22**, 415–427 (1958).

HUTCHINSON,G.E.: The Ecological Theater and the Evolutionary Play. New Haven: Yale Univ. Press 1965.

HUWALD,K.: Resistenzentwicklung bei Tetranychus urticae Koch in Abhängigkeit von der Vorgeschichte der Population und dem physiologischen Zustand der Wirtspflanze. Z. Angew. Entom. **56**, 1–40 (1965).

HUXLEY,J.: Towards the new systematic. In: The New Systematics, pp.1–46 (HUXLEY,J., ed.). Oxford (1940).

HUXLEY,J.: Evolution, the Modern Synthesis. New York and London (1942).

HUXLEY,J.S.: Morphism as a clude for the study of population dynamics. Proc. Roy. Soc. B. **145**, 319–322 (1956).

IMAM,A.G., ALLARD,R.W.: Population studies in predominantly selfpollinated species. VI. Genetic variability between and within natural populations in wild oats from differing habitats in California. Genetics **51**, 49–62 (1965).

IYAMABO,D.E.: Species introduction and growth in African Savanna. In: Second World Consultation on Forest Tree Breeding, vol.1, pp.81–99. Rome: FAO 1969.

JAIN,S.K., ALLARD,R.W.: The nature and stability of equilibria under optimizing selection. Proc. Nat. Ac. Sci. Wash. **54**, 1436–1443 (1965).

JAIN,S.K., BRADSHAW,A.D.: Evolutionary divergence among adjacent plant populations. I. The evidence and its theoretical analysis. Heredity **21**, 407–441 (1966).

JAIN,S.K., MARSHALL,D.R.: Genetic changes in a barley population analyzed in terms of some life cycle components of selection. Genetica **38**, 355–374 (1967).

JAIN,S.K., WORKMAN,P.L.: Generalized F-statistics and the theory of inbreeding and selection. Nature **214**, 674–680 (1967).

JARVIS, J. M., STENEKER, G. A., WALDRON, R. M., LEES, J. C.: Review of Silvicultural Research. Departmental Publ. no. 1156. Canada Department of Forestry (1966).

JEFFERS, J. N., BLACK, T. M.: An analysis of variability in Pinus contorta. Forestry 36, 199–218 (1963).

JEFFREY, W. W.: Foresters and the challenge of integrated resource management. For. Chron. 46, 196–199 (1970).

JOHNSSON, H.: Avkommepröving av björk. For. Skogsträdsförädling Årsbok, 1966 (1967).

JOHNSSON, H.: Cone and seed production from Pine seed orchards over a period of years. Årsb. För. Skogstr. Förädl., Uppsala (1968).

JONEBORG, S.: Monströs kottebildning hos granen. Vegetativt skott eller kotte? Sv. Skogsv. Tidskr. 43, 453–462 (1945).

JONES, D. A.: On the polymorphism of cyanogenesis in Lotus corniculatus. I. Selection by animals. Can. J. Gen. Cyt. 8, 351–368 (1966).

JONES, D. A.: On the polymorphism of cyanogenesis in Lotus corniculatus L. II. The interaction with Trifolium repens L., Heredity 23, 453–455 (1968).

JONES, D. F.: Unisexual maize plants and their bearing on sex differentiation in other plants and animals. Genetics 19, 552–564 (1934).

JONES, E. W.: The structure and reproduction of the virgin forest of the north temperate zone. New Phytol. 44, 130–148 (1945).

JONES, E. W.: Ecological studies on the rain forest of Southern Nigeria. I. J. Ecol. 43, 56 (1955).

JONES, E. W.: Ecological studies on the rain forest of Southern Nigeria. II. J. Ecol. 44, 483–504 (1956).

JOSEPHSON, H. R., HEIR, D.: The United States. In: A World Geography of Forest Resources, pp. 149–182 (HADEN-GUEST, S., WRIGHT, J. K., TECLAFF, E. M., eds.). New York: The Ronald Press 1956.

JOSHI, B. C., JAIN, S. K.: Clinal variation in natural populations of Justicia simplex. Amer. Nat. 98, 123–125 (1964).

JOWETT, D.: Populations of Agrostis ssp. tolerant to heavy metals. Nature 182, 816–817 (1958).

JOWETT, D.: Population studies on lead-tolerant Agrostis tenuis. Evolution 18, 70–80 (1964).

KARLIN, S.: Equilibrium Behavior of Population Genetic Models with Non-random Mating. Stanford, Calif.: Standford Univ. Press 1969.

KEAY, R. W.: Wind-dispersed species in a Nigerian forest. J. Ecol. 45, 471–478 (1957).

KEDHARNATH, S., UPHADAY, L. P.: Chiasma frequency in Pinus roxburhii Sarg. and P. elliottii Engelm. Silvae Gen. 16, 112–113 (1967).

KELLISON, R. C.: Establishment and management of clonal seed orchards of pine. In: Second World Consultation on Forest Tree Breeding, pp. 1355–1366. Rome: FAO 1969.

KEMP, R. H.: Trials of exotic tree species in the savanna region of Nigeria. Research Paper no. 4. Savanna Forestry Research Station, Ministry of Agriculture and Natural Resources, Nigeria (1969).

KENNEDY, L. L.: Seed — from collection to established tree. For. Chron. 45, 421–427 (1969).

KERNER, MARILAUN, v.: Pflanzenleben. Leipzig (1891).

KERSHAW, K. A.: Quantitative and Dynamic Ecology. London: Edw. Arnold 1964.

KERSTER, H. W., LEVIN, D. A.: Neighborhood size in Lithospermium caroliniense. Genetica 60, 577–587 (1948).

KESSLER, S.: Selection for and against ethological isolation between Drosophila pseudoobscura and Drosophila persimilis. Evolution 20, 634–645 (1966).

KETTLEWELL, H. B. D.: Industrial melanism in Lepidoptera and its contribution to our knowledge of evolution, Prov. 10th Congr. Ent. 2, 831–841 (1956).

KETTLEWELL, H. B. D., BERRY, R. J.: The study of a cline. Amathes glareosa Esp. and its melanic f. edda Staud (Lep.) in Shetland. Heredity 16, 403–414 (1961).

KETTLEWELL, H. B. D., BERRY, R. J.: Gene flow in a cline. Amathes glareosa Esp. and its melanic f. edda Staud (Lep.) in Shotland. Heredity 24, 1–14 (1969 a).

KETTLEWELL, H. B. D., BERRY, R. J., CADBURY, C. J., PHILLIPS, C. G.: Differences in behaviour dominance and survival within a cline. Amathes glareosa Esp. (Lep.) and its melanic f. edda Staud. in Shetland. Heredity 24, 15–26 (1969 b).

KIILS, A. D., CHROSCIEWICZ, Z.: Prescribed fire — its place in reforestation. For. Chron. 46, 448–451 (1970).

KIMURA, M.: Rules for testing stability of a selective polymorphism. Proc. Nat. Ac. Sci. Wash. 42, 336–340 (1956).

KIMURA, M.: A model of a genetic system which leads to close linkage by natural selection. Evolution 10, 278–287 (1956).

KIMURA, M.: Natural selection as the process of accumulating genetic information in adaptive evolution. Gen. Res. 2, 127–140 (1961).

KIMURA, M.: Diffusion models in population genetics. J. Appl. Prob. 1, 177–232 (1964).

KIMURA, M.: Stochastic processes in population genetics with special reference to distribution of gene frequencies and gene fixation. In: Math. Topics in Pop. Gen., Biom. 1, pp. 178–209. Heidelberg: Springer 1970.

KIMURA, M., CROW, J. E.: The measurement of effective population number. Evolution 17, 279–288 (1963).

KIMURA, M., WEISS, G. H.: The stepping stone model of population structure and the decrease of genetic correlation with distance. Genetics 49, 561–576 (1964).

KING, J. P., JEFFERS, R. M., NIENSTAEDT, H.: Effects of varying proportions of self-pollen on seed yield, seed quality and seedling development in Picea glauca. Proc. Work. Group IUFRO Sect. 22 on Sex. Repr. of For. Trees, Finland (1970).

KING, K. F. S.: Agri-Silviculture (The Taungya System). Bulletin no. 1. Dept. of Forestry, Univ. of Ibadan (1968).

KINLOCH, B. B., PARKS, G. K., FOWLER, C. W.: White pine blister rust: simply inherited resistance in sugar pine. Science 167, 193–194 (1970).

KIRK, M. D.: How to save Southern Ontario's woodlots. For. Chron. 46, 26–30 (1970).

KITAGAWA, O.: Genetic divergence in M. Vetukhivs populations of Drosophila pseudoobscura. Gen. Res. 10, 303–312 (1967).

KLOPFER, P. H.: Environmental determinants of faunal diversity. Amer. Nat. 93, 337–342 (1959).

KNABE, W.: Man-made forests on man-made ground. In: FAO World Symposium on Man-Made Forests and Their Industrial Importance, pp. 1165–1173. Rome: FAO 1967.

KNIGHT, G. R., ROBERTSON, A., WADDINGTON, C. H.: Selection for sexual isolation within a species. Evolution 10, 14–22 (1956).

KOEHN, R. K.: The component of selection in the maintenance of a serum esterase polymorphism. Proc. 12. Int. Congr. Gen. 1, 227 (1968).

KOHLER, A.: Zum Pflanzengeographischen der Robinie in Deutschland. Beitr. Nat. Forsch. Südw. Deutschl. 22, 3–18 (1963).

KOJIMA, K.: Stable equilibria for the optimum model. Proc. Nat. Ac. Sci. Wash. 45, 989–993 (1959).

KOJIMA, K. I.: Is there a constant fitness value for a given phenotype? No! Evolution 25, 281–285 (1971).

KOJIMA, K. I., TOBARI, Y. N.: The pattern of viability changes associated with genotype frequency at the alcohol dehydrogenase locus in a population of Drosophila melanogaster. Genetics 61, 201–209 (1969).

KOJIMA, K., LEWONTIN, R. C.: Evolutionary significance of linkage and epistasis. In: Math. Topics in Pop. Gen., Biom. 1, pp. 337–366. Heidelberg: Springer 1970.

KOLLER, D.: The survival value of germination-regulating mechanisms in the field. Herbage Abst. 34, 1–7 (1964).

KOOPMAN, K. F.: Natural selection for sexual isolation between Drosophila pseudoobscura and Drosophila persimilis. Evolution **4**, 135–148 (1950).

KOPECKY, F.: Pollen flight examinations performed with inactive marking. Proc. Sec. 22 IUFRO, Hungary 1966, **1**, 94–96 (1966).

KOSKI, V.: Pollen dispersal and its significance in genetics and silviculture. Helsinki (1967).

KOSKI, V.: A study of pollen dispersal as a mechanism of gene flow in conifers. Comm. Inst. For. Fenn. **70**, 1–78 (1970).

KOSKI, V.: Measuring the catch of forest tree pollen. Proc. Work. Group IUFRO Sec. 22 on Sex. Repr. of For. Trees, Finland (1970).

KOSKI, V.: Embryonic lethals and empty seeds in Picea abies and Pinus silvestris. Paper presented to IUFRO Sec. 22, Work. Group on Repr. of For. Trees, Gainesville (1971).

KÖSTLER, J. N., BRÜCKNER, E., BIBELRITHER, H.: Die Wurzeln der Waldbäume. Hamburg: Parey 1968.

KOZUBOW, G. M.: Eine Form von Pinus silvestris mit roten Antheren. Bot. Z. **47**, 276–280 (1962).

KRAMER, P. J.: Some effects of various combinations of day and night temperatures and photoperiod on the height growth of loblolly pine seedlings. For. Sci. **3**, 45–55 (1957).

KRIEBEL, H. B.: Patterns of genetic variation in sugar maple. Ohio Agr. Exp. Sta. Wooster, Ohio, Res. Bull. no. 791, 47 (1957).

KRUCKEBERG, A. R.: The ecology of serpentine soils: a symposium. III. Plant species in relation to serpentine soils. Ecology **35**, 267–274 (1954).

KRUG, H. P.: Planning of afforestation and planting in Brazil. In: FAO World Symposium on Man-Made Forests, pp. 1219–1235. Rome: FAO 1967.

KRUGMAN, S. L.: Incompatibility and inviability systems among some Western North American pines. Proc. Work. Group IUFRO Sect. 22 on Sex. Repr. of For. Trees, Finland (1970).

KUGLER, H.: Blütenökologie. Stuttgart: Fischer 1970.

KUHNHOLTZ-LORDAT, G.: La Terre incendiée: Essai d'agronomie comparée. Nimes: Editions de la Maison Carée, Ateliers Brugier (1939).

KUMLER, M. L.: Two edaphic races of Senecio silvaticus. Bot. Gaz. **130**, 187–191 (1969).

KUNTZ, J. E., RIKER, A. J.: The use of radioactive isotope to ascertain the role of root grafting in the translocations of water, nutrients and disease inducing organisms among forest trees. Proc. Int. Conf. Peaceful Uses Atom. Energy (1955).

LACAZE, J. F.: Le choix des provenances d'Abies grandies. Rev. For. Fran. 613–624 (1967).

LACAZE, J. F.: Comportement de diverses provenances d'epicea de Sitka en France. Rev. For. Franz. **22**, 45–54 (1970).

LAESSLE, A. M.: Spacing and competition in natural stands of sand pine. Ecology **46**, 65–72 (1965).

LAMB, A. F. A.: Choice of pines for lowland tropical sites. In: FAO Symposium on Man-Made Forests and Their Industrial Importance, vol. II, pp. 1009–1029. Rome: FAO 1967 a.

LAMB, A. F. A.: A review of natural regeneration techniques in some tropical lowland forests. Sylva. **48**, 27–29 (1967 b).

LAMB, A. F. A.: Gmelina arborea. Fast Growing Timber Trees of the Lowland Tropics no. 1. Commonw. For. Inst. Dept. of Forestry, Univ. of Oxford (1968 a).

LAMB, A. F. A.: Cedrela odorata. Fast Growing Timber Trees of the Lowland Tropics no. 2. Commonwealth Forestry Inst. Dept. of Forestry, Univ. of Oxford (1968 b).

LAMB, A. F. A.: Enrichment planting in English speaking countries of the tropics. In: Report of the Session, FAO Committee on Forest Development in the Tropics, pp. 44–56. Rome: FAO 1969 a.

LAMB, A. F. A.: Artificial regeneration within the humid lowland tropical forest. Common. For. Rev. **48**, 41–53 (1969 b).

LAMB, A. F. A., COOLING, E. N. G.: Exploration, utilization and conservation of low altitude tropical pine gene resources. Common. Fore. Rev. **49**, 41–51 (1970).

LAMB, A. F. A., NTIMA, O. O.: Terminalia ivorensis. Fast growing Timber Trees of the Lowland Tropics. no. 5. Common. Forestry Inst. Dept. of Forestry Univ. of Oxford (1971).

LAND, S. B.: Intraspecific variation in sea-water tolerance of Loblolly pine (Pinus taeda L.). M.S. Thesis, N.C. State Univ., Raleigh (1967).

LANDRIGE, J.: A genetic and molecular bais for heterosis in Arabidopsis and Drosophila. Amer. Nat. **96**, 5–27 (1962).

LANE, C., ROTHSCHILDT, M.: Observation on colonies of the Narrow-bordered Five-spot Burnet (Zygaena lonicerae v. Schwer) near Bicester. Entomologist **94**, 79–81 (1961).

LANGLET, O.: Om variationen hos tallen (Pinus silvestris L.) och dess samband med klimatet. Svenska Skogsv. För. Tidskrift **32**, 131–143 (1934).

LANGLET, O.: Till frågan om sambandet mellan temperatur och växtgränser. Medd. St. Skogsf. Inst. **28**, 299–412 (1935).

LANGLET, O.: Norrlandstallens praktiska och systematiska avgränsning. Sv. Skogsv. För. Tidskr. 425–436 (1959).

LANGLET, O.: Mellaneuropeiska granprovenienser i svensk skogsbruk. Kungl. Lant- och Skogsbr. Ak. Tidskr. 259–329 (1960).

LANGLET, O.: Patterns and terms of intra-specific ecological variability. Nature **200**, 347–348 (1963).

LANGLET, O.: Tvåhundra år genekologi. Svensk Bot. Tidskr. **58**, 273–308 (1964).

LANGER, W.: Selbstfertilität und Inzucht bei Picea omorica. Silv. Gen. **8**, 69–104 (1959).

LANGER, W., STERN, K.: Untersuchungen über den Austriebstermin von Fichten und dessen Beziehung zu anderen Merkmalen. Allg. Forst- u. Jagdztg. **135**, 53–60 (1964).

LANNER, R. M.: Needed: a new approach to the study of pollen dispersal. Silv. Gen. **15**, 50–52 (1966).

LA RUE, C. D.: Root graftings in trees. Amer. J. Bot. **21**, 121–126 (1934).

LATTER, B. D. H.: The evolution of non-additive genetic variance under artificial selection I. Modification of dominance at a single autosomal locus. Austr. J. Biol. Sci. **17**, 427–435 (1964).

LAWRENCE, W. H., REDISKE, J. H.: Fate of sown Douglas-Fir seed. For. Sci. **8**, 210–219 (1962).

LEDIG, F. T., PERRY, T. O.: Variation in photosynthesis and respiration among Loblolly pine progenies. Proc. 9th South Conf. For. Tree Impr. 120–128 (1967).

LEDIG, F. T., FRYER, J. H.: The serotinous cone habit in Pinus rigida as related to selection in introgression. Paper IUFRO Congr. Gainesville 1971, Work. Group Quant. Genetics (1971).

LEE, J. A.: A study of plant competition in relation to development. Evolution **14**, 18–28 (1960).

LEFEBVRE, C.: Self-fertility in maritime and zinc-mine populations of Armeria maritima (Mill.) Willd. Evolution **24**, 571–577 (1970).

LERNER, I. M.: Ecological genetics: Synthesis. Gen. Roday, Proc. 11. Int. Congr. Gen. **2**, 488–494 (1963).

LERNER, M.: Genetic Homeostasis. New York: Yohn Wiley 1954.

LESSMANN, D.: Ein Beitrag zur Verbreitung und Lebensweise von Megastigmus spermotrophus Wachtl und Megastigmus bipunctatus Swederus. Diss. Forstl. Fak. Univ. Göttingen (1971).

LEVENE, H.: Genetic equilibrium when more than one ecological niche is available. Amer. Nat. **87**, 331–333 (1953).

LEVIN, B. R.: A model for selection in systems of species competition. In: Concepts and models in Biomath., pp. 237–275. New York: Dekker 1969.

LEVIN, B. R.: The operation of selection in situations of interspecific competition. Evolution **25**, 249–264 (1971).

LEVIN, D. A., KERSTER, H. W.: Natural selection for reproductive isolation in Phlox. Evolution **21**, 679–687 (1967).

LEVIN, D. A., KERSTER, H. W.: The dependence of bee-mediated pollen and gene dispersal upon density. Evolution **23**, 560–571 (1969).

LEVINS, R.: Mendelian species as adaptive systems. Gen. Sys. **6**, 33–39 (1961).

LEVINS, R.: Theory of fitness in a heterogeneous environment. I. Fitness set and adaptive function. Amer. Nat. **96**, 361–373 (1962).

LEVINS, R.: Theory of fitness in a heterogeneous environment II. Developmental flexibility and niche selection. Amer. Nat. **97**, 75–90 (1963).

LEVINS, R.: Theory of fitness in a heterogeneous environment. III. The response to selection. J. Theor. Biol. **7**, 224–240 (1964).

LEVINS, R.: The theory of fitness in a heterogenous environment. IV. The adaptive significance in a heterogeneous environment. V. Optimal genetic systems. Genetics **52**, 891–905 (1965 b).

LEVINS, R.: The strategy of model building in population biology. Amer. Sci. **54**, 421–431 (1966).

LEVINS, R.: Evolutionary consequences of flexibility. In: Pop. Biol. and Evol., pp. 67–70. Syracuse, N.Y.: Syracuse Univ. Press 1967.

LEVINS, R.: Evolution in changing environments. Monogr. Pop. Biol. 2. Princeton, N.J.: Princeton Univ. Press 1968 a.

LEVINS, R.: Towards an evolutionary theory of the niche. In: Evolution and Environment, pp. 325–340. New Haven: Yale Univ. Press 1968 b.

LEVINS, R.: Fitness and optimization. In: Math. Topics in Pop. Gen., Biom. 1, pp. 389–400. Heidelberg: Springer 1970.

LEVINS, R., MACARTHUR, R.: The maintenance of genetic polymorphism in a spatially heterogeneous environment. Amer. Nat. **100**, 585–589 (1966).

LEWIS, H.: Catastrophic selection as a factor in speciation. Evolution **16**, 257–271 (1962).

LEWIS, N. B.: Regeneration of man-made forests. In: FAO Symposium on Man-Made Forests and Their Industrial Importance, vol. I., pp. 321–343. Rome: FAO 1967.

LEWONTIN, R. C.: Evolution and theory of games. J. Theor. Biol. **1**, 382–403 (1961).

LEWONTIN, R. C.: Interdeme selection controlling a polymorphism in the house mouse. Amer. Nat. **96**, 65–78 (1962).

LEWONTIN, R. C.: Selection for colonizing ability. In: Genetics of Col. Species, pp. 77–94. New York: Academic Press 1964 a.

LEWONTIN, R. C.: The interaction of selection and linkage II. Optimum models. Genetics **50**, 757–782 (1964 b).

LEWONTIN, R. C.: Selection in and of populations. In: Ideas in Modern Biol., Proc. 16th. Int. Congr. Zool. 6. Garden City, N.Y.: Nat. Hist. Press 1965.

LEWONTIN, R. C.: Introduction. In: Pop. Biol. and Evol. Syracuse, N.Y.: Syracuse Univ. Press 1967.

LEWONTIN, R. C., KOJIMA, K. I.: The evolutionary dynamics of complex polymorphism. Evolution **14**, 458–472 (1960).

LEWONTIN, R. C., WHITE, M. J. D.: Interaction between inversion polymorphisms of two chromosom pairs in the grasshopper Moraba scurra. Evolution **14**, 116–129 (1960).

LI, C. C., HORVITZ, D. G.: Some methods of estimating the inbreeding coefficient. Amer. J. Hum. Gen. **5**, 107–117 (1953).

LIBBY, W. J., STETTLER, R. F., SETZ, F. W.: Forest genetics and forest tree breeding. Ann. Rev. Gen. **3**, 469–494 (1969).

LIETH, H.: Phenology in productivity studies. In: Ecological Studies. I: Temperate Forest Ecosystems, pp. 29–46 (REICHLE, D. E., ed.). New York, Heidelberg, Berlin: Springer Verlag 1970.

LINES, R.: Standardization of methods for provenance research and testing. In: Proc. XIX IUFRO Congr. Munich pp. 672–718 (1967).

LINES, R., MITCHELI, A. F.: Differences in phenology of Sitka spruce provenances. Brit. For. Comp. Rep. 173–184 (1964).

LITTLE, S., DOOLITTLE, W. T.: Natural hybrids among pond, loblolly and pitch pines. U.S. For. Serv. Res. Pap. NE-67, 1–22 (1967).

LOUCKS, O. L.: Evolution of diversity, efficiency and community stability. Amer. Zool. 10, 17–25 (1970).

LUDWIG, W.: Zur Theorie der Konkurrenz. Neue Ergebnisse und Probleme der Zoologie. Klatt Festschrift 516–537 (1950).

LUNDEGÅRDH, H.: Environment and Plant Development. London: Arnold 1931.

LUSH, J. L.: Rates of genetic changes in populations of farm animals. Proc. IX. Int. Congr. Gen. 1, 589–599 (1953).

LÜTTGE, U.: Über die Zusammensetzung des Nektars und den Mechanismus seiner Sekretion I. Planta 56, 189–212 (1961).

LUTZ, H. J.: Fire as an ecological factor in the boreal forest of Alaska. Jo. For. 58, 454–460 (1960).

LYR, H., POLSTER, H., FIEDLER, H.-J.: Gehölzphysiologie. Jena: VEB Gustav Fischer Verlag 1967.

MACARTHUR, R. H.: On relative abundance of bird species. Proc. Nat. Ac. Sci. 43, 293–295 (1957).

MACARTHUR, R. H.: On the relative abundance of species. Amer. Nat. 94, 25–34 (1960).

MACARTHUR, R.: The theory of the niche. In: Pop. Biol. and Evol., pp. 159–176. Syracuse, N.Y.: Syracuse Univ. Press 1967.

MACARTHUR, R. H.: Species packing and competition equilibrium for many species. Theoret. Pop. Biol. 1, 1–11 (1970).

MACARTHUR, R. H., LEVINS, R.: Competition, habitat selection and character displacement in a patchy environment. Proc. Nat. Ac. Sci. Wash. 51, 1207–1210 (1965).

MACARTHUR, R. H., WILSON, E. O.: The theory of island biogeography. Princeton, N. J.: Princeton Univ. Press. 1967.

MACATEE, W. L.: Distribution of seeds by birds. Amer. Midl. Nat. 38, 214–223 (1947).

MALOGOLOWKIN, Ch., SIMMONS, A. S., LEVENE, H.: A study of sexual isolation between certain strains of Drosophila paulistorum. Evolution 19, 95–103 (1965).

MAJOR, J., BAMBERG, S. A.: Some cordillerean plant species new for the Sierra Nevada of California. Madrono 17, 93–109 (1963).

MARCET, E.: Pollenuntersuchungen an Föhren (Pinus silvestris L.) verschiedener Provenienz. Mitt. Schweiz. Anst. Forstl. Versuchsw. 27, 1–45 (1951).

MARSHALL, J. T.: Interrelations of Abert and Brown towhees. Condor 62, 49–64 (1960).

MARTIN, N. D.: An analysis of forest succession in Algonquin Park, Ontario. Ec. Monogr. 29, 187–218 (1959).

MATHENY, W. A.: Seed dispersal. Ithaca: Slingerland 1931.

MATHER, K.: Polymorphism as an outcome of disruptive selection. Evolution 9, 52–61 (1955).

MAYNARD SMITH, J.: Disruptive selection, polymorphism and sympatric speciation. Natura 195, 60–62 (1962).

MAYNARD SMITH, J.: Sympatric speciation. Amer. Nat. 100, 637–650 (1966).

MAYR, E.: Animal species and evolution. Cambridge, Mass.: Harvard Univ. Press 1963.

McCOMB, A. L., JACKSON, J. K.: The role of tree plantations in savanna development. Unasylva 23, 8–18 (1969).

McELWEE, R. L.: Radioactive tracer techniques for Pine pollen flight studies and an analysis of short-range pollen behavior. Ph.D. Thesis, N.C. State Univ., Raleigh, Dept. of Forestry (1970).

McNAUGHTON, S. J., WOLF, L. L.: Dominance and the niche in ecological systems. Science 167, 131–139 (1970).

MCNEILLY, T.: Evolution in closely adjacent plant populations III. Agrostis tenuis on a small copper mine. Heredity **23**, 99–108 (1968).

MCNEILLY, T., ANTONOVICS, J.: Evolution in closely adjacent plant populations IV. Barriers to gene flow. Heredity **23**, 205–218 (1968).

MCNEILLY, T., BRADSHAW, A. D.: Evolutionary processes in populations of copper tolerant Agrostis tenuis. Evolution **22**, 108–118 (1968).

MCWILLIAM, J. R., GRIFFING, B.: Temperature dependent heterosis in maize. Austr. J. Biol. Sci. **18**, 569–583 (1965).

MEEUSE, A. J. D.: A possible case of interdependence between a mammal and a higher plant. Arch. Need. Zool. **13**, 314–318 (1958).

MEEUSE, A. J. D.: Fundamentals of Phytomorphology. New York: Ronald Press 1966.

MEHRA, P. N., BAWA, K. S.: Chromosomal evolution in tropical hardwoods. Evolution **23**, 466–481 (1969).

MELCHERS, G.: Genetik und Evolution. Ztschr. Ind. Abstr. u. Vererb. Lehre **76**, 229–240 (1939).

MELECHOW, J. S.: Zur Frage der natürlichen Verjüngung der Fichte auf Brandflächen. Forstwiss. Centralbl. **78**, 47–85 (1934).

MELVILLE, R.: The application of biometrical methods for the study of elms. Proc. Linn. Soc. London **151**, 152–166 (1939).

MELVILLE, R.: Plant conservation and the Red Data Book. Biol. Cons. **2**, 185–188 (1970).

MELVILLE, R.: Red Data Book, Vol. 5. International Union for the conservation of Nature and National Resources, Morges, Switzerland (1971).

MERGEN, F.: A toxic principle in the leaves of Ailanthus. Bot. Gaz. a1, 32–36 (1959).

MERREL, D. J.: Measurement of sexual isolation and selective mating. Evolution **4**, 326–331 (1950).

MERREL, D. J.: Selective mating as a cause of gene frequency changes in laboratory populations of Drosophila melanogaster. Evolution **7**, 287–296 (1953).

MESAROVIC, M. D., ed.: Systems theory and biology. Heidelberg, New York: Springer 1968.

METTLER, L. E., GREGG, T. G.: Populations genetics and evolution. Englewood Cliffs, N. J.: Prentice-Hall 1969.

MEYER, H.: Über das Umsetzen 30/40jähriger Fichten im Verlauf einer 10jährigen Zuwachsperiode. Archiv f. Forstw. **14**, 299–311 (1965).

MICHAELIS, P.: Cytoplasmatic inheritance in Epilobium and its theoretical significance. Adv. Gen. **6**, 287–298 (1954).

MIKOLA, P.: Special techniques for poorly drained sites including peat bogs, swamps etc. In: FAO World Symposium on Man-Made Forests and Their Industrial Importance, pp. 367–382. Rome: FAO 1967.

MIKOLA, P.: Afforestation of treeless areas. Unasylva **23**, 35–48 (1969).

MILLICENT, E., THODAY, J. M.: Gene flow and divergence under disruptive selection. Science **131**, 311–312 (1960).

MISRA, R. K.: Vectorial analysis for genetic clines in body dimensions in populations of Drosophila subobscura Coll. and a comparison with D. robusta Sturt. Biometrics **22**, 469–487 (1966).

MITSCHERLICH, G.: Wald, Wachstum und Umwelt. D. Sauerländer, Frankfurt a. M., Bd. 1 (1970).

MÖBIUS, M.: Die vegetative Vermehrung der Pflanzen. Jena: G. Fischer 1940.

MODE, C. J.: A mathematical model for the co-evolution of obligate parasites and their hosts. Evolution **12**, 158–165 (1958).

MODE, C. J.: A model of a host-pathogen system with particular reference to the rusts of cereals. In: Biometrical Genetics, pp. 84–100 (KEMPTHORNE, P., ed.). New York: Pergamon Press 1960.

MODE, C. J.: A generalized model of a host-pathogen system. Biometrics **17**, 386–404 (1961).

MOFFET, A. A.: Genetical studies in Acacias III. Chlorosis in interspecific hybrids. Heredity 20, 609–620 (1965).

MOLL, R. H., LANNQUIST, J. H., FORTUNO, V. J., JOHNSSON, E. C.: The relationship of heterosis and genetic diversity in maize. Genetics 52, 139–144 (1965).

MOONEY, H. A., HARLAN, J. R.: Convergent evolution of mediterraneanclimate evergreen sclerophyll shrubs. Evolution 24, 292–303 (1970).

MOORE, C. W. E.: Interaction of species and soil in relation to the distribution of eucalypts. Ecology 40, 734–735 (1959).

MOORE, J. A.: Competition between Drosophila melanogaster and D. simulans. I. Population cage experiments. Evolution 6, 407–426 (1952).

MORAN, O. A. P.: Balanced polymorphisms with unlinked loci. Austr. J. Biol. Sci. 16, 1–6 (1963).

MORAN, P. A. P.: On the measurement of natural selection dependent on several loci. Evolution 17, 182–186 (1963).

MOREE, R.: An effect of negative assortative mating on gene frequency. Science 118, 600–601 (1953).

MORGENSTERN, E. K.: Genetic variation in seedlings of Picea mariana (Mill.) BSP I. Correlation with ecological factors. II. Variation patterns. Silv. Gen. 18, 151–161, 161–167 (1969).

MORGENSTERN, E. K., FARRAR, J. L.: Introgressive hybridization in red spruce and black spruce. Tech. Rep. no. 4, Faculty of Forestry, Univ. of Toronto (1964).

MORK, E.: Temperaturen som Föryngringsfaktor i de trönderske granskoger. Medd. Norsk. Skogsförsöksv. 5, no. 16 (1933).

MORLEY, F. W. H.: The inheritance and ecological significance of seed dormany in subterranean clover (Trifolium subterraneum L.). Austr. J. Biol. Sc. 11, 261–274 (1958).

MORLEY, F. H. W.: Natural selection and variation in plants. Cold Spring Harbor Symp. Quant. Biol. 24, 47–56 (1959).

MOSTYN, H. P.: The role of plantations in the industrial economy of Zambia. In: FAO Symposium on Man-Made Forests and Their Industrial Importance, vol. II, pp. 945–963. Rome: FAO 1967.

MÜLDER, D.: Die Disposition der Kiefer für den Kienzopfbefall als Kernproblem waldbautechnischer Abwehr. Schriftenr. Forstl. Fak. Göttingen 10 (1953).

MÜLDER, D.: Beitrag zur Individualauslese bei der Blasenrostresistenzzüchtung mit Pinus strobus. Ztschr. Forstgen. u. Forstpflanzen 4, 89–99 (1955).

MÜLLER, C. H.: Ecological control of hybridization in Quercus: A factor in the mechanism of evolution. Evolution 6, 147–161 (1952).

MÜLLER, P.: Verbreitungsbiologie der Blütenpflanzen. Verh. Geob. Inst. Zürich, 30, Bern (1955).

MURTY, B. R., ARUNACHALAM, V., JAIN, O. P.: Factor analysis in relation to breeding system. Genetica 41, 179–189 (1970).

MUTCH, R. W.: Wild fires and ecosystems — a hypothesis. Ecology 51, 1047–1051 (1970).

NAMKOONG, G.: Statistical analysis of introgression. Biometrics 22, 488–502 (1966).

NAMKOONG, G.: Nonoptimal local races. Proc. 10th South. Conf. For. Tree Impr. 149–153 (1969).

NAMVAR, K.: Versuche über den Aufbau einer genetischen Isolierungsbarriere durch disruptive Auslese bei Drosophila melanogaster. Diss. Forstl. Fak. Univ. Göttingen (1971).

NARISE, T.: The effect of relative frequency of species in competition. Evolution 19, 350–354 (1965).

NEI, M., MURATA, M.: Effective population size when fertility is inherited. Gen. Res. 8, 257–260 (1966).

NEI, M., IMAIZUMI, Y.: Efficiency of selection for increased or decreased recombination. Amer. Nat. 102, 90–93 (1968).

NEUMANN,F.P., SCHANTZ-HANSEN,T., REES,L.W.: Cone scale movements of jack pine (Pinus banksiana Lamb.). Minn. For. Notes, no. 142 (1964).

NICHOLSON,E.M.: Handbook to the conservation section of the International Biological Program. I.B.P. Handbook no. 5. Oxford, Edinburgh: Blackwells Scientific Publications 1968.

NIENSTAEDT,H.: The ecotype concept in forest tree genetics. Sta. Pa. Lake States For. Exp. St. St. Paul, Minn. no. 81, 14–24 (1961).

NIENSTAEDT,H.: Variation and adaptive stability of white spruce (Moench) Voss seed sources. In: Proceedings of the Eleventh Meeting of the Committee on Forest Tree Breeding in Canada, pp. 183–194 (1968).

NITSCH,J.P.: Réactions photopéripdiques chez les plantes ligneuses. Bull. Soc. Bot. Franc. **106**, 259–287 (1959).

NTIMA,O.O.: The Araucarias. Fast Growing Timber Trees of the Lowland Tropics. Commonwealth Forestry Inst. of Forestry, Univ. of Oxford (1968).

NYE,P.H., GRRENLAND,D.J.: The soil under shifting cultivation. Technical Communication no. 51. Commonwealth Bureau of Soils, Harpenden, no. 51 (1960).

NYGREN,A.: Apomixis in the angiosperms. In: Handbuch der Pflanzenphysiologie, Band 18, S. 531–596. Heidelberg, New York, Berlin: Springer 1967.

O'DONALD,P.: Measuring the intensity of natural selection. Nature **220**, 197–198 (1968).

O'DONALD,P.: Models of the evolution of dominance. Proc. Roy. Soc. Ser. B, **171**, 127–143 (1968).

ODUM,E.P.: The strategy of ecosystem development. Science **164**, 262–270 (1969).

ODUM,G.: Germination of ancient seeds. Dansk. Bot. Arch. **24** 2 (1965).

OKABE,S., HASHIGICHI,S.: An application of games theory to estimation of optimum mixtures of resistances genes in multiline varieties for disease resistance in crop breeding. Proc. 12. Int. Congr. Gen. **1**, 248 (1968).

OLSSON,G.: Self-incompatibility and outcrossing in rape and white mustard. Hereditas **46**, 241–252 (1960).

ORNDUFF,R.: The reproductive system of Jepsonia hererandra. Evolution **25**, 300–311 (1971).

ORR-EWING,A.L.: A cytological study of the effects of self-pollination on Pseudotsuga menziesii (Mirb.) Franco. Silv. Gen. **6**, 179–185 (1957).

OSMASTON,H.A.: Pollen and seed dispersal in Chlorophora and Parkia. Commw. For. Rev. **44**, 97–105 (1965).

OWEN,A.R.G.: A genetical system allowing of two stable equilibria. Nature **170**, 1127–1128 (1952).

PALMBLAD,I.G.: Competition in experimental populations of weeds with emphasis on the regulation of population size. Ecology **49**, 26–33 (1968).

PANIN,W.A.: Frühe und späte Formen der Fichte. Lesn. Chosjaistwo **13**, 17–21 (1960).

PARROT,L.: Le climat, facteur sélectif, et l'adaptation génétique de Juglans nigra L., espèce exotique au Quebec, Canada. Silv. Gen. **20**, 1–9 (1971).

PARSONS,P.A.: Polymorphism and the balanced polygenic combination. Evolution **17**, 564–574 (1963a).

PARSONS,P.A.: Migration as a factor in natural selection. Genetica **33**, 184–206 (1963b).

PARSONS,P.A., ALLARD,R.W.: Seasonal variation in Lima bean seed size: An example of genotypic environmental interaction. Heredity **14**, 115–123 (1960).

PARSONS,P.A., BODMER,W.F.: The evolution of overdominance: Natural selection and heterozygote advantage. Nature **190**, 7–12 (1961).

PATERNIANI,E.: Selection for reproductive isolation between two populations of maize, Zea maize L. Evolution **23**, 534–548 (1969).

PAXMAN,G.J.: The maximum likelihood estimation of the number of self-sterility alleles in a population. Genetics **48**, 1029–1032 (1963).

PEACOCK, J. T., MCMILLAN, C.: The photoperiodic response of American Prosopis and Acacia from a broad latitudinal distribution. Amer. J. Bot. 55, 153–159 (1968).

PENFOLD, A. R., WILLIS, J. L.: The Eucalypts. London, New York: Leonard Hill (Books) Ltd. 1961.

PENGELLEY, E.: Differential development patterns and their adaptive value in various species of the genus Citellus. Growth 30, 137–142 (1966).

PERCIVAL, M. S.: Types of nectar in angiosperms. New Phytol. 60, 235–281 (1961).

PERRY, T. O.: Dormancy of trees in winter. Science 171, 29–36 (1971).

PERSSON, A.: Frequenzen von Kiefernpollen in Südschweden. Ztschr. Forstgen. 4, 129–137 (1955).

PERSSON, C.: Gene-for-gene Relationships in host-parasite systems. Can. J. Bot. 37, 1111–1130 (1959).

PERSSON, C.: Genetic polymorphism in parasitic systems. Nature 212, 266–267 (1966).

PERSSON, C.: Genetic aspects of parasitism. Can. J. Bot. 45, 1193–1204 (1967).

PFAHLER, P. L.: Fitness and variability in the cultivated species of Avena. Crop. Sci. 4, 29–31 (1964).

PHARIS, R. P., FERRELL, W. K.: Difference in drought resistance between coastal and inland sources of Douglas fir. Can. J. Bot. 44, 1651–1659 (1966).

PHILIPS, J. F. V.: General biology of the flowers, fruits and young regeneration of the more important species of the Knyska forest. South. Agr. J. Sci. 33, 366–417 (1926).

PIANKA, E. R.: Latitudinal gradients in species diversity: A review of concepts. Amer. Nat. 100, 33–46 (1966).

PIELOU, E. C.: An Introduction of Mathematical Ecology. New York, London: John Wiley 1969.

PIGOTT, C. D., WALTERS, S. M.: On the interpretation of the discontinuous distribution shown by certain British species of open habitats. J. Ecol. 42, 95–116 (1954).

PIMENTEL, D.: Population ecology and the genetic feedback mechanism. Proc. XI. Int. Congr. Gen. Den Haag 2, 483–488 (1965).

PIMENTEL, D.: Population regulation and genetic feedback. Science 159, 1432–37 (1968).

PIMENTEL, D., NAGEL, W. B., MADDEN, L. J.: Space-time structure of the environment and the survival of parasitehost systems. Amer. Nat. 97, 141–168 (1963).

PIMENTEL, D., AL-HAFIDH, R.: Ecological control of a parasite population by genetic evolution in the parasite-host system. Ann. Ent. Soc. Amer. 58, 1–6 (1965).

PIMENTEL, D., FEINBERG, E. H., WOOD, P. W., HAYES, J. T.: Selection spatial distribution and the coexistence of competing fly species. Amer. Nat. 99, 97–109 (1965).

POHL, F.: Untersuchungen über die Bestäubungsverhältnisse der Traubeneiche. Beih. Bot. Centralbl. 51, Abt. I (1933).

POHL, F.: Ein Fall zweckloser Protandrie. Ber. Dt. Bot. Ges. 53, 17–28 (1935).

POHL, F.: Die Pollenerzeugung der Windblütler. Beih. Bot. Centralbl. 57, Abt. A (1937a).

POHL, F.: Die Pollenkorngewichte einiger windblütiger Pflanzen und ihre ökologische Bedeutung. Beih. Bot. Centralbl. 57, Abt. A (1937b).

POLK, R. B.: Reproductive phenology and precocity as factors in seed orchard development. Proc. 5, Centr. Stat. For. Tree Impr. Conf. 13–21 (1966).

POLLACK, E., ROBINSON, H. F., COMSTOCK, R. E.: Inter-population hybrids in open pollinated varieties of maize. Amer. Nat. 91, 387–391 (1957).

PORSCH, O.: Windpollen und Blumeninsekt. Oesterr. Bot. Ztschr. 103, 1–18 (1956).

POTTER, L. D., ROWLEY, J.: Pollen rain and vegetation, San Augustin Plains, New Mexico. Bot. Gaz. 122, 1–25 (1960).

PRAKASH, S. R. C., LEWONTIN, R. C., HUBBY, J. L.: A molecular approach to the study of genic heterozigosity in natural populations IV. Patterns of genic variation in central, marginal and isolated populations of Drosophila pseudoobscura. Genetics 61, 841–858 (1969).

PRAVDIN, L. F.: Die Kiefer, ihre Variabilität, intraspezifische Taxonomie und Züchtung. Nauka, Moskau 191, 1964.

PRESTON, F. W.: The canonical distribution of commonness and rarity: Part. I. Ecology 43, 185–215 (1962a).

PRIEHÄUSSER, G.: Die Variabilität der Fichte in systematischer Hinsicht. Ber. Bayer. Bot. Ges. 25, 100–135 (1962).

PROUT, T.: The effects of stabilizing selection on the time of development in Drosophila melanogaster. Gen. Res. a3, 364–382 (1962).

PROUT, T.: The estimation of fitness from genotypic frequencies. Evolution 19, 546–551 (1965).

PRYOR, L. D.: Evolution in eucalyptus. Austr. J. Sci. 22, 2–48 (1959).

PRYOR, L. D.: Eucalyptus in plantations — present and future. In: FAO Symposium on Man-Made Forests and Their Industrial Importance, pp. 993–1008. Rome: FAO 1967.

PUGSLEY, A. T.: Inheritance of a correlated day-length response in spring wheat. Natur 207, 108 (1965).

PURSEGLOVE, J. W.: The spread of tropical crops. In: Gen. of Col. Spec., pp. 375–390. New York: Academic Press 1964.

PUTWAIN, P. D., HARPER, J. L.: Studies in the dynamics of plant populations II. Components and regulation of a natural population of Rumex acetosella L. J. Ecol. 35–47 (1968).

RAMIREZ, B. W.: Host specifity of fig-wasps (Agaonidae). Evolution 24, 680–691 (1970).

TAO, S. V., DE BACH, P.: Experimental studies on hybridization and sexual isolation between some Aphytis species (Hymenoptera: Aphelinidae). III. The significance of reproductive isolation between interspecific hybrids and parental species. Evolution 23, 525–533 (1969).

RAPER, J. R., KRONGELB, G. S., BAXTER, M. G.: Number and distribution of incompatibility factors in Schizophyllum. Amer. Nat. 92, 221–232 (1958).

RATTRAY, G.: Notes on the pollination of South African cycads. Trans. Roy. Soc. S. Afr. 3, 259–270 (1913).

RAVEN, P. H.: Catastrophic selection and edaphic endemism. Evolution 18, 336–338 (1964).

REINIG, W. E.: Elimination und Selektion. Jena 1937.

REMPE, H.: Untersuchungen über die Verbreitung des Blütenstaubs durch die Luftströmungen. Planta 27, 13–27 (1937).

RENDEL, J. M.: Selection for canalization of the acute phenotype in Drosophila melanogaster. Austr. J. Biol. Sci. 13, 36–47 (1960).

RESCIGNO, A., RICHARDSON, I. W.: On the competitive exclusion principle. Bull. Math. Bioph. 27, 85–89 (1965).

RICHARDS, P. W.: The Tropical Rain Forest. Cambridge, England: Cambridge Univ. Press 1952.

RICHARDS, P. W.: The Tropical Rain Forest. Cambridge, England: Cambridge Univ. Press 1964.

RICHARDS, P. W.: Speciation in the tropical rain forest and the concept of the niche. Biol. J. Linn. Soc. London 1, 149–153 (1969).

RICHARDSON, R. H.: Models and analyses of dispersal patterns. In: Math. Tropics in Pop. Gen., Biom. 1, pp. 79–103. Heidelberg: Springer 1970.

RICHARDSON, S. D.: Training for forest industries and timber marketing. Unasylva 23, 15–23 (1969).

RICHARDSON, S. D.: The end of forestry in Great Britain. Commonw. For. Rev. 49, 324–335 (1970).

RIDLEY, H. N.: The Dispersal of Plants Throughout the World. Reeve, Ashford 1930.

RIEGER, R., MICHAELIS, A., GREEN, M. M.: A Glossary of Genetics and Cytogenetics. Heidelberg, Berlin, New York: Springer 1968.

ROANE, C. W., STAKMAN, E. C., LOEGERING, W. C., STEWART, D. M., WATSON, W. M.: Survival of physiological races of Puccinia graminis var. tritice on wheat near Barberry bushes. Phytopathology 50, 40–44 (1960).

ROBERTSON, A.: Selection in animals: synthesis. Cold Spring Harbor Symp. Quant. Biol. **20**, 225–229 (1955).

ROBERTSON, F. W.: A test of sexual isolation in Drosophila. Gen. Res. **8**, 181–187 (1966).

ROCHE, L.: The value of short term studies in provenance research. Commonw. For. Rev. **47**, 14–26 (1968).

ROCHE, L.: A genecological study of the genus Picea in British Columbia. New Phytol. **68**, 505–554 (1969).

ROCHE, L.: The conservation of forest gene resources in Canada. For. Chron. **47**, 215–217 (1971).

ROCHE, L.: Variation, selection and breeding of coniferous tree species: an introduction. Laur. For. Res. Centre Quebec, Inf. Rep. QX 22 (1971).

ROEMER, H.: Über die Reichweite des Pollens beim Roggen. Z. f. Züchtung 17–28 (1931).

ROSENZWEIG, M. L., MACARTHUR, R. H.: Graphical representation and stability conditions of predator-prey interactions. Amer. Nat. **97**, 209–224 (1963).

ROWLANDS, D. G.: Genetic control of flowering in Pisum sativum L. Genetica's Gravenhage **35**, 75–94 (1964).

RUBNER, K.: Keimung von grün- und rotzapfigen Fichten. Tharandter Forstl. Jahrb. **89**, 247–251 (1938).

RUBNER, K.: Kiefernrassenstudien in der deutschen Bundesrepublik. Forstarchiv 30, 165–174, 205–213 (1959).

RUDLOFF, SCHANDERL: Befruchtungsbiologische Studien an Zwetschen, Pflaumen, Mirabellen und Reineclauden. Gartenbauw. **10**, 23–34 (1937).

RUDOLF, P. O.: Collecting and handling seeds of forest trees. Yearb. Agr., USDA, Wash. D.C., 211–232 (1961).

SAKAI, K. I.: Contributions to the problem of species colonization from the viewpoint of competition and migration. In: Gen. of Col. Spec., pp. 215–244. New York: Academic Press 1964.

SAKAI, K. I., MIYAZAKI, Y.: Genetic studies in natural populations of forest trees II. Family analysis: A new method for quantitative genetic studies. Paper for IUFRO Group Quant. Gen. Meeting at Brno (1970).

SAKAI, K. I., PARCK, Y. G.: Genetic studies in natural populations of forest trees III. Genetic differences within a forest of Cryptomeria japonica. Theor. Appl. Gen. **41**, 13–17 (1971).

SALISBURY, E. J.: The oak-hornbeam wood of Hartfordshire. J. Ecol. **6**, 14–52 (1919).

SALISBURY, E. J.: Ecological aspects of plant taxonomy. In: The New Systematics, pp. 329–340 (HUXLEY, J., ed.), Oxford (1940).

SALISBURY, E. J.: The Reproductive Capacity of Plants. London: Bell 1942.

SAMMETA, K. P. V., LEVINS, R.: Genetics and ecology. Ann. Rev. Gen. **4**, 469–488 (1970).

SANDERS, M.: Soil survey and site selection in Zambia. In: FAO World Symposium on Man-Made Forests and Their Industrial Importance, pp. 1415–1428. Rome: FAO 1967.

SARGENT, A. B.: Vitality of seeds of Pinus contorta. Bot. Gen. **5**, 54–60 (1880).

SARSON, B., MURTY, B. R.: Effect of disruptive selection on productivity in Brassica campestris. Proc. 12. Int. Congr. Gen. **1**, 271 (1968).

SARTORIUS, P., HENLE, H.: Forestry and Economic Development. New York, Washington, London: Frederick A. Praeger 1968.

SARVAS, R.: On the flowering of birch and the quality of seed crop. Comm. Inst. For. Fenn. no. 38 (1952).

SARVAS, R.: Investigations into the flowering and seed quality of forest trees. Comm. Inst. For. Fenn. no. 45 (1955).

SARVAS, R.: Studies on the seed setting of Norway spruce. Medd. Norske Skogsforsv. no. 48 (1957).

SARVAS, R.: Investgations on the flowering and seed crop of Pinus silvestris. Comm. Inst. For. Fenn. **53**, 1–198 (1962).

SARVAS,R.: The annual period of development of forest trees. Proc. Finnish Aca. Sci. Lett. 211–231 (1965).

SARVAS,R.: Temperature sum as a restricting factor in the development of forest in the sub-Arctic. Unesco, Nat. Res. Res. Org. Symp. on Ecol. of Sub-Arctic regions, Paper no.27 (1966).

SARVAS,R.: Pollen dispersal within and between subpopulations: Role of isolation and migration in microevolution of forest trees. In: Proc. 14. IUFRO Congr. München, vol.3, 332–345 (1967).

SARVAS,R.: Investigations on the flowering and seed crop of Picea abies. Comm. For. Fenn. 67, 1–84 (1968).

SATCHELL,J.E.: Resistance in oak (Quercus ssp.) to defoliation by Tortrix viridana L. in Roudsea Wood National Nature Reserve. Ann. Appl. Biol. 50, 431–442 (1962).

SAUNIER,R.E., WAGLE,R.F.: Root grafting in Quercus turbinella Greene. Ecology 46, 749–750 (1965).

SAX,H.: Chromosome pairing in Larix species. J. Arn. Arb. 13, 368–374 (1932).

SAX,H.J.: Chromosome number and morphology in the conifers. J. Arn. Arb. 14, 356–375 (1933).

SAYLOR,L.C., SMITH,B.W.: Meiotic irregularity in species and interspecific hybrids in Pinus. Amer. J. Bot. 53, 453–468 (1966).

SCHAFFALITZKY DE MUCKADELL,M.: Investigations on aging of apical meristems in woody plants and its importance in silviculture. Det Forstl. Forsv. i Danmark 25, 309–455 (1959).

SCHANDERL,H.: Untersuchungen über die Befruchtungsverhältnisse bei Stein- und Kernobst in Westdeutschland. Gartenbauw. 6, 55–63 (1932).

SCHANDERL,H.: Befruchtungsbiologische Studien an Birnen. Gartenbauw. 11, 168–179 (1937).

SCHANDERL,H.: Variationsstudien über den Eintritt des Blühens bei Apfelsämlingen. Tagungsber. Dt. Ak. Landw. Wiss. no.35, 111–120 (1962).

SCHANDERL,H.: Untersuchungen über die Blütenökologie und die Embryonenbildung bei Juglans regia. Biol. Centralbl. 83, 71–103 (1964).

SCHARLOO,W.: The effect of disruptive and stabilizing selection on the expression of a cubitus interreptus mutant in Drosophila. Genetics 50, 553–562 (1964).

SCHARLOO,W., HOOGMOED,M.S., TER KUILE,A.: Stabilizing and disruptive selection on a mutant character in Drosophila I. The phenotypic variance and its components. Genetics 56, 709–726 (1967).

SCHIMPER,A.F.W.: Die Indo-Malayische Strandflora. Jena: Fischer 1891.

SCHMALHAUSEN,I.I.: Natural selection and information. Izv. Ak. Nauk SSSR, Ser. Biol. 25, 19–38 (1960a).

SCHMALHAUSEN,I.I.: Evolution and cybernetics. Evolution 14, 509–524 (1960b).

SCHMIDT,H.: Versuche über die Pollenverteilung in einem Kiefernbestand. Diss. Forstl. Fak. Univ. Göttingen (1970).

SCHMIDT,W.: Der Massenaustausch in freier Luft und verwandte Erscheinungen. Probleme der Geophysik. Hamburg (1925).

SCHMIDT-VOGT,H.: Die Verzweigungstypen der Fichte (Picea abies L.) und ihre Bedeutung für die forstliche Pflanzenzüchtung. Ztschr. Forstgenetik 1, 81–91 (1952).

SCHMIDT-VOGT,H.: Die Zapfen- und Samenreifung im Hochgebirge. In: Forstsamengewinnung und Pflanzenzucht für das Hochgebirge. BLV München (1964).

SCHMIDT-VOGT,H.: Anzucht von Pflanzen für das Hochgebirge. In: Forstsamengewinnung und Pflanzenanzucht für das Hochgebirge. BLV München, Basel, Wien, 214–217 (1965).

SCHMIDT-VOGT,H.: Wachstum und Wurzelentwicklung von Schwarzerlen verschiedener Herkunft. Allg. Forst- u. Jagdztg. 142, 149–155 (1971).

SCHMUCKER,Th., STERN,K.: Breeding systems and population structure in Gymnosperms. Int. Bot. Congr. Seattle, Symp. on Breeding Structures (1969).

SCHNELL, F. W.: The covariance between relatives in the presence of linkage. Stat. Gen. and Plant Breed., Nat. Sci. — Nat. Res. Counc. Publ. 982, 468–486 (1963).

SCHOENIKE, R. W., RUDOLF, T. D., SCHANTZ-HANSEN, T.: Cone characteristics in a Jack Pine seed sources plantation. Minnesota For. Notes 76 (1959).

SCHOLZ, E.: Die Braunmaserbirke. Forst und Jagd Sonderheft Samenplantagen 28–36 (1960).

v. SCHÖNBORN, A.: Gibt es Bodenrassen bei den Waldbäumen? Mitt. Staatsforstverw. Bayern 36, 289–297 (1967).

SCHOPF, T. J. M., GOOCH, J. L.: Gene frequencies in a marine Ectoproct: A cline in natural populations related to sea temperature. Evolution 25, 286–289 (1971).

SCHRÖCK, O.: Die Abhängigkeit der phototropischen Reaktion bei Pinus silvestris von der Breiten-, Längen- und Höhenlage des Herkunftsortes. Flora 157, 13–24 (1967 a).

SCHRÖCK, O.: Herkunfts- und Eignungsfrühtestung bei Forstsaatgut durch Keimlingsuntersuchungen des individuellen Wachstumsganges und der phototropischen Reaktion. Forstsamen und Forstpflanzen 5–11 (1967 b).

SCHULTZ, A. M.: A study of an ecosystem: the arctic tundra. In: The Ecosystem Concept in Natural Resources Management, pp. 77–94 (VAN DYNE, G. M., ed.). New York, London: Academic Press 1969.

SCHULTZ, R. P., WOODS, F. W.: The frequency and implications of root-grafting in loblolly pine. For. Sci. 13, 226–239 (1967).

SCHUSTER, L.: Über den Sammeltrieb des Eichelhähers (Garrulus). Vogelwelt 71, 9–17 (1950).

SCHÜTZ, W. M., USANIS, S. A.: Intergenotypic competition in plant populations II. Maintenance of allelic polymorphisms with frequency-dependent selection and mixed selfing and random mating. Genetics 61, 857–891 (1969).

SCHWERDTFEGER, F.: Oekologie der Tiere I, Autökologie. Hamburg, Berlin: Parey 1963.

SCHWERDTFEGER, F.: Oekologie der Tiere II. Demökologie. Hamburg, Berlin: Parey 1968.

SCOTT, C. W.: Pinus radiata. FAO Forestry and Forest Products Studies no. 14 (1960).

SERNANDER, R.: Den skandinaviska vegetationens spridningsbiologie, Uppsala (1901).

SHANTZ, H. L.: An estimate of the shrinkage of Africa's tropical forests. Unasylva 2, 66–67 (1948).

SHANTZ, H. L.: In: TURNER, B. L., (Ed.): Vegetational changes in Africa. Rep. no. 169. College of Agriculture, Univ. of Arizona (1958).

SHEPPARD, P. M.: Natural Selection and Heredity. London: Hutchinson 1958.

SHUTZ, W. M., BRIM, C. A.: Inter-genotypic competition in soybeans I. Evaluation of effects and proposed field plot design. Crop Sci. 7, 371–376 (1967).

SILEN, R.: Pollen dispersal considerations for Douglas fir. J. For. 60, 790–795 (1962).

SILVERSIDES, C. R.: Developments in logging mechanization in eastern Canada. H.R. MacMillan Lectureship in Forestry, 1964. Univ. of British Columbia (1964).

SIMAK, M., GUSTAFSSON, Å.: Röntenanalys och det norrländska tallfröets grobarhet. Sv. Skogs. för Tidskr. 57, 475–486 (1959).

SIMPSON, G. G.: Rates of evolution in animals. In: Genetics, Paleontology and Evolution (JEPSEN, G. I., ed.). Princeton, N. J.: Princeton Univ. Press 1949.

SINGER, M. C.: Evolution of food-plant preference in the butterfly Euphydras editha. Evolution 25, 383–389 (1971).

SINGH, M., LEWONTIN, R. C.: Stable equilibria under optimizing selection. Proc. Nat. Ac. Sci. Wash. 56, 1345–1348 (1966).

SINSKAJA, E. N.: The species problem in modern botanical literature. Usp. Sovj. Biol. 15, 326–359 (1942).

SIREN, G.: Mechanizing forest regeneration in Sweden. In: FAO World Symposium on Man-Made Forests and Their Industrial Importance, pp. 1069–1086. Rome: FAO 1967.

SKELLAM, J. G.: Random dispersal in theoretical populations. Biometrika 38, 196–218 (1951).

SLOBODKIN, L. B.: Growth and Regulation of Animal Populations. New York (1961).

SLOBODKIN, L. B.: The strategy of evolution. Amer. Sci. **52**, 342–357 (1964).

SLOBODKIN, L. B.: Toward a predictive theory of evolution. In: Pop. Biol. and Evol., pp. 187–205. Syracuse, N.Y.: Syracuse Univ. Press 1967.

SLUDER, E. R.: Gene flow patterns in forest tree species and implications for tree breeding. Proc. Sec. World Cons. Tree Breed. 1141–1150 (1969).

SNAYDON, R. W.: Competitive ability of natural populations of Trifolium repens and its relation to differential response to soil factors. Heredity **16**, 522–534 (1961).

SNAYDON, R. W.: The growth and competitive ability of contrasting natural populations of Trifolium repens L. on calcareous and acid soils. J. Ecol. **50**, 439–447 (1962).

SNAYDON, R. W.: Population differentiation on a mosaic environment I. The response of Anthoxanthum odoratum populations to soils. Evolution **24**, 257–269 (1970).

SNAYDON, R. W., BRADSHAW, A. D.: Differential response to calcium within the species Festuca ovina L. New Phytol. **60**, 219–234 (1961).

SNAYDON, R. W., BRADSHAW, A. D.: The performance and survival of contrasting natural populations of white clover when planted into an upland Festuca/Agrostis sward. J. Brit. Grassland Soc. **17**, 113–118 (1962 a).

SNAYDON, R. W., BRADSHAW, A. D.: Differences between natural populations of Trifolium repens L. in response to mineral nutrients. J. Exp. Bot. **13**, 422–434 (1962 b).

SNOW, D. W.: A possible factor in the evolution of fruiting seasons in tropical forests. Oikos **15**, 274–281 (1965).

SNYDER, E. B., WAKELEY, P. C., WELLS, O. O.: Slash pine provenance tests. J. For. **65**, 414–420 (1967).

SOBOLEFF, A. N., FOMITZEFF, A. W.: Der Samenertrag von Waldbeständen. Beilage zu Band 18 Mittl. Kaiserl. Forstinst. Petersburg (1908).

SOEGAARD, B.: Variation and inheritance of resistance to attack by Didymascella thujina in Western Red Cedar and related species. In: Breeding Pest Resistant Trees, pp. 83–88. New York: Pergamon Press 1966.

SÖRENSEN, F.: Embryonic genetic load in coastal Douglas Fir, Pseudotsuga menziesii var. Menziesii. Amer. Nat. **103**, 389–398 (1969).

SPAGER, W.: Über das Vorkommen von Lokalrassen des kleinen Frostspanners (Cheimatobia brumata C.). Arb. Phys. Angew. Ent., Berlin Dahlem **5**, 50–76 (1938).

SPERLICH, D.: Chromosomale Strukturanalysen an einer Marginalpopulation von Drosophila subobscura. Z. Vererbungsl. **95**, 73–81 (1964).

SPERLICH, D.: Equilibria for inversion by X-rays in isogenic strains of Drosophila pseudoobscura. Genetics **53**, 835–842 (1966).

SPURR, S. H.: Role of introduced species in forest genetics. Sta. Pap. Lake States For. Exp. Sta., St. Paul, Minn. no. 81, 10–14 (1960).

SQUILLACE, A. E.: Racial variation in slash pine as affected by climatic factors. USDA For. Serv., Southeastern For. Exp. Stat. Ashville, Paper no. E-21 (1966 a).

SQUILLACE, A. E.: Geographic variation in slash pine. For. Sci. Mon. 10 (1966 b).

SQUILLACE, A. E.: Racial patterns for monoterpens in cortical oleoresin of Slash Pine. XV. Congr. IUFRO, Gainesville Working Group on Quant. Gen. (1971).

SQUILLACE, A. E., KRAUS, J. F.: Effects of inbreeding on seedling yield, germination, rate of germination and seedling growth in Slash Pine. Proc. For. Gen. Workshop Macon, Ga. (1962).

SQUILLACE, A. E., SILEN, R. R.: Racial variation in Ponderosa pine. For. Sci. Mon. 2, pp. 27 (1962).

STADT, G.: Die Genetik und Evolution der Heterözie in der Gattung Fragaria I. Untersuchungen an Fragaria orientalis. Ztschr. f. Pflanzenz. **58**, 245–277 (1967).

STADT, G.: Die Genetik und Evolution der Heterözie in der Gattung Fragaria III. Untersuchungen an hexa- und oktoploiden Arten. Ztschr. f. Pflanzenz. **59**, 83–102 (1968).

STAIRS, G. R.: Cold Hardiness in intraspecific hybrids of Tulip Poplar. Proc. 15. Northeast For. Tree Impr. Conf., pp 6 (1967).

STAKMAN, E. C., HARRAR, J. G.: Principles of Plant Pathology. New York: Ronald Press 1957.

STALKER, H. D.: Sexual isolation studies in the species complex Drosophila virilis. Genetics 27, 238–257 (1942).

STANEK, W.: A forest drainage experiment in northern Ontario. Pulp and Paper Mag. Can. 69, 58–62 (1968).

STASZKIEWICZ, J.: Cone variation in Picea abies (L.) Karst from Poland. Zaklad Dend. I Arbor. Korn. 9–18 (1967).

STEBBING, E. P.: The Forests of West Africa and the Sahara. London, Edinburgh: W. and R. Chambers Ltd. 1937.

STEBBINS, G. L.: Rates of evolution in plants. In: Gen. Pal. and Evolution, pp. 229–242 (JEPSEN, G., ed.). Princeton, N.Y.: Princeton Univ. Press 1949.

STEBBINS, G. L.: Variation and Evolution in Plants. New York: Columbia Univ. Press 1950.

STEBBINS, G. L.: Self fertilization and population variability in the higher plants. Amer. Nat. 91, 337–354 (1957).

STEBBINS, G. L.: Longevity, habitat and release of genetic variability. Cold Spring Harbor Symp. Quant. Biol. 23, 365–378 (1958).

STEBBINS, G. L.: The inviability, weakness and sterility of interspecific hybrids. Adv. Gen. 9, 147–215 (1958).

STEBBINS, G. L.: The comparative Evolution of genetic systems. In: The Evolution of Life, pp. 197–226. Chicago: Univ. of Chicago Press 1960.

STEBBINS, G. L.: Chromosomal Evolution of Higher Plants. London: Edward Arnold 1971.

STEBBINS, G. L., MATZKE, E. B., EPLING, C.: Hybridization in a population of Quercus marilandica and Quercus ilicifolia. Evolution 1, 79–88 (1947).

STEENBERG, B. K.: Report of the Second Session, FAO Committee on Forest Development in the Tropics, pp. 7–11. Rome: FAO 1969.

STEHR, G.: The determination of sex and polymorphism in microevolution. Can. Ent. 96, 418–428 (1964).

STEPHENS, S. G.: Canalization of gene action of Gossypium leaf shape and its bearing on certain evolutionary mechanism. J. Gen. 46, 346–357 (1945).

STERN, K.: Über einen grundsätzlichen Unterschied der forstlichen Saat- und Pflanzengutgesetzgebung in der Schweiz und der Bundesrepublik Deutschland. Schweiz. Ztschr. f. Forstw. 145–163 (1960).

STERN, K.: Über den Erfolg einer über drei Generationen geführten Auslese auf frühes Blühen bei Betula verrucosa. Silv. Gen. 10, 53–64 (1961 a).

STERN, K.: Plusbäume und Samenplantagen. Sauerländer, Frankfurt a. M. (1961 b).

STERN, K.: Über einige Kreuzungsversuche zur Frage des Vorkommens von Arthybriden Betula verrucosa × Betula pubescens. Deutsche Baumschule 15, 1–10 (1963).

STERN, K.: Über die Abhängigkeit des Blühens der Sandbirke von Erbgut und Umwelt. Silv. Gen. 12, 26–30 (1963 a).

STERN, K.: Versuche über die Selbststerilität bei der Sandbirke. Silv. Gen. 12, 80–82 (1963 b).

STERN, K.: Herkunftsversuche für die Forstpflanzenzüchtung — erläutert am Beispiel zweier Herkunftsversuche mit Birken. Der Züchter 34, 181–219 (1964).

STERN, K.: Vollständige Varianzen und Kovarianzen in Pflanzenbeständen I. Ein Modell für Konkurrenz zwischen Genotypen. Silv. Gen. 14, 87–91 (1965).

STERN, K.: Die Bewertung des Merkmals Austriebstermin in einem Züchtungsvorhaben mit Fichten in Schleswig-Holstein. Forstarchiv 37, 70–74 (1966).

STERN, K.: Minimum standards for provenance testing and progeny testing for certification purposes. In: Second World Consultation on Forest Tree Breeding, pp. 1447–1452. Rome: FAO 1969 a.

STERN, K.: Einige Beiträge genetischer Forschung zum Problem der Konkurrenz in Pflanzenbeständen. Allg. Forst- u. Jagdztg. **140**, 253–262 (1969b).

STERN, K.: Population structure of forest tree species, pp. 109–113. In: Genetic Resources in Plants (FRANKEL, O. H., BENNETT, E., eds.). Oxford, Edinburgh: Blackwells Scientific Publication 1970.

STERN, K.: Neuere Ergebnisse von Forstgenetik und Forstpflanzenzüchtung. Allg. Forstzeitschr. **26**, 47–50 (1971).

STERN, K.: The concept of genetic gain in breeding rust resistant trees. In: Biology of rust resistance in forest trees (BINGHAM, R. T., HOFF, R. J., and MCDONALD, edrs.). Misc. Public. 1221, U.S. Dept. Agric., Forest Service, pp. 299–307. Washington, D.C.: 1972.

STERN, K.: Über die Ergebnisse einiger Versuche zur räumlichen und zeitlichen Verteilung des Pollens einzelner Kiefern. Ztschr. f. Pflanzenz. **67**, 313–326 (1972a).

STERN, K.: Über eine Methode zur Schätzung der Pollen-Immigration in Beständen und Samenplantagen von Nadelbäumen. Forstpflanzen–Forstsamen **3** (1972b).

STERN, K.: Labeling of seeds of seed-dispersal studies in conifers. Silv. Gen. **21**, 61 (1972c).

STERN, K., GREGORIUS, H. R.: Schätzungen der effektiven Populationsgröße bei Pinus silvestris. Theor. Appl. Gen. **42**, 107–110 (1972).

STEWART, O. C.: Fire as the first great force employed by man, pp. 115–133. In: Man's Role in Changing the Face of the Earth (THOMAS, JR., W. L., ed.). Chicago: Univ. of Chicago Press 1956.

STONE, D. E.: A unique balanced breeding system in the vernal pool mouse-tail. Evolution **13**, 151–174 (1959).

STONE, E. C., GOOR, A. Y.: Afforestation techniques for arid countries. In: FAO World Symposium on Man-Made Forests and Their Industrial Importance. pp. 345–366. Rome: FAO 1967.

STOPP, K.: Botanische Analyse des Driftgutes vom Mittellauf des Kongoflusses mit kritischen Bemerkungen über die Bedeutung fluviatiler Hydratochorie. Beitr. Biol. Pflanzen **32**, 427–449 (1956).

STRAND, L.: Pollen dispersal. Silv. Gen. **6**, 129–136 (1957).

STRITZKE, S.: Untersuchungen über die befruchtungsbiologischen Verhältnisse bei Haselnußsorten unter besonderer Berücksichtigung ökologischer Verhältnisse. Arch. Gartenbau **10**, 573–608 (1962).

STUTZ, H. C., THOMAS, L. K.: Hybridization and introgression in Cowania and Purshia. Evolution **18**, 183–185 (1964).

SWEENEY, J. R.: Responses of vegetation to fire. Univ. Cal. Publ. Bot. **28**, 143–250 (1956).

SYLVEN, N.: The influence of climatic conditions on type composition. Imp. Bur. Plt. Gent. Herb. Bull. 21–38 (1937).

TAFT, K. A.: The effect of controlled pollination and honeybees on seed quality of yellow poplar (Liriodendron tulipifera L.) as assessed by X-ray photography. Tech. Rep. 13, N.C. State Univ., School of For., Raleigh (1962).

TAKKTAJAN, A.: Evolution der Angiospermen. Jena: Fischer 1959.

TAMARIN, R. H., KREBS, C. J.: Microtus population biology II. Genetic changes at the transferrin locus in fluctuating populations of the vole species. Evolution **23**, 183–211 (1969).

TANSLEY, A. G.: On competition between Galium saxatile and G. silvestris on different types of soils. J. Ecol. **5**, 177–191 (1917).

TANSLEY, A. G.: The use and abuse of vegetational concepts and terms. Ecology **16**, 284–307 (1935).

TAUBER, H.: Investigations of the mode of pollen transfer in forested areas. Rev. Paleobot. Palynol. **3**, 277–286 (1967).

TEICH, A. H.: Research on the genetic basis of white spruce improvement. Petawawa 1968–70. In: Proceedings of the 12th meeting of the Committee on Forest Tree Breeding in Canada, pp. 95–99 (1970).

THODAY, J. M.: Components of fitness. Symp. Soc. Exp. Bio. Cambr. 7, 96–113 (1953).

THODAY, J. M.: Effects of disruptive selection: the experimental production of a polymorphic population. Nature 181, 1124–1125 (1958).

THODAY, J. M.: Effects of disruptive selection. I. Genetic flexibility. Heredity 13, 187–203 (1959).

THODAY, J. M.: Effects of disruptive selection. III. Coupling and repulsion. Heredity 14, 35–49 (1960).

THODAY, J. M.: Effects of selection for genetic diversity. Genetics Today, Proc. XI. Int. Congr. Gen. 533–540 (1964).

THODAY, J. M.: Genetics and the integration of reproductive systems. Insect. Reproduction (Symp. 2, Roy. Ent. Soc.) 108–120 (1964).

THODAY, J. M.: The general importance of disruptive selection. Gen. Res. 9, 119–120 (1967).

THODAY, J. M., BOAM, T. B.: Effects of disruptive selection. II. Polymorphism and divergence without isolation. Heredity 13, 205–218 (1959).

THODAY, J. M., GIBSON, J. B.: Isolation by disruptive selection. Nature 163, 1164–1166 (1960).

THULIN, I. J.: Breeding of Pinus radiata through seed improvement and clonal afforestation. In: Second World Consultation on Forest Tree Breeding, pp. 1109–1118. Rome: FAO 1969.

TIGERSTEDT, P.: Experiments on selection for development rate in Drosophila melanogaster. Ann. Ac. Sci. Fenn., Ser. A, 148 (1969).

TIMOFEEF, V. P.: 100 Jahre Versuchsrevier der Landwirtschaftlichen Timirjasew Akademie. Forstwirtsch. Verlag Moskau (1965).

TIREN, L.: Om granens kottsättning, des periodicitet och samband med temperatur och nederbörd. Medd. Stat. Skogsf. Anst. 28, 413–524 (1935).

TOOLE, E. H., BROWN, E.: Final results of the Duvel buried seed experiment. J. Agr. Res. 72, 201–210 (1946).

TROUP, R. S.: The Silviculture of Indian Trees, vol. II. Oxford: The Clarendon Press 1921.

TUNSTELL, G.: Canada. In: World Geography of Forest Resources, pp. 127–147 (HADEN-GUEST, S., WRIGHT, J. K., TECLAFF, E. M., eds.). New York: The Ronald Press 1956.

TURESSON, G.: The scope and import of genecology. Hereditas 4, 171–176 (1923).

TURNER, J. R. G.: Why does the genotype not congeal? Evolution 21, 645–656 (1967).

TURNER, J. R. G.: On supergenes II. The estimation of genetic excess in natural populations. Genetics 39, 82–93 (1968).

TURNER, J. R. G.: Changes in mean fitness under natural selection. In: Math. Topics in Pop. Gen., Biom. 1, pp. 32–78. Heidelberg: Springer 1970.

TÜXEN, R. (Ed.): Anthropogene Vegetation. Int. Symp. Stolzenau 1961, Junk, Den Haag (1966).

UNDERWOOD, G.: Categories of adaptation. Evolution 8, 365–377 (1954).

UPHOF, J. C.: Ecological relations of plants with ants and termites. Bot. Rev. 8, 563–598 (1942).

VAARAMA, A., VALANNE, T.: Induced mutations and polyploidy in Birch. Betula ssp. Final Rep. Proj. E-FS-47, Dep. Bot. Univ. Turku (1970).

VACLAV, E.: Verbreitung, Standortansprüche und Wachstum der Braunmaserbirke in Europa. Mitt. Forstw. Inst. Hochschule für Landw. Prag 6, 217–237 (1963).

VAN ANSCHL, E. P., RIKER, A. J., KOUBA, T. F., SNOMI, V. E., BOYSEN, R. A.: The climatic distribution of blister rust on White Pine in Wisconsin. Lake St. For. Exp. Sta., USDA For. Service, Sta. Paper no. 87 (1961).

VAN BUJTENEN, J. P.: Testing loblolly pine for drought resistance. Tech. Rep. 13, Texas Forest Service. 15 pp. (1966).

VAN BUJTENEN, J. P., STERN, K.: Marginal populations and provenance research. Proc. XIV. IUFRO Congr. München 3, 319–331 (1967).

VAN DER PIJL, L.: The dispersal of plants by bats. Act. Bot. Neerl. **6**, 291–315 (1957).

VAN DER PIJL, L.: Flowers free from the environment? Blumea **4**, 32–38 (1958).

VAN DER PIJL, L.: Ecological aspects of flower evolution. Evolution **14** (**15**), 403–416 (44–59) (1960/61).

VAN DER PIJL, L.: Ecological aspects of fruit evolution. Proc. Ned. Ac. Wet. **69**, 587–640 (1966).

VAN DER PIJL, L.: Principles of Dispersal in Higher Plants. Berlin: Springer Verlag 1969.

VAN DER PLANCK, J. E.: Disease Resistance in Plants. New York: Academic Press 1968.

VAN DYNE, G. M., ed.: The Ecosystem Concept in Natural Resource Management. New York, London: Academic Press 1969.

VAN MIEGROET, M.: La définition du but de la sylviculture. In: FAO Symposium on Man-Made Forests and Their Industrial Importance, pp. 1371–1393. Rome: FAO 1967.

VAN MIEGROET, M.: Natural and artificial regeneration. In: FAO Symposium on Man-Made Forests and Their Industrial Importance, pp. 1713–1734. Rome: FAO 1967.

VAN VALEN, L.: Nonadaptive aspects of evolution. Amer. Nat. **96**, 305–308 (1960).

VAN VALEN, L.: Introgression on laboratory populations of Drosophila persimilis and Drosophila pseudoobscura. Heredity **18**, 205–214 (1963).

VAN VALEN, L.: Selection in natural populations. III. Measurement and estimation. Evolution **19**, 514–528 (1965).

VAN VALEN, L.: Selection in natural populations. VI. Variation genetics and more graphs for estimation. Evolution **21**, 402–405 (1967).

VASEK, F. G.: Outcrossing in natural populations. I. The Breckenridge Mountain population of Clarkia exilis. Evolution **18**, 213–218 (1964).

VASEK, F. G.: The evolution of Clarkia ungulata derivatives adapted to relatively xeric environments. Evolution **18**, 26–42 (1964).

VECCHI, P.: Evolution and importance of land races in breeding. In: Second World Consultation on Forest Tree Breeding, pp. 1263–1278. Rome: FAO 1969.

VENKATESH, C. S.: Cleistogamy in Eucalyptus terticornis. Paper for Work. Group Rep. For. Trees, Int. IUFRO Congr. Gainesville (1971).

VOGL, R. J.: One hundred and thirty years of plant succession in a southeastern Wisconsin Lowland. Ecology **50**, 248–255 (1969).

VOGL, M., SCHÖNBACH, H., HAEDICKE, E.: Experimentelle Untersuchungen zur relativen Rauchhärte im Rahmen eines Provenienzversuchs mit der japanischen Lärche. Arch. f. Forstw. **17**, 101–115 (1968).

VOGT, W.: Latin America Timber Ltd. Unasylva **2**, 19–25 (1948).

VOLTERRA, V.: Variazione e fluttazione del numero XX d'individui in specie animali conviventi. Mem. Ac. Naz. Lincei **2**, 31–113 (1926).

VON FRISCH, K.: Aus dem Leben der Bienen. Berlin, Göttingen, Heidelberg: Springer Verlag 1953.

VUKOREP, I.: Beziehungen zwischen chemischen Bodeneigenschaften und dem Zuwachs von Schwarzpappeln. Göttinger Bodenkundl. Ber. **15**, 1–108 (1970).

WADDINGTON, C. H.: The strategy of the genes. Allen and Unwin, London (1957).

WADDINGTON, C. H.: Experiments on canalizing selection. Gen. Res. **1**, 140–150 (1960).

WADDINGTON, C. H.: Introduction to the symposium. In: Genetics of Col. Spec. pp. 1–6. New York: Academic Press 1964.

WADDINGTON, C. H.: The paradigm for the evolutionary process. In: Pop. Biol. and Evol., pp. 37–45. Syracuse, N. Y.: Syracuse Univ. Press 1967.

WADDINGTON, C. H., ROBERTSON, E.: Selection for developmental canalization. Gen. Res. **7**, 303–312 (1966).

WAHLENBERG, W. G.: Loblolly pine. The School of Forestry, Duke University (1960).

WAHRBURTON, F. E.: A model of natural selection based on a theory of guessing games. J. Theor. Biol. **16**, 78–96 (1967).

WAKELEY, P. C.: Results of the south-wide pine seed source study through 1960–61. Proc. VI. South Conf. Tree Impr. Gainesville, pp. 10–24 (1961).

WALLACE, B.: The estimation of adaptive values of experimental populations. Evolution 6, 333–341 (1952).

WALLACE, B.: Genetic divergence of isolated populations of Drosophila melanogaster. Proc. IX. Int. Congr. Gen. 761–764 (1954).

WALLACE, B.: Modes of reproduction and their genetic consequences. Stat. Gen. and Plant. Breed., Nat. Ac. Sci. — Nat. Res. Counc., Publ. 982, 3–20 (1963).

WALTER, H.: Die Vegetation der Erde in ökologischer Betrachtung. Bd. 1. Jena: Gustav Fischer 1962.

WALTER, H.: Die Vegetation der Erde in öko-physiologischer Betrachtung. Bd. II. Jena: Gustav Fischer 1968.

WANG, B. S. P., SZIKLAI, O.: A review of forest tree seed certification. For. Chron. 45, 378–385 (1969).

WANG, C. W., PERRY, T. C.: The ecotypic variation of dormancy, chilling requirements and photoperiodic response in Betula species. Proc. X. Int. Congr. Gen. Montreal 307 (1958).

WASSER, R. G.: Flowering phenology study. Virginia Div. of For. Occ. Rep. no. 28 (1967).

WATKINS, R., SPANGELO, L. P. S.: Components of genetic variance in the cultivated Strawberry. Genetics 59, 93–103 (1968).

WATT, K. E. F., ed.: Systems Analysis in Ecology. New York, London: Academic Press 1966.

WEBB, L. J., TRACEY, J. G., HADDOCK, K. P.: A factor toxic to seedlings of the same species associated with living roots of the non-gregarious subtropical rain-forest tree Grevillea robusta. J. App. Ecol. 4, 13–25 (1967).

WEBBER, B., ARNOTT, J. T., WEETMAN, G. F., CROOME, G. C. R.: Advance growth destruction, slash coverage and ground conditions in logging operations in eastern Canada. Woodlands Paper no. 8, PPRIC, Pointe Claire, Quebec (1969).

WEBER, H.: Vegetative Fortpflanzung bei Spermatophyten. In: Handbuch der Pflanzenphysiologie, Bd. 18. Heidelberg, London, New York: Springer 1967.

WECK, J.: Vom Umsetzen unserer Waldbäume. Allg. Forstzeitschr. 13, 717–720 (1958).

WEETMAN, G. F.: The need to establish a national system of natural forested areas. For. Chron. 46, 31–33 (1970).

WELLS, O. O.: Geographic variation in ponderosa pine. I. The ecotypes and their distribution. II. Correlations between progeny performance and characteristics. Silv. Gen. 13, 89–103, 125–132 (1964).

WENT, F. W.: Ecology of desert plants. II. Ecology 30, 1–13 (1949).

WERFT, R.: Über die Lebensdauer der Pollenkörner in der freien Atmosphäre. Biol. Centralbl. 70, 131–137 (1951).

WESTERGAARD, M.: The mechanism of sex determination in dioecious flowering plants. Adv. Gen. 9, 217–235 (1958).

WHITE, M. J. D.: Animal cytology and evolution, 2nd ed. Cambridge, England: Cambridge Univ. Press 1954.

WHITEHEAD, D. R.: Wind pollination in the Angiosperms: evolutionary and environmental considerations. Evolution 23, 28–36 (1969).

WHITTAKER, R. H.: Communities and Ecosystems. Toronto: Collier-Macmillan 1970.

WILKINS, D. A.: Sampling for genecology. Scot. Plant. Breed, Stat. Rep. 92–96 (1959).

WILKINSON, R. C., HANOVER, J. W., WRIGHT, J. W., FLAKE, R. H.: Genetic variation in the monoterpene composition of White Spruce. For. Sci. 17, 83–90 (1971).

WILLIAMS, C. B.: Patterns in the Balance of Nature. New York, London: Academic Press 1964.

WILLIAMS, W.: Genetical Principles and Plant Breeding. Oxford: Blackwells Scientific Publications 1964.

WILLIAMSON, M. H.: An elementary theory of interspecific competition. Nature **180**, 42–425 (1957).

WILLIAMSON, M. H.: Selection, controlling factors and polymorphism. Amer. Nat. **92**, 329–335 (1958).

WILSON, E. O.: The challenge from related species. In: Gen. of Col. Spec., pp. 7–24. New York: Academic Press 1964.

WILSON, F.: Biological control and the genetics of colonizing species. In: Gen. of. Col. Spec., pp. 307–329. New York: Academic Press 1964.

WILSON, J.: Experimental determination of fitness interactions in Drosophila melanogaster by the method of marginal populations. Genetics **59**, 501–511 (1968).

WINSTEAD, J. E.: Population differences in seed germination and stratification requirements of sweet gum. For. Sci. **17**, 34–36 (1971).

WISEMAN, J. D. H.: The determination and significance of past temperature changes in the upper layer of the equatorial Atlantic Ocean. Proc. Roy. Soc. Lond., Ser. A**222**, 296–323 (1954).

WOLSTENHOLME, D. R., THODAY, J. M.: Effects of disruptive selection. VII. A third chromosome polymorphism. Heredity **18**, 413–432 (1963).

WOODSON, E.: The geography of flower color in Butterfly weed. Evolution **18**, 143–163 (1964).

WORRALL, J., MERGEN, F.: Environmental and genetic control of dormancy in Picea abies. Phys. Plant. **20**, 733–745 (1967).

WRIGHT, J. W.: Species crossability in spruce in relation to distribution and taxonomy. For. Sci. **1**, 319–349 (1955).

WRIGHT, S.: Evolution in mendelian populations. Genetics **16**, 97–159 (1931).

WRIGHT, S.: Size of population and breeding structure in relation to evolution. Science **87**, 430–431 (1938).

WRIGHT, S.: Isolation by distance. Genetics **31**, 336–352 (1946).

WRIGHT, S.: The theoretical variance within and among subdivisions of a population that is in a steady state. Genetics **37**, 312–321 (1952).

WRIGHT, S.: Modes of selection. Amer. Nat. **90**, 5–24 (1956).

WRIGHT, S.: The interpretation of population structure by F-statistics with special regard to systems of mating. Evolution **19**, 395–420 (1965).

WRIGHT, S.: Factor interaction and linkage in evolution. Proc. Roy. Soc. Ser. B**162**, 80–104 (1965).

WRIGHT, S.: Random drift and the shifting balance theory of evolution. In: Mathematical topics in population genetics. Biom. 1, pp. 1–31. Heidelberg: Springer 1970.

YARBROUGH, K., KOJIMA, K. I.: The mode of selection at the polymorphic esterase-6 locus in cage populations of Drosophila melanogaster. Genetics **57**, 667–686 (1967).

YLI-VAKKURI, P.: Untersuchungen über organische Wurzelverbindungen zwischen Bäumen in Kiefernbeständen. Acta. For. Fenn. **60**, 1–117 (1953).

YOSHIMICHI, K., KURABAYASHI, M.: Evolution and variation in Trillium VII. Migration between northern and eastern population groups of T. kamtschaticum. Evolution **14**, 232–237 (1960).

ZOBEL, B.: Gene preservation by means of a tree improvement program. Proc. 13th Meeting Comm. on For. Tree Breed. Can. (in press) 1972.

ZOBEL, B. J., BARBER, J., BROWN, C. L., PERRY, T. O.: Seed orchards, their concept and management. J. For. **56**, 815–825 (1958).

ZOHARY, M.: Die verbreitungsökologischen Verhältnisse der Pflanzen Palästinas. I. Beih. Bot. Zbl. **56**, 1–155 (1937).

Species and Family Index

Page numbers *in italics* refer to figures or tables.

Subject Index

Standard roman type numbers refer to the text, numbers in **bold type** refer to definitions, numbers *in italics* indicate citation in figures or tables.

Ecological Studies

Analysis and Synthesis

Editors: J. Jacobs,
O. L. Lange, J. S. Olson,
W. Wieser

Vol. 1: Analysis of Temperate Forest Ecosystems

Editor: D. E. Reichle
91 figs. XII, 304 pages. 1970
Cloth DM 52,—; US $20.10
ISBN 3-540-04793-X

Contents: Analysis of an
Ecosystem. — Primary
Producers. — Consumer
Organisms. — Decomposer
Populations. — Nutrient
Cycling. — Hydrologic Cycles.

Vol. 2: Integrated Experimental Ecology

Methods and Results of Ecosystem Research in the German Solling Project

Editor: H. Ellenberg
53 figs. XX, 214 pages. 1971
Cloth DM 58,—; US $22.40
ISBN 3-540-05074-4

Contents: Introductory
Survey. — Primary Production.
— Secondary Production.—
Environmental Conditions. —
Range of Validity of the
Results. —Editorial Note.

Vol. 3: The Biology of the Indian Ocean

Editor: B. Zeitzschel
in cooperation with
S. A. Gerlach
286 figs. XIII, 549 pages. 1973
Cloth DM 123,—; US $55.40
ISBN 3-540-06004-9

The present volume contains
much new information and
some conclusions regarding
the functioning and organization of the ecosystems of
the Indian Ocean.

Vol. 4: Physical Aspects of Soil Water and Salts in Ecosystems

Editors: A. Hadas;
D. Swartzendruber;
P. E. Rijtema; M. Fuchs;
B. Yaron
221 figs. 61 tab. XVI,
460 pages. 1973
Cloth DM 94,—; US $36.20
ISBN 3-540-06109-6

Contents: Water Status
and Flow in Soils:
Water Movement in Soils.
Energy of Soil Water and
Soil-Water Interactions. —
Evapotranspiration and
Crop-Water Requirements:
Evaporation from Soils
and Plants. Crop-Water
Requirements. — Salinity
Control.

Vol. 5: Arid Zone Irrigation

Editors: B. Yaron;
E. Danfors; Y. Vaadia
Approx. 180 figs.
Approx. 500 pages. 1973
Cloth DM 94,—; US $36.20
ISBN 3-540-06206-8

Contents: Arid Zone
Environment. — Water
Resources. — Water Transport
in Soil-Plant-Atmosphere
Continuum. — Chemistry
of Irrigated Soils-Theory and
Application. — Measurements
for Irrigation Design and
Control. — Salinity and
Irrigation. — Irrigation
Technology. — Crop Water
Requirement.

Vol. 6: K. Stern, L. Roche; Genetics of Forest Ecosystems

Approx. 155 figs.
Approx. 540 pages. 1973
ISBN 3-540-06095-2
In preparation

Vol. 7: Mediterranean Type Ecosystems

Origin and Structure

Editors: F. di Castri;
H. A. Mooney.
Approx. 88 figs.
Approx. 500 pages. 1973
Cloth DM 78,—; US $30.10
ISBN 3-540-06106-1

Contents: Convergence
in Ecosystems. — Physical
Geography of Lands with
Mediterranean Climates. —
Vegetation in Regions
of Mediterranean Climate. —
Soil Systems in Regions
of Mediterranean Climate. —
Plant Biogeography. —
Animal Biogeography
and Ecological Niche. —
Human Activities Affecting
Mediterranean Ecosystems.

Distribution rights for
U. K. Commonwealth,
and the Traditional British
Market (excluding Canada):
Chapman & Hall Ltd., London

Prices are subject to change
without notice

**Springer-Verlag
Berlin
Heidelberg
New York**

München Johannesburg
London Madrid New Delhi
Paris Rio de Janeiro
Sydney Tokyo Utrecht Wien

Trees: Structure and Function

By **Martin H. Zimmermann,**
Charles Bullard Professor
of Forestry and Director
of Harvard Forest,
Harvard University

Claud L. Brown,
Professor of Forestry and
Botany, University
of Georgia

with a chapter by
Melvin T. Tyree,
Botany School, University
of Cambridge, England

111 figs. XIV, 336 pp. 1971
Cloth DM 72,—; US $27.80
ISBN 3-540-05367-0

Prices are subject to change
without notice

Trees have the distinction of being the largest and oldest living organisms on earth. Although the herbaceous habit has made unprecedented evolutionary gains, trees are still the most conspicuous plants covering the habitable land surface of the earth. Trees have always been of much interest to botanists, and many of the early investigations concerning the structure and function of plants were conducted with trees. Not only do trees perform all the cellular activities of most unicellular and herbaceous plants . . . they perform a good deal more.

This book is devoted largely to those aspects of tree physiology which are peculiar to tall woody plants. Throughout the book the emphasis is on function as it relates to structure. The authors describe how trees grow, develop and operate, not how these functions are modified in different types of environment. The approach is functional rather than ecological . . . how trees work, not how they behave in various habitats.

The text contains many literature citations without being encyclopedic.

Springer-Verlag
Berlin Heidelberg New York
München Johannesburg London Madrid
New Delhi Paris Rio de Janeiro Sydney
Tokyo Utrecht Wien